A HISTORY OF
NEAR-EARTH
OBJECTS
RESEARCH

A HISTORY OF
NEAR-EARTH
OBJECTS
RESEARCH

Erik M. Conway
Donald K. Yeomans
Meg Rosenburg

National Aeronautics and Space Administration

Office of Communications
NASA History Division
Washington, DC 20546

NASA SP-2022-4235

Library of Congress Cataloging-in-Publication Data

Names: Conway, Erik M., 1965– author. | Yeomans, Donald K., author. | Rosenburg, Meg, author. | United States. NASA History Division (2021–), issuing body.

Title: A history of near-earth objects research / Erik M. Conway, Donald K. Yeomans, Meg Rosenburg.

Other titles: NASA SP (Series) ; 4235.

Description: Washington, DC : National Aeronautics and Space Administration, Office of Communications, NASA History Division, 2022. | Series: SP ; 2022-4235 | Includes bibliographical references and index. | Summary: "In 2016, NASA took on a new responsibility: Planetary Defense Coordination. That event reflected a growing interest in, and concern about, the threat of celestial impacts. In ancient times, the solar system's small bodies— asteroids and comets—were sometimes seen as ill omens, warnings from the gods. In modern times, they've come to be seen as the solar system's rubble, leftovers from its formation, but still largely ignored until the late 20th century. Increasingly, they've been seen by scientists as objects worthy of study; by the general public, and the US government, as potential threats to be mitigated; and by space advocates as future resources. This book tells the story of those re-interpretations and NASA's role in them" –Provided by publisher.

Identifiers: LCCN 2021035177 (print) | LCCN 2021035178 (ebook) | ISBN 9781626830684 (paperback) | ISBN 9781626830691 (Adobe pdf)

Subjects: LCSH: Near-Earth objects–Research. | Near-earth asteroids–Research. | Collisions (Astrophysics)–Research. | Asteroids–Collisions with Earth. | Comets–Collisions with Earth. | Impact craters–History. | Astronomy–History. | Hazard mitigation–International cooperation. | United States. National Aeronautics and Space Administration–Research.

Classification: LCC QB651.C66 2022 (print) | LCC QB651 (ebook) | DDC 523.44–dc23/eng/20211209 | SUDOC NAS 1.21:2022-4235

LC record available at *https://lccn.loc.gov/2021035177*

LC ebook record available at *https://lccn.loc.gov/2021035178*

This publication is available as a free download at
http://www.nasa.gov/ebooks.

ISBN 978-1-62683-068-4

9 781626 830684

CONTENTS

	Acknowledgments	*vii*
	List of Figures and Tables	*ix*
	Introduction	1
CHAPTER 1	Small Solar System Bodies Before the 20th Century	13
CHAPTER 2	Recognizing Impact as a Geological Process	31
CHAPTER 3	Finding and Characterizing Near-Earth Objects Through 1990	51
CHAPTER 4	Cosmic Impacts and Life on Earth	79
CHAPTER 5	Recognizing Cosmic Hazard	105
CHAPTER 6	Automating Near-Earth Object Astronomy	141
CHAPTER 7	Studying the Rubble of the Solar System	175
CHAPTER 8	Impacts as Natural Hazards	213
CHAPTER 9	Asteroids as Steppingstones and Resources	245
CHAPTER 10	Organizing Planetary Defense	273
CHAPTER 11	Conclusion	307

Appendix 1
Asteroid and Comet Designations, Numbers, and Names — *317*

Appendix 2
Heliocentric Orbit Classifications for Near-Earth Objects — *321*

Glossary	*325*
Acronyms and Abbreviations	*329*
Select Bibliography	*335*
About the Authors	*343*
The NASA History Series	*345*
Index	*361*

ACKNOWLEDGMENTS

This history was funded by the Planetary Defense Coordination Office at NASA Headquarters, via a task order to the Jet Propulsion Laboratory (JPL). The requirement was to produce a manuscript in two years, and to manage that, Erik Conway asked Meg Rosenburg, who had previously done research on impact science, and Don Yeomans, who was a central player in the development of NASA's Near-Earth Object (NEO) research programs, to coauthor this work. In our original schema, Rosenburg was to write two chapters on the history of impact science and on impact mitigation research; Yeomans was to write chapters on premodern and modern NEO discovery and modern characterization research; and Conway was to write chapters on modern search systems, automation, and policy development. However, after a number of rounds of reviews and revisions, there is no longer a simple way to map chapter to author.

In addition to the more than 40 interviewees, who selflessly provided their remembrances and insights, we'd like to thank David Morrison, Clark Chapman, Andrew Butrica, Rob Landis, Kelly Fast, Lindley Johnson, Steve Garber, James Anderson, Brian Odom, and the NASA History Division's anonymous reviewers for their thorough reviews and extensive comments on various iterations of this manuscript. We are indebted to the NASA History Division's former archivists, Robyn Rodgers, Liz Suckow, and Colin Fries, for research support. Molly Stothert-Maurer and Rachel Castro of the University of Arizona's Special Collections provided access to Tom Gehrels's papers. At the JPL library, Mickey Honchell provided considerable interlibrary loan support. Robert McMillan of SpaceWatch contributed documents and an extensive set of interviews during Conway's research visit to Tucson. Steven Larson of the Catalina Sky Survey (CSS) gave a very informative tour of the survey's observatories as well as interviews on CSS's origin. Nick Kaiser, Richard Wainscoat, and David Tholen of the University of Hawai'i Institute for Astronomy were particularly helpful during Conway's visit there. David

Portree, of the U.S. Geological Survey's Astrogeology Science Center, and Lauren Amundson, of the Lowell Observatory, aided Rosenburg's research visit to Flagstaff, Arizona. Yeomans's interview with Ed Tagliaferi provided insights into the initial congressional interest in NEOs as well as the little-known Earth orbital observations of atmospheric impacts by detectors designed to monitor nuclear weapons testing.

Yeomans provided documents and journals from his career for this work, as did JPL's Paul Chodas, Alan W. Harris, Kenneth Lawrence, and Amy Mainzer. In addition to providing five oral interviews, Lindley Johnson of NASA Headquarters gave Conway access to his files and to material turned over to him by his predecessor, Tom Morgan of Goddard Space Flight Center.

Several online documentation efforts are worth calling attention to. The personal home page of Clark Chapman of the Southwest Research Institute contains numerous documents and essays reflecting his own career and the growth of NEO astronomy. The National Space Society's online document library was invaluable for digital scans of some older documents. The NASA Planetary Defense Coordination Office maintains an online document library, though because it is a new office, the collection is biased toward the recent. NASA's Small Bodies Assessment Group maintains a public document collection and places all of its meeting agendas, minutes, and recommendations online too. It is an excellent example of public disclosure and governance. Readers of this history will also see frequent references to the International Astronomical Union's Minor Planet Center website and JPL's Center for Near-Earth Object Studies website, both of which contain valuable information about NEOs and NEO research.

Many thanks to the outstanding production team in the Communications Support Services Center (CSSC). Lisa Jirousek exhibited her customary attention to detail in meticulously copyediting the manuscript. Andrew Cooke performed careful quality assurance. Michele Ostovar most skillfully laid out the attractive book design. Heidi Blough expertly prepared the index. Tun Hla professionally oversaw the printing. Thank you all for your skill, professionalism, and customer-friendly attitudes in bringing this book to fruition.

This research was carried out at the Jet Propulsion Laboratory, California Institute of Technology, under a contract with the National Aeronautics and Space Administration.

LIST OF FIGURES AND TABLES

BOX 1-1   Bode's Law.

FIGURE 1-1   Giuseppe Piazzi, discoverer of dwarf planet Ceres.

FIGURE 1-2   Harvard College Observatory (HCO) plate I-10321, an image from 27 December 1893.

FIGURE 2-1   Gene Shoemaker's comparison of Meteor Crater to two nuclear test craters in Nevada, Teapot Ess and Jangle U, 1959.

FIGURE 3-1   Eleanor Helin at the Mount Palomar 18-inch Schmidt telescope.

FIGURE 3-2   Gene and Carolyn Shoemaker.

FIGURE 3-3   Tom Gehrels in his office in August 1997.

FIGURE 4-1   Geologist's view of time since the beginning of the Mesozoic Era.

FIGURE 4-2   Map of the Chicxulub location.

BOX 5-1   Key concepts from the unpublished 1981 Shoemaker report.

FIGURE 5-1   Size-frequency-consequence chart from the unpublished 1981 Shoemaker report.

FIGURE 5-2   Comet Shoemaker-Levy 9 discovery image.

FIGURE 5-3   Uncertainty in 1997 XF11's position for its 2028 close approach to Earth.

FIGURE 6-1   Robert S. McMillan in the Spacewatch control center.

FIGURE 6-2   The Yarkovsky effect.

FIGURE 6-3   The YORP effect.

FIGURE 6-4   The Torino Scale.

FIGURE 7-1   Bus-DeMeo asteroid spectral classification system (taxonomy).

TABLE 7-1   Comet and asteroid mission timelines.

FIGURE 7-2   Steven Ostro of JPL and Alan W. Harris of the German Aerospace Center (DLR).

FIGURE 7-3   433 Eros.

FIGURE 7-4   Comet Tempel 1.

FIGURE 7-5   Comet Churyumov-Gerasimenko.

FIGURE 8-1   Overall hazard from NEO impacts by diameter.

FIGURE 8-2   Remaining hazard after completion of the "Spaceguard Survey" goal.

FIGURE 8-3   Earth-Sun Lagrange points.

FIGURE 8-4   Hazard reduction from sub-global impactors as a function of survey completeness.

FIGURE 8-5   Lindley N. Johnson and Eleanor Helin.

FIGURE 8-6   NEO discovery statistics as of December 2019.

FIGURE 8-7   NEOCam detector development.

FIGURE 9-1   NASA historical budget and projection from 2009 to 2014.

FIGURE 9-2   Asteroid Redirect Mission Option A concept.

FIGURE 10-1   Impact trajectory of 2008 TC3.

FIGURE 10-2   Fireballs detected by U.S. government sensors, 1988–2019.

FIGURE 10-3   Orbit of 2012 DA14 compared to the impact trajectory of the Chelyabinsk bolide.

TABLE A1-1   Naming conventions for asteroids and comets.

FIGURE A2-1   Orbit classes.

INTRODUCTION

On 7 January 2016, the National Aeronautics and Space Administration (NASA) established the Planetary Defense Coordination Office. Built on the foundation of NASA's older Near-Earth Objects Observations Program, this new organization is intended to continue funding and coordinating ongoing search and research efforts for near-Earth objects (generally short-handed as "NEOs"), identify strategies to reduce the risk of future large impacts by NEOs (called "mitigation"), and coordinate emergency response actions with other federal agencies and international partners. The new office's initial budget was set at $40 million per year, up from about $4 million in 2010.[1]

A number of highly public recent events preceded this action. Most prominently, an object about 20 meters in diameter had impacted Earth's atmosphere and exploded at an altitude of about 23 kilometers near the Russian city of Chelyabinsk on 15 February 2013, releasing the energy equivalent of about 500 kilotons of TNT. The explosion was high enough above the surface that it did only modest structural damage, but even so, there were over 1,500 injuries, mostly from glass shattered by the violent shock wave initiated in the blast. By chance, the fireball was recorded on dashboard cameras installed by Russian citizens to record accidents and police encounters, resulting in an unprecedented circumstance: unlike every other known NEO impact, the Chelyabinsk airburst was a visual event that could be widely shared—and broadcast worldwide.[2]

---

1. Anon., "NASA Office To Coordinate Asteroid Detection, Hazard Mitigation," NASA, 17 January 2016, *http://www.nasa.gov/feature/nasa-office-to-coordinate-asteroid-detection-hazard-mitigation*.

2. P. G. Brown et al., "A 500-Kiloton Airburst over Chelyabinsk and an Enhanced Hazard from Small Impactors," *Nature* 503, no. 7475 (14 November 2013): 238–241, doi:10.1038/nature12741; Olga P. Popova, Peter Jenniskens, Vacheslav Emel'yanenko, Anna Kartashova, Eugeny Biryukov, Sergey Khaibrakhmanov, Valery Shuvalov, et al., "Chelyabinsk Airburst, Damage Assessment, Meteorite Recovery,

The Chelyabinsk event was the largest NEO airblast since the Russian Tunguska event of 1908. The energy of that explosion, though poorly recorded due to the location's remoteness in a Siberian forest and the lack of communications technologies, is gauged to have been about an order of magnitude greater than Chelyabinsk's. Estimated to have been around 40 meters in diameter, the Tunguska impactor flattened about 2,200 square kilometers of forest.[3] For comparison, the largest recorded California wildfire as of September 2021, ignited by lightning and known as the August Complex fire of August 2020, burned nearly 4,200 square kilometers.[4]

These extraterrestrial visitations have always happened. Earth experiences annual "meteor showers" of objects that are nearly always too small to reach the surface. These clouds of near-Earth objects have names—the "Perseids" occur in August, while the "Geminids" occur in December. They are the result of Earth's passage through the debris trails of comets or asteroids, with the Perseids, for example, originating with Comet Swift-Tuttle and the Geminids with asteroid 3200 Phaethon. And there are other such showers, too. Ancient civilizations knew of them, though they generally interpreted them as signs from their gods, and not as we do: as relics of the solar system formation process. Interestingly, for most of human history, these apparitions were often imbued with malign intent—they were, in other words, seen as portending ill.

The Idea of Cosmic Risk

The classical astronomer who probably did the most to categorize the variety of damage foretold by comet appearances was Claudius Ptolemaeus (AD 100–175), usually called Ptolemy. Ptolemy is mostly known for his *Almagest*, the treatise in which he developed the theory of epicycles to describe the motions of the heavenly bodies. In Ptolemy's cosmology, the heavens rotated around a spot slightly off-center from a stationary Earth. Although incorrect, Ptolemy's theory predicted the motion of the known planets and stars quite accurately, well within the measurement capabilities of astronomers for another

and Characterization," *Science* 342, no. 6162 (29 November 2013): 1069–1073, doi:10.1126/science.1242642.

3.   Donald K. Yeomans, *Near-Earth Objects: Finding Them Before They Find Us* (Princeton, NJ: Princeton University Press, 2013), p. 119; Popova et al., "Chelyabinsk Airburst."

4.   See "Top 20 Largest California Wildfires," *https://www.fire.ca.gov/media/4jandlhh/top20_acres.pdf* (accessed 8 September 2021).

16 centuries. Because of this, Ptolemy remained a central authority in astronomy until the Renaissance.

Ptolemy's *Almagest* did not include comets because he, like the influential Greek philosopher Aristotle (384–322 BC) before him, did not consider them heavenly bodies. Aristotle had considered them emanations arising from Earth itself, while Ptolemy only mentioned comets once, in his astrological work, the *Tetrabiblos*. In this work, he described in some detail how one could "read" the omen represented by a comet's apparition: its position with respect to the Sun indicated the timing of the coming disaster, while its tail noted the location. The shape of the comet was supposed to foretell the nature of the disaster.[5] The *Tetrabiblos* was just as influential as the *Almagest*, though in a different way: it was embedded in Catholic theology during the many centuries of the European Middle Ages.

If classical astronomy can be seen as an effort to understand and predict the motions of heavenly bodies, classical astrology can be interpreted as astronomers' efforts to predict the influence of those bodies on humans and on human events. Ptolemy was not alone in practicing astronomy and astrology simultaneously (though he placed them in different books); most astronomers prior to the modern era made their livings as astrologers. Society's wealthier members were much more interested in predictions relevant to everyday life than they were in the mere motion of the heavens. This was equally true in ancient China, where court astronomers had been making detailed observations of the heavens, and of comets, for centuries before the *Almagest* in order to provide astrological advice to their rulers.[6]

Finally, it is important to note that while premodern civilizations saw comets as omens of ill fortune, they did not see comets as the *cause* of that ill fortune. Fickle, or vengeful, gods and goddesses were the actual agents of destruction; comets were merely warnings.

The *Almagest*'s influence gradually broke down among Western astronomers during the 17th century, rooted as it was in the failing notion of geocentrism. (It is less clear that the influence of the *Tetrabiblos* ever did fade completely, although it is no longer part of Christian doctrine.) Comet

---

5.   Donald K. Yeomans, *Comets: A Chronological History of Observation, Science, Myth, and Folklore* (New York: John Wiley & Sons, Inc., 1991), pp. 14–15.

6.   Chinese astronomers were the first to notice the anti-solar direction of comet tails, for example. See Yeomans, *Comets: A Chronological History*, pp. 42–48.

apparitions of 1607 and 1618 once more sparked debate among astronomers over their nature. The 1618 comet was the first to be observed with a telescope, and astronomers began to understand them as physical objects, rather than as omens sent by deities. But comets were also not quite fully accepted as permanent features of the cosmos. Johannes Kepler, who in this era demonstrated that planetary orbits were calculable as ellipses, believed comets were ephemeral and moved in straight lines. Other period astronomers believed them eternal, moving in orbits, though not necessarily centered on the Sun.

Edmond Halley's 1705 treatise on comets began to solidify elite opinion around the idea that at least some comets were repeat visitors. Using a technique developed by Isaac Newton, he computed orbits for 24 comets that had appeared between 1337 and 1698, and he contended that a single comet actually accounted for three of those visits. It had appeared in 1531, 1607, and 1682. He also predicted its return in late 1758 or early 1759.[7] Though Halley did not live to see his comet's next reappearance, it did return. A German amateur made the first recorded observation on Christmas evening, 1758.[8] Halley's Comet, as we now name it, settled the question about cometary longevity and rendered these objects relatively predictable members of the cosmos. If one could track a comet for the few days or weeks of an apparition, one could generate its orbital characteristics and predict its future motion.

Halley also had the foresight to wonder what would happen should a comet strike Earth. But he did not strongly pursue the question—perhaps wisely, for the sake of his reputation. The same century in which Halley's comet prediction was made and confirmed also witnessed the development of the doctrine of uniformitarianism within the natural sciences. Uniformitarianism taught (in various formulations) that "the present is the key to the past." It was articulated in 1795 by James Hutton and subsequently popularized (and purged of its religious overtones) by John Playfair's 1802 revival, *Illustrations of the Huttonian Theory of the Earth*.[9] Uniformitarianism proposed that the same natural laws and processes operating today have always operated throughout Earth history, at the same rate and across the entire planet:

---

7. Ibid., pp. 118–119.
8. Ibid., p. 132.
9. Martin J. S. Rudwick, *Bursting the Limits of Time: The Reconstruction of Geohistory in the Age of Revolution* (Chicago: University of Chicago Press, 2007), p. 465.

…no powers are to be employed that are not natural to the globe, no action to be admitted of except those of which we know the principle, and no extraordinary events to be alleged in order to explain a common appearance.[10]

Hutton's uniformitarianism was adopted by Charles Lyell, whose *Principles of Geology*, published between 1830 and 1833, established it as a central tenet of geological thinking in the 19th century. In a prior statement, Hutton illustrated the distasteful connotation attached to the catastrophist side, rejecting appeals "to any destructive accident in nature, or to the agency of preternatural cause, in explaining what actually appears."[11]

For both Hutton and Lyell, changes on Earth, like the gradual cycles of erosion or uplift, occurred over an inconceivably vast time span, providing no hope of deducing either a beginning or an end.[12] Intended to remove natural science from theology—for example, the biblical flood would be catastrophic, and thus unacceptable as an explanation under uniformitarian ideals—it also led scientists to reject natural cataclysms as an explanation for Earth's features. Extraterrestrial impacts could not be shapers of Earth. Thus one interesting effect of uniformitarianism on the Earth sciences was the separation of Earth from its cosmos. Only the Sun and Moon were permitted to have influence on Earth. By the end of the 19th century, not only were comets no longer widely seen as omens of ill events; they were no longer seen as relevant to human, or Earthly, affairs at all.

Rocks from Space

Comets were but one type of occasional celestial visitor. Meteors—and the mysterious remnants they occasionally left behind—were another. Reports of meteorite falls date back nearly as far as the dawn of written history, with accounts preserved in ancient Egyptian, Chinese, Greek, and Roman literature. Myths and religious writings record still more ancient events, such as

---

10. James Hutton, *Theory of the Earth: With Proofs and Illustrations*, vol. 2 (printed for Cadell and Davies, London, and William Creech, Edinburgh, 1793), p. 547.

11. James Hutton, "X. Theory of the Earth; or, an Investigation of the Laws Observable in the Composition, Dissolution, and Restoration of Land upon the Globe," *Transactions of the Royal Society of Edinburgh* 1, no. 2 (1788): 285.

12. Rudwick, *Bursting the Limits of Time*, p. 169; Joe D. Burchfield, *Lord Kelvin and the Age of the Earth* (Chicago: University of Chicago Press, 1990), p. 9.

a set of traditions related to the Campo del Cielo fall in northern Argentina 4,000–5,000 years ago, the Bible passage referring to the Temple of Artemis in Ephesus as "the image that fell down from Jupiter," the Phoenician legend of Astarte bringing "a star fallen from the sky" to the island of Tyre to be worshipped, and the disputed meteoric origin of the Black Stone of the Kaaba in Mecca. Tools and items of religious significance made of meteoric iron have also been discovered at many archeological sites. The oldest known worked iron artifacts are a collection of Egyptian beads dated to 3200 BC, made of hammered and rolled meteoric iron, and the extraterrestrial signature of a knife taken from the tomb of King Tutankhamun in 1925 has recently been confirmed by chemical analysis. Many Native American artifacts have been discovered at burial sites across North America, and several Chinese Bronze Age axe-heads containing meteoric iron blades have been documented.[13]

Despite these widespread accounts and the demonstrated influence of meteorite falls on cultural myths and traditions, they were not generally accepted as "rocks from space" until well into the 19th century. The absence of known source bodies, pressure to find explanations in known terrestrial phenomena, and skepticism of eyewitness accounts all played a role in fostering doubts as to the extraterrestrial origin of meteorites. In the 1700s, the conception of the divinely maintained clockwork solar system established by Isaac Newton was no place for stray objects. Even when observations of comets and stellar novae established that the universe does indeed change over time, the existence of cosmic debris remained difficult to accept.

---

13. G. Brent Dalrymple, *The Age of the Earth* (Palo Alto, CA: Stanford University Press, 1994), p. 258; W. B. Masse, M. J. Masse, "Myth and Catastrophic Reality: Using Myth To Identify Cosmic Impacts and Massive Plinian Eruptions in Holocene South America," in *Myth and Geology*, Special Publication no. 273, ed. Luigi Piccardi and W. Bruce Masse (London: Geological Society of London, 2007), p. 177; C. C. Wylie and J. R. Naiden, "The Image Which Fell Down from Jupiter," *Popular Astronomy* 44 (1936): 514; John G. Burke, *Cosmic Debris: Meteorites in History* (Berkeley, CA: University of California Press, 1991), pp. 221–225; Thilo Rehren et al., "5,000 Years Old Egyptian Iron Beads Made from Hammered Meteoritic Iron," *Journal of Archaeological Science* 40, no. 12 (2013): 4785–4792; Daniela Comelli et al., "The Meteoritic Origin of Tutankhamun's Iron Dagger Blade," *Meteoritics & Planetary Science* (2016); Rutherford J. Gettens, Roy S. Clarke, Jr., and William Thomas Chase, "Two Early Chinese Bronze Weapons with Meteoritic Iron Blades," in *Occasional Papers*, vol. 4, no. 1, (Washington, DC: Smithsonian Institution, 1971); J. Hua, "Mining and Metallurgical Technology," in *A History of Chinese Science and Technology*, vol. 3, ed. Yongxiang Lu (Berlin: Springer Berlin, 2015), pp. 239–240.

This long insistence on an empty universe, as well as the tinge of superstition attached to reports of falls and fireballs, fueled attempts to find explanations for these phenomena in more familiar, terrestrial experiences. One such explanation involved the condensation—often in conjunction with a lightning strike—of minerals within the atmosphere. This idea, bolstered by contemporary enthusiasm for recently discovered electrical phenomena, retained support well into the 19th century. By that time, however, chemical analyses had revealed that the majority of fallen and found stones are both remarkably similar to each other in texture and composition and markedly distinct from the typical rocks of Earth. This observation pointed to a single, external origin for meteorites, and the Moon was an obvious candidate. Fanned by several dramatic eruptions of Vesuvius and Etna starting in the 1760s and followed by alleged observations of lunar eruptions, support for a lunar origin for meteorites persisted into the middle of the 19th century, eventually to be defeated by dynamical considerations.[14]

While the atmospheric and lunar volcanism theories for the origin of meteorites declined, a cosmic origin seemed increasingly possible. In the 1790s, E. E. F. Chladni drew an explicit and unprecedented connection between meteors, fireballs, and "native irons," proposing an extraterrestrial origin for these three phenomena, previously considered distinct. He proposed that meteorites were fragments of matter that either had never consolidated into a planet or had emerged from the violent disruption of a planet by a collision or internal explosion of some kind. Chladni's first report was followed over the next several years by an entirely coincidental spate of highly prominent falls, culminating in the spectacular fall of thousands of stones at L'Aigle, France, in 1803. The widespread publication of these events, including a comprehensive report on the L'Aigle fall completed by Jean-Baptiste Biot on behalf of France's Minister of the Interior, did much to establish that rocks do indeed sometimes fall from the sky.[15]

---

14. John G. Burke, *Cosmic Debris: Meteorites in History* (Berkeley, CA: University of California Press, 1991); Ursula B. Marvin, "Meteorites in History: An Overview from the Renaissance to the 20th Century," in *The History of Meteoritics and Key Meteorite Collections: Fireballs, Falls and Finds*, Special Publication no. 256.1, ed. G. J. H. McCall et al. (London: Geological Society of London, 2006), pp. 15–71.

15. Ernst Florens Friedrich Chladni, *Ueber den Ursprung der von Pallas gefundenen und anderer ihr ähnlicher Eisenmassen (etc.)* (Riga: Johann Friedrich Hartknoch, 1794); Burke, *Cosmic Debris*; Marvin, "Meteorites in History."

Discovering Near-Earth Objects

The discovery of Ceres in 1801 provided a new hypothesis for the origins of these rocks, although other, more local explanations retained their currency until about 1860. While conducting observations for what would become a famous star catalog, Italian astronomer Giuseppe Piazzi discovered a new object—Ceres—in an orbit that appeared more planet-like than comet-like. Ceres also fit what is known as the Titius-Bode law. Published in 1766, this rule inferred from the spacing of the inner planets that there should be a planet between Mars and Jupiter; and another beyond Saturn. The 1781 discovery of Uranus approximately where the Titius-Bode law had forecast seemed a confirmation of its accuracy; Ceres, too, was in the right place. Piazzi first thought it might be a comet, but other astronomers doubted that. After the 1802 discovery of Pallas, traveling in a similar orbit, William Herschel declared the two objects to be a new class of celestial body: asteroids.[16]

The asteroids occupying orbital space between Mars and Jupiter are known collectively today as "main-belt asteroids," to mark a distinction between them and the many other varieties of asteroid orbits discovered since. They are still the largest known asteroid population, and one, Ceres itself, was later reclassified as a "dwarf planet" by vote of the International Astronomical Union in 2006.[17] Main-belt asteroids' orbits generally keep them far away from Earth, and they are not classed as near-Earth objects. But their orbits can be perturbed, and over the eons some have been shifted into orbits that take them into the inner solar system.

The first such near-Earth asteroid to be discovered was 433 Eros, found in 1898 by German astronomer Gustav Witt.[18] Since the discovery of Eros, scientists (and many devoted amateur astronomers) have identified many thousands of near-Earth asteroids. But far from the uniformitarian view that extraterrestrial impacts could have no influence on Earth, a very large impact is now blamed for the demise of the dinosaurs, and numerous craters associated

---

16. Giorgia Foderà Serio, Alessandro Manara, and Piero Sicoli, *Giuseppe Piazzi and the Discovery of Ceres* (Tucson: University of Arizona Press, 2002).

17. International Astronomical Union, "IAU 2006 General Assembly: Result of the IAU Resolution Votes," 24 August 2006, *https://www.iau.org/news/pressreleases/detail/iau0603/* (accessed 30 October 2020).

18. Joseph Veverka, "Eros: Special Among the Asteroids," in *Asteroid Rendezvous*, eds. Jim Bell and Jacqueline Mitton (NY: Cambridge University Press, 2002), pp. 1–9.

with large extraterrestrial impacts have been identified on, or under, Earth's surface. These impact craters are known as "astroblemes." Bringing the geological community around to the idea that asteroids and comets have in fact left marks on Earth's surface was the life's work of Eugene Shoemaker, who also trained the Apollo astronauts in field geology and whose wife, Carolyn, was one of the most prolific discoverers of comets during the era when film was the discovery medium. The recognition that large impacts could influence the evolution of life itself is an extension of what is known as the Alvarez hypothesis, named for a famous 1980 paper by Luis and Walter Alvarez.[19] This posits that an impact at the end of the Cretaceous period forced the dinosaurs into extinction. Acceptance of the idea that large impacts have transformed both the surface of Earth and life itself represents a rejection of strict uniformitarianism, or what Stephen J. Gould referred to as "substantive uniformitarianism" in a 1965 essay.[20]

Three events in close succession brought the potential destructiveness of cosmic impacts to public attention and began moving NEOs into policy salience. The first was the unexpected close flyby of Earth by 1989 FC, discovered after it had already passed by. Discovery of the enormous, buried crater left by the end-Cretaceous impactor in 1990 (or, as we will see in chapter 4, reinterpretation of an already-known structure), was the second. The spectacular collision of Comet Shoemaker-Levy 9 with Jupiter in July 1994, witnessed by ground observatories, by the Hubble Space Telescope, and by the Galileo spacecraft en route to Jupiter, was the third, and most dramatic, event. These led to Congress directing NASA to find at least 90 percent of the 1-kilometer-diameter or larger near-Earth asteroids within 10 years (see chapter 5). And NASA currently has a directive to discover and track 90 percent of all the near-Earth asteroids larger than 140 meters—those about seven times the size of the Chelyabinsk impactor—which could cause regional damage on Earth in the event of an impact.[21]

---

19. Luis W. Alvarez, Walter Alvarez, Frank Asaro, and Helen V. Michel, "Extraterrestrial Cause for the Cretaceous-Tertiary Extinction," *Science* 208 (6 June 1980).

20. James Lawrence Powell, *Night Comes to the Cretaceous: Dinosaur Extinction and the Transformation of Modern Geology* (New York: W. H. Freeman and Company, 1998); S. J. Gould, "Is Uniformitarianism Necessary?" *American Journal of Science* 263 (1965): 223–228.

21. Yeomans, *Near-Earth Objects*, pp. 68–69.

Once near-Earth objects became objects of policy concern, they also became hazards to be mitigated. The discovery of NEOs and determination of their trajectories was one component of mitigation, but as policy measures started being discussed, other aspects emerged. Deflecting specific threats was one recurrent theme of discussion; another was civil defense. The first decade of the 21st century witnessed the first impact-related civil defense exercises in the United States, and both NASA and the European Space Agency had plans for deflection demonstration missions to be carried out in the 2020s. At the international level, advocates of planetary defense had established an International Asteroid Warning Network to coordinate observations as well as communications about particular asteroid risks, along with a separate planning group for space-based deflection missions should one be necessary. This also reflects the gradual internationalization of NEO concern.

In a recent article, Valerie Olson has argued that asteroids and comets have come to be seen as environmental objects, as distinct from astronomical ones.[22] They are no longer merely intellectual curiosities or research objects for scientists. They are natural hazards, like flooding or wildfires, and are increasingly treated as risks subject to cost-benefit analysis. As we will see in later chapters, astronomers used this tool of the administrative state as part of their advocacy for NEO discovery surveys and planetary defense efforts.

Most near-Earth asteroid discoveries have happened since 1990. In large part, that is due to a technological change in the way astronomers look for them. As we will detail in later chapters, for most of the 20th century, astronomers interested in near-Earth objects (there were not many) had to compare two pieces of film, taken of the same piece of sky but at different times, to find whatever had moved between exposures. It was a painstaking, slow process. During the 1970s, though, Jet Propulsion Laboratory researchers worked to improve charge-coupled detector (CCD) technology for eventual use in space-borne cameras. A prototype CCD-based camera was used by James Westphal and James Gunn at Palomar Observatory in 1976.[23] By 1981, University of

---

22. Valerie A. Olson, "NEOecology: The Solar System's Emerging Environmental History and Politics," in *New Natures: Joining Environmental History with Science and Technology Studies* (Pittsburgh, PA: University of Pittsburgh Press, 2013), pp. 195–211.

23. James E. Gunn and James A. Westphal, "Care, Feeding, and Use of Charge-Coupled Device Imagers at Palomar Observatory," *SPIE Solid State Imagers for Astronomy* 290 (1981): 16–23.

Arizona astronomer Tom Gehrels was trying to fund a CCD-based telescope dedicated to hunting near-Earth objects at Kitt Peak Observatory; failing to raise enough money, he and a small team developed the "Spacewatch camera," which was attached to an existing 0.9-meter telescope to produce its first electronic images late in 1983. CCDs lent themselves easily to computer-assisted analysis, and by the mid-1990s, other teams were also striving to automate the process of NEO discovery, orbit determinations, and risk-to-Earth assessments. By the 2000s, hundreds of near-Earth objects were being discovered per year with the aid of automation; by the time these words were written in 2019, more than 20,000 were known.[24]

24. See *http://neo.jpl.nasa.gov/stats/* and *https://cneos.jpl.nasa.gov/stats/totals.html.*

CHAPTER 1

SMALL SOLAR SYSTEM BODIES BEFORE THE 20TH CENTURY

As we saw in the introduction, comets were well known to celestial observers in the ancient world, if also widely misunderstood. In this chapter, we will expand on pre-20th-century comet observations and examine the first discoveries of what we currently call asteroids (often also called "minor planets").

These small solar system bodies were largely the province of astronomy until the late 19th century, when they began to spark interest among a small handful of geologists. This set off a controversy lasting a few decades over whether asteroids and comets could be geological agents.

Ceres—The First Asteroid Discovery

The discovery of the first asteroid, named Ceres, took place after a well-organized search by several European astronomers had been initiated to find a missing planet between the orbits of Mars and Jupiter. The actual discovery of Ceres itself, however, is credited to an astronomer who knew nothing of this organized search effort. The fortuitous discovery of asteroid Ceres took place on the first evening of the 19th century (1 January 1801) by Gioacchino Giuseppe Maria Ubaldo Nicolò Piazzi, a Catholic priest and director of the Palermo Observatory in Sicily.[1]

---

1. Much of this section on the discovery of Ceres is based upon the work of Serio and others: G. F. Serio, A. Manara, and P. Sicoli, "Giuseppe Piazzi and the Discovery of Ceres," in *Asteroids III*, ed. W. F. Bottke, Jr., A. Cellino, P. Paolicchi, and R. P. Binzel (Tucson, AZ: University of Arizona Press, 2002), pp. 17–24.

The notable, but untimely, organized search for the missing planet has its roots in the so-called Titius-Bode law that purported to explain the relative distances of the planets from the Sun. Often called simply Bode's law, it is today thought of as little more than a curious mathematical relationship, but toward the end of the 18th century, it was taken seriously—especially when the 1781 discovery of Uranus at a distance of about 19 au seemed consistent with this pattern. If this relationship is assumed to be valid, the argument went, then there could well be an undiscovered planet in the gap between Mars and Jupiter at a heliocentric distance of 2.8 au.

Baron Franz Xaver von Zach, director of the Seeberg, or Gotha, Observatory in Germany, organized a search for the missing planet. He intended to have each of 24 astronomers agree to search a particular region of the zodiacal sky by comparing the observed stars to an accurate catalog of known stars.[2] Each observing zone would be 15 degrees in ecliptic longitude and

Box 1-1. Bode's Law.

The Bode's law relationship denotes the relative heliocentric distances of the planets as follows:

The Sun-Saturn distance is taken as 100 units.

Mercury's distance is then 4 units.
Venus is 4+ 3 = 7 units.
Earth is 4 + 6 = 10 units.
Mars is 4 + 12 = 16 units.
Gap = 4 + 24 = 28 units.
Jupiter = 4 + 48 = 52 units.
Saturn = 4 + 96 = 100 units.
Uranus = 4 + 192 = 196 units.

Thus, if Earth has a heliocentric distance of 1 astronomical unit (1 au), then the first planet Mercury (n=0) is at a distance of 0.4 au and subsequent planetary distances are given by a = 0.3 n + 0.4, where a is the distance to the Sun (in au) and n = 0, 1, 2, 4, 8, 16, 32, 64, and so on, with the values of n doubling as planets are added.

---

2.   Baron von Zach planned to limit the search to the zodiacal regions of the sky, where the plane of ecliptic contains the paths of the Sun, Moon, and planets, and where a new planet would be most likely to reside.

7–8 degrees latitude above and below the ecliptic plane. Any observed star not in the existing catalog would therefore be a candidate for the missing planet. Notwithstanding the fact that not all of the 24 invited astronomers actually participated (and some were not invited to be a part of the search in a timely manner), the observers agreeing to this plan became known as the "Lilienthal Society" or the "Himmelspolizei" (Celestial Police).

Giuseppe Piazzi, who would discover the searched-for planet, by chance never received his invitation; apparently, Barnaba Oriani in Milan was asked by von Zach to invite Piazzi to join but

Figure 1-1. Giuseppe Piazzi, discoverer of dwarf planet Ceres.

never issued the request.[3] Completely unaware of the efforts of the Celestial Police, he was pursuing his own effort to construct a reference star catalog.[4] Observing with the Palermo Observatory's vertical circle instrument made by Jesse Ramsden, with its modest aperture of 7.5 centimeters, he needed to observe each star for at least four nights before its catalog position could be established. On the evening of 1 January 1801, he detected an "unknown tiny star" in the shoulder of the constellation Taurus. Piazzi reobserved the new object the following night on 2 January 1801 and noted its apparent motion with respect to neighboring stars. By 4 January, he was convinced that he had discovered a new planet, or perhaps a comet that lacked nebulosity. He contacted the press, and soon news of Piazzi's discovery was spreading throughout Europe. On 24 January, after collecting position observations for a total of 14 nights, Piazzi wrote of his find to his friend Barnaba Oriani—the same astronomer who was to have invited him to join the Celestial Police—and to Johann Elert Bode in Berlin. In his correspondence, Piazzi included celestial

---

3.   Serio et al., "Giuseppe Piazzi and the Discovery of Ceres," p. 19.
4.   Cunningham notes that Piazzi's observing assistant, Niccolò Cacciatore, should also get credit for the discovery of Ceres. In 1817, Cacciatore succeeded Piazzi as the director of the Palermo Observatory. Clifford J. Cunningham, *Discovery of the First Asteroid, Ceres* (New York: Springer-Verlag, 2015).

positions for 1 January and 23 January and noted that the object's motion had changed from retrograde westward to direct eastward on 11 January.[5]

For the new object, Bode assumed a circular orbit with a diameter consistent with the Titius-Bode law. He pointed out that Piazzi's observations were consistent with these assumptions. Bode announced a new planet to the press in Berlin, Hamburg, and Jena and named it "Juno." Von Zach was more in favor of naming it "Hera," and this name, at least for a short period of time, was widely used in Germany. Piazzi himself initially called his planet "Ceres Ferdinandea" in honor of the patron goddess of Sicily and King Ferdinand of Bourbon. In the end, the name Ceres was accepted by the astronomical community.[6]

Piazzi's 22 observations covered the interval from 1 January through 11 February 1801, but additional observations could not be taken for several months thereafter because the position of Ceres in the sky was too close to the Sun as it moved from the evening sky to the morning sky. Without an accurate orbit and ephemeris (predicted positions at specific times), attempts to reobserve the new planet in August proved unsuccessful.[7]

Piazzi's complete set of observations was finally published in the September 1801 issue of the *Monatliche Correspondenz*, a German scientific journal. Piazzi received some criticism for not publishing his observations more promptly;

---

5.   Serio et al., "Giuseppe Piazzi and the Discovery of Ceres," p. 19.

6.   Ibid., p. 20.

7.   An asteroid orbit, for a given instantaneous epoch, is represented by six parameters that define the size, shape, and orientation of the object's path about the Sun, along with the object's position along this path. Of particular interest are the eccentricity (e), which is zero (or one) for a circular (or parabolic) path about the Sun and the semimajor axis (a), which is half the distance between the object's closest distance to the Sun (perihelion, or q) and the object's furthest distance from the Sun (aphelion or Q). So, $q = a (1 - e)$ and $Q = a (1 + e)$. Early orbit determination techniques were quite difficult, and oftentimes the task was made easier by assuming that the eccentricity was zero for a planet or asteroid while the eccentricity of a comet was often taken to be one. One of the authors, Don Yeomans, is old enough to have started out doing comet orbits using only logarithms, pencil, and paper in the 1960s. Then, what a treat to get the use of a mechanical (not electronic) Friden calculator, then the use of an IBM 360 mainframe computer with punch card inputs—and finally a (blissfully easy) laptop computer. An object's ephemeris can be thought of as a table of future (or past) times and the associated, predicted celestial positions for the object. In general, the celestial positions are the object's right ascension, or angular distance measured eastward along the celestial equator from the intersection of the celestial equator and the ecliptic plane (vernal equinox), and its declination, or its angular distance above or below the celestial equator.

he had delayed with the hope of computing an orbit himself. However, he encountered tremendous difficulties, even assuming a circular orbit. Orbit determination techniques of the day normally assumed that the object's orbital path was either circular (eccentricity = 0) or parabolic (eccentricity = 1). It was not clear that the orbit of Ceres fell into either of these categories. What was needed was a general orbit determination technique that made no initial assumptions about the object's orbital eccentricity or heliocentric distance.

The German mathematician Carl Friedrich Gauss, only 24 years old at the time, seized upon this challenge and, within little more than a month, had developed the required orbit determination process. Using three of the 22 observations that Piazzi provided (1 January, 21 January, and 11 February 1801), Gauss initially computed the orbital parameters of Ceres. Subsequent refinements of his orbit determination process iteratively improved the match between the observed and computed positions using a so-called "least squares technique"—another of his inventions.[8] Thus Gauss was able to adjust the initial orbital parameters to fit all of the valid observations. Gauss was widely acknowledged as the greatest mathematician of his time, and his least squares technique, as well as a modified version of his initial orbit determination process, is still in use today.

In September of 1801, von Zach published several forecasts of the prospective orbit, including his own and that of Gauss, which was markedly different from the others (and would turn out to be correct). Using an ephemeris for Ceres provided by Gauss, von Zach observed Ceres on 7 December and, after bad weather cleared, again on 31 December 1801 and 11 January 1802. Using Gauss's ephemeris, Wilhelm Olbers also observed Ceres from Bremen on 2 January 1802.[9] The first asteroid had been discovered—along with the means to compute orbits and ephemerides for it, as well as all asteroids discovered thereafter.

Credit for inventing the term "asteroid" often falls to William Herschel, but attribution should actually go to the English music historian Dr. Charles Burney, Sr., and his son Charles, Jr.[10] In May of 1802, Herschel asked the

---

8. Walter Kaufmann Bühler, *Gauss: A Biographical Study* (NY: Springer Science & Business Media, 2012); Donald Teets and Karen Whitehead, "The Discovery of Ceres: How Gauss Became Famous," *Mathematics Magazine* 72, no. 2 (1999): 83–93.

9. Serio et al., "Giuseppe Piazzi and the Discovery of Ceres," p. 21.

10. Clifford J. Cunningham, "The First Four Asteroids: A History of Their Impact on English Astronomy in the Early Nineteenth Century" (Ph.D. diss., University of

senior Burney if he would furnish a Latin or Greek name for the small "stars" that had been lately found. In a subsequent letter to his son, Burney suggested the Greek word "aster" to denote "star-like," and the younger Burney suggested that "oid" be added to denote resemblance and form "asteroid." Herschel first used the term on 6 May 1802, in his memoir presented to the Royal Society entitled "Observations of the Two Lately Discovered Celestial Bodies," but his choice was not immediately greeted with great enthusiasm. Indeed, Herschel was accused of purposefully applying a lesser designation for these objects so as not to detract from his own discovery of the planet Uranus in 1781, and the English author John Corry even accused Hershel of "philosophical quackery."[11] Piazzi himself also rejected the term "asteroid," perhaps because his pride would not allow his discovery to be known as anything other than a primary planet. Even so, by 1830, the term "asteroid" was commonly used in England, and the American astronomer Benjamin Apthorp Gould, founder of the *Astronomical Journal* in 1849, gave the term his stamp of approval in 1848.[12]

The second asteroid discovery was made on 28 March 1802 by Wilhelm Olbers in Bremen. Olbers made his discovery of the seventh-magnitude asteroid—named Pallas—while observing stars in the constellation of Virgo in order to more easily establish the positions of Ceres, which was then residing in the neighboring constellation of Coma Berenices.[13] Olbers also suggested that, since the orbital distances of both Ceres and Pallas were in general agreement with Bode's law, these two objects might be the fragments of a once larger planet between Mars and Jupiter and that this fragmentation could have been initiated by an internal force or possibly have resulted from an impact with a comet. At least for some time after the breakup, fragments would be expected to have orbits that intersect in the same region of the sky, so if Olbers's conjecture was correct, there could be more asteroids where the

---

Southern Queensland, 2014), pp. 46–97.

11. Ibid., p. 62.

12. Ibid., p. 96.

13. An asteroid's magnitude is a logarithmic measure of its apparent observed brightness. An asteroid's apparent magnitude depends upon its distance from the Sun and Earth, its phase angle (Sun-asteroid-Earth angle), and its absolute magnitude (H), which is defined as its magnitude when both the distances are equal to one au and the phase angle is zero. An object is barely naked-eye visible at sixth magnitude, and for every magnitude it increases, its apparent brightness decreases by a factor of about 2.5.

fragmentation event had occurred or where the orbits of Ceres and Pallas intersected. It was by observing in and near these regions of sky in Cetus and Virgo that Karl Ludwig Harding at Lilienthal made the third asteroid discovery (Juno) on 1 September 1804. Then, on 29 March 1807, Olbers made his second asteroid discovery (Vesta) while searching in the same regions of sky where Ceres, Pallas, and Juno had been discovered. These new additions to the asteroid family provided two more objects to test the planetary fragmentation hypothesis. The mathematician Joseph-Louis Lagrange examined the consequences of such an event in 1812 and reached the conclusion that Olbers's hypothesis was extraordinary but not unlikely.[14]

There was then a rather long gap between the discovery of Vesta in 1807 and the discovery of a fifth asteroid (Astraea) in 1845, likely because Astraea was not located in the same region of sky as the others, and suitable reference star maps for these other regions were not readily available. By October 1857, there were 50 known asteroids, and Olbers's fragmentation hypothesis was beginning to seem untenable. French mathematician François Arago noted that a large number of the asteroids discovered had orbits that did not intersect in a manner that would support the fragmentation hypothesis.[15]

In 1867, the American astronomer Daniel Kirkwood hypothesized that the asteroids had originated from the uniting of particles within rings of nebular material between the orbits of Mars and Jupiter. These rings were separated by gaps (now known as the Kirkwood gaps) kept clear by the perturbing effects of Jupiter.[16] This view of the asteroid belt's structure is favored by modern astronomers.

Eros—The First Near-Earth Asteroid

For most of the 19th century, asteroid discoveries were made with (often inadvertent) visual observations of moving starlike points of light in the neighborhood of fixed stars. There is an oft-repeated anecdote that because asteroids could be misleadingly stellar in appearance, astronomers trying to observe

---

14. Joseph-Louis Lagrange, "Sur l'Origine des Comètes," *Connaissance des Tems ou des Mouvemens Célestes a li Usage des Astronomes et des Navigateurs pour l'An 1814* (Paris: Bureau des Longitudes, April 1812), pp. 211–218.
15. F. Arago, *Astronomie Populaire*, vol. 4 (Paris: Gide, 1857), pp. 173–180.
16. Daniel Kirkwood, *Meteoric Astronomy: A Treatise on Shooting-Stars, Fire-Balls, and Aerolites* (Philadelphia: J. B. Lippincott & Co., 1867), p. 110.

true stars would refer to the asteroids as "vermin of the skies."[17] The introduction of photographic methods brought with it the first detection of a near-Earth asteroid, Eros.

Realizing that low-quality ephemerides were making it difficult to find already-observed asteroids (known as "recovery"), Max Wolf at Heidelberg was the first to develop techniques for discovering minor planets photographically. He affixed a camera to the telescope tube, and while the telescope was used to identify the appropriate star field and track at a sidereal rate, the much wider field of view of the attached camera allowed the identification of trailed images after 2- to 3-hour exposures. However, the low precision of these photographic images was not suitable for position measurements. On the following evening, or as soon as possible, a precise reobservation of the recovered asteroid was usually possible using naked-eye micrometer readings in conjunction with a long-focal-length refractor telescope. The observer could communicate the photographic position to other observers by telegraph if weather conditions prevented an observational attempt on the next night. The first successful photographic discovery of an asteroid, 323 Brucia, was made by Wolf on 20 December 1891.[18] Thus initially, the photographic recovery of an asteroid involved identifying the object first using a camera affixed to the telescope tube and then—armed with an approximate photographic position—later determining a more precise position using telescopic, naked-eye micrometer observations.[19] Later on, of course, cameras would be located directly at the telescope's focal plane, rather than affixed to the telescope tube.

---

17. According to Prof. David Hughes, the term "vermin of the skies" was first used by the Austrian astronomer Edmund Weiss (1837–1917), the director of the Vienna Observatory, who sometimes used the term in conversation. See Frederick H. Seares, "Address of the Retiring President in Awarding the Bruce Medal to Professor Max Wolf" in the year 1930, *Publications of the Astronomical Society of the Pacific* 42 (1930): 5–22.

18. Asteroid 323 Brucia was named after the American patron of astronomy, Catherine Wolfe Bruce.

19. Hans Scholl and Lutz Schmadel provide an in-depth review for the discovery of 433 Eros, the first near-Earth asteroid to be discovered, and B. G. Marsden provides a detailed chronological account for the early naked-eye and photographic discoveries of comets and asteroids. Hans Scholl and Lutz D. Schmadel, "Discovery Circumstances of the First Near-Earth Asteroid 433 Eros," in *Beiträge zur Astronomiegeschichte*, vol. 5, ed. Wolfgang R. Dick and Jürgen Hamel (Frankfurt: Harri Deutsch, 2002), pp. 210–220; B. G. Marsden, "Comets and Asteroids: Searches and Scares," *Advances in Space Research* 33 (2004): 1514–1523.

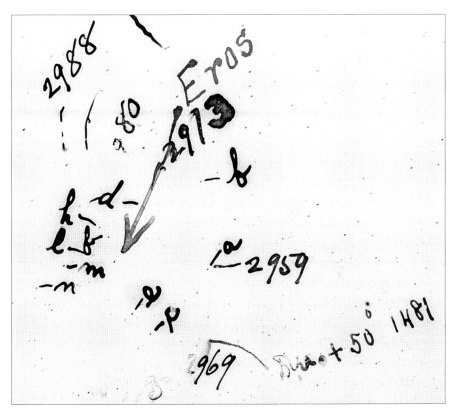

Figure 1-2. Harvard College Observatory (HCO) plate I-10321, an image from 27 December 1893. This plate, taken before 433 Eros's discovery, is a "precovery" image identified after Eros's discovery during a search of the plate archives by Williamina Fleming as part of an effort to prepare for a 1901 observing campaign. Precovery images were, and are, used to improve understanding of orbits. (Courtesy of Harvard College Observatory)

Credit for the discovery of the first near-Earth asteroid, 433 Eros, goes to Gustav Witt, along with his observing assistant Felix Linke, who helped Witt in guiding the telescope-camera combination during long exposures. No stranger to finding new objects, Witt had made his first photographic discovery of a main-belt asteroid, 422 Berolina, two years earlier in 1896. The discovery observations of Eros were made with the aid of a 6-inch, f/3.5 portrait lens camera on Saturday, 13 August 1898, at the Urania-Sternwarte in Berlin.[20] On that evening, Witt and Linke were attempting to rediscover the long-lost asteroid 185 Eunike in order to provide the precise positions required

---

20. Witt did not provide low-precision celestial positions for the discovery on 13 April, but rather on the next night, when he made naked-eye micrometer observations of Eros

to improve knowledge of this object's orbit. Their photographic plate not only recovered 185 Eunike, but also 119 Althaea, as well as the 0.4-millimeter discovery trail of Eros.

An independent photographic discovery of Eros on 13 August 1898 is sometimes attributed to Auguste Charlois at Nice Observatory in France. At the time, Charlois was one of the leading discoverers of asteroids, but because of poor weather on 14–15 August and the fact that 14 August was a Sunday and 15 August was a French holiday, Charlois did not follow up his photographic discovery of Eros with more precise naked-eye micrometer measurements until Tuesday, 16 August 1898. He may have waited until Tuesday to develop the plate that was taken the previous Saturday, or, because of some telescope tracking problems, perhaps he did not initially notice the Eros trail on his photographic plate.[21] Later, Julius Bauschinger, director of the Königliches Astronomisches Rechen-Institut in Berlin and editor of the Berliner *Astronomisches Jahrbuch*, a contemporary scientific journal, noted that Eros was likely one of many with orbits between Earth and the asteroid belt.[22]

As the number of known asteroids continued to grow and the first near-Earth asteroids (NEAs)[23] were added to the inventory, another debate began to smolder over what might happen if one of these stray bodies were to strike Earth—or what we might find if one had done so in the past.

Meteor Crater and the Moon

In November of 1891, Grove Karl Gilbert, chief geologist for the United States Geological Survey,[24] set out on what he described as a peculiar errand:

---

using the 12-inch aperture refractor made by Carl Bamberg at Berlin, then the largest telescope in Prussia.

21. Scholl and Schmadel, "Discovery Circumstances of the First Near-Earth Asteroid 433 Eros."

22. Ibid.

23. See appendix 2. The terms "near-Earth asteroid" and "near-Earth object" were not widely in use until the late 20th century.

24. Having served on several surveys of the American west, Grove Karl Gilbert joined the United States Geological Survey (USGS) when it formed in 1879 and had already published several pioneering field reports by the time he was appointed chief geologist in 1888.

"I am going to hunt a star."[25] He was referring to the buried remnant of an extraterrestrial body that—perhaps—had left its mark in the layers of sandstone and limestone of northern Arizona: a bowl-shaped depression known today as Barringer Meteorite Crater (or, more commonly, Meteor Crater). The relatively new idea that this particular crater might have been formed in an impact event had taken hold of Gilbert's imagination a few months earlier at the 1891 annual meeting of the American Association for the Advancement of Science (AAAS), where he heard the eminent mineralogist A. E. Foote describe meteoric iron fragments that had been discovered at the site. In the discussion that followed, Gilbert "suggested that this so-called 'crater' was like the depressions on the surface of the Moon produced by the impact of an enormous meteoric mass."[26]

This brief comment illuminates the many questions and lines of reasoning that intersect in any discussion of NEOs as historicized objects. Three months after the AAAS meeting, he would survey the crater himself and eventually set his sights on the lunar landscape, his thoughts on the population of bodies that might have left their distinctive marks there. In setting out for Arizona with the Moon in the back of his mind, Gilbert was wading into an already-longstanding debate over the origin of the lunar surface features: had they been formed by internal forces, like the relatively familiar action of volcanoes on our own planet, or were they scars left by foreign objects that had careened into the lunar surface from without? And if the latter case could be believed, what were these impacting bodies, and why were they seemingly absent from the skies today? This debate was inextricably tied to the even older problem concerning the origin of meteorites, which, as we saw in the introduction, hinged on the eventual acceptance that extraterrestrial objects, however small, could make their way through space to intercept Earth.

After hearing about the meteoric iron fragments, Gilbert sent a colleague to assess a possible impact origin for Meteor Crater, but the study resulted in no satisfying conclusion, and Gilbert arrived at the site himself at the end of

---

25. William Morris Davis, *Biographical Memoir: Grove Karl Gilbert* (Washington, DC: U.S. Government Printing Office, 1926), p. 183.

26. A. E. Foote, "Geologic Features of the Meteoric Iron Locality in Arizona," in *Proceedings of the Academy of Natural Sciences of Philadelphia*, vol. 43 (Philadelphia: Academy of Natural Sciences, 1891), p. 407. Until well into the 20th century, the word "crater" to describe a landform strongly implied a volcanic origin.

October in 1891.[27] With his USGS colleague Marcus Baker, he carried out two tests to attempt to distinguish between the two outstanding hypotheses for crater formation, steam explosion versus meteor impact. Both of these tests crucially relied upon the (now known to be erroneous) assumptions that, given an impact origin, the mass of the impacting body would be very large, made of iron (hence magnetic), and buried beneath the crater floor. Under these circumstances, the volume of the crater rim should exceed that of the cavity, part of that volume being occupied by the buried impactor. A detailed topographic survey could compare these volumes, and this was the first of Gilbert's tests. The other relied on a magnetic survey of the region to look for a buried magnetized mass. It only took a few days to see that neither test was looking favorable for the impact hypothesis, and Gilbert ultimately concluded, "It follows, under my postulates, that the great meteor is not in the hole...chief attention should be given to other explanations of the crater."[28]

In retrospect, it might seem obvious that Gilbert's two crucial tests would fail, not because the impact hypothesis was wrong, but because his assumptions about the buried "star" were incorrect. Today, Meteor Crater is known to have been formed by an impact about 50,000 years ago, in which a 50-meter nickel-iron body struck Earth while traveling at about 12 kilometers per second, impacting, and mostly vaporizing altogether. Under these circumstances, there could be no iron mass to be found beneath the crater floor. The developing understanding of impact physics as an essentially explosive process would make this clear by the middle of the 20th century, but for Gilbert, the evidence unequivocally ruled out the hypothesis he had originally favored, and he had to abandon it.[29]

This episode left such an impression on Gilbert that when he did eventually publish his results several years later, it was in the context of a philosophical argument. In his outgoing address as the president of the Geological Society of America in 1895, he related his theory of "The Origin of Hypotheses: Illustrated by the Discussion of a Topographic Problem," in which he presented the puzzle

27. Grove Karl Gilbert, "The Origin of Hypotheses, Illustrated by the Discussion of a Topographic Problem," *Science III*, 53 (1 January 1896): 5.

28. Grove Karl Gilbert, "Notes Made in Arizona, Oct. 22–Nov. 19, 1891," Index No. 51, Accession No. 3448, U.S. Geological Survey Field Records File, U.S. National Archives, hereafter cited as Notebook 51, 14 November 1891.

29. H. J. Melosh and G. S. Collins, "Meteor Crater Formed by Low-Velocity Impact," *Nature* 434, no. 7030 (2005): 157.

of the Arizona crater, the two rival hypotheses, his crucial tests, and the eventual conclusion as a lesson in how to properly address a geological problem.[30] In the case of Gilbert's "little limestone crater," he had gone to Arizona with a preferred theory in mind (impact) and by careful and correct scientific reasoning had completed the process of elimination, leaving with an altogether different conclusion, that the crater was a product of a steam explosion.[31] The entire episode amounted to an example of proper scientific conduct.

As we have seen, the very first time Gilbert heard of the Arizona crater, the lunar surface had leapt to his mind, and Meteor Crater has been linked to the Moon ever since. By Gilbert's time, Earth's Moon had become a planetary body with a surface that could be resolved in ever more detail, making it available for analysis using geological methods. Cooperation among and tensions between these disciplines touch on every aspect of near-Earth object studies and persist to the present day.[32]

After his surveys at Meteor Crater, Gilbert investigated several nearby volcanic features before returning to Washington. As he wrote in the Survey's *Annual Report*, his investigations of potential mechanisms for volcanic steam explosions led him "to give attention to the crateriform hollows of the moon, which have been ascribed by some writers to the impact of meteoric masses falling to its surface."[33] As he saw it, his experience as a geologist—"one who has given much thought to the origin of the forms of terrestrial topography"[34]— was perfectly suited to addressing the puzzle of the lunar craters.

Using projectile experiments in the laboratory, telescopic observations of the Moon, and analysis of lunar photographs,[35] Gilbert, with his geologist's

---

30. Gilbert, "The Origin of Hypotheses," p. 5.

31. Ibid., p. 10.

32. For example: "Interview of Eugene Shoemaker, June 27, 1994," transcript of proceedings before the U.S. National Aeronautics and Space Administration, Washington, DC, p. 6.

33. G. K. Gilbert, *14th Annual Report* (Washington, DC: U.S. Geological Survey, 1892–93), p. 187.

34. Grove Karl Gilbert, *The Moon's Face: A Study of the Origin of Its Features* (Washington, DC: Philosophical Society of Washington, 1893), p. 242.

35. Ibid., pp. 242–248. Taken at the Lick Observatory, these photographs were provided by the Smithsonian Institution, with further aid from George E. Hale at the Kenwood Physical Observatory in Chicago. Gilbert's own telescopic observations were made in August, September, and October of 1892 at the U.S. Naval Observatory, using the 26.5-inch refractor with a magnifying power of 400. This time period, which afforded 18 clear nights, allowed him to view the lunar surface through two full

eye, scrupulously inspected the lunar topography and assessed each of the major theories of crater formation. He opened the resulting study with a survey of key aspects of crater appearance and distribution, including a description of their varying abundance by size, uneven distribution between the rough highlands and smoother maria, and overlapping relationships. He also noted the differences in morphology between small craters and large ones, and described qualitatively the scaling of depth with diameter in each of the two size classes.[36] These observations provided evidence for or against the volcanic and meteoric hypotheses for crater origins.

Whereas his Arizona study had led Gilbert to consider a volcanic origin for Meteor Crater, this analysis convinced him of the opposite for the craters on the Moon. To begin with, he noted that the sizes of lunar craters vastly overshadow those of terrestrial volcanoes, and the enormous range of crater sizes is entirely unlike the distribution of Earth's features. An even bigger problem, Gilbert argued, was the vast disparity in form. The most common type of terrestrial volcano, the stratovolcano (e.g., Mount Vesuvius), with its tall sloping sides, relatively shallow caldera, and misshapen or altogether absent central cones, looks very different from the typical crater described at the beginning of his paper. "Thus," he concluded, "through the expression of every feature the lunar crater emphatically denies kinship with the ordinary volcanoes of the earth."[37]

Turning to the meteoric hypothesis, Gilbert cited Richard A. Proctor's *The Moon* (1873) as the oldest developed theory he could find, also mentioning an 1882 article by A. Meydenbauer and another theory "said to be contained in *Die Physiognomie de Monde*, by 'Asterios,' Nordlingen, 1879."[38] These theories faced difficult obstacles in Gilbert's view. First, the lunar craters were far too large to have been formed by any meteors that had ever been observed to

---

lunations, during which the terminator—where the illumination angle is lowest and the topography most exaggerated—passed over every part of the lunar disk.

36. Ibid., pp. 243–248.

37. Ibid., pp. 250–252. Gilbert also considered shield volcanoes and maars but found only the latter to be a plausible analog, and only then in the case of the smallest lunar craters. He also dismissed a tidal origin for the craters and a hypothesis involving ice buildup.

38. Ibid., p. 257; Richard Anthony Proctor, *The Moon: Her Motions, Aspect, Scenery and Physical Condition* (London: Longmans, Green, 1873); A. Meydenbauer, "Die Gebilde der Mondoberfläche," *Sirius* 15 (1882): 59–64; W. Thiersch and A. Thiersch ("Asterios"), *Die Physiognomie des Mondes* (Nördlingen: Beck, 1879).

strike Earth. While this difficulty could be resolved by supposing that the Moon's bombardment had occurred in the distant past, it hinted at a scale of catastrophe beyond imagining and required the additional assumption that any similar craters formed on Earth had been subsequently weathered away.

Another difficulty concerned the material properties of the surface and their behavior in a crater-forming collision, since lunar surface materials must somehow have been simultaneously plastic enough for huge cavities to be sculpted and rigid enough to support and retain the resulting landforms. Gilbert dismissed this issue by invoking the difference in scale between geological processes and laboratory experiments, noting that "under sufficient strains great bodies of rock both bend and flow."[39]

The final two obstacles faced by meteoric theories concerned the expected shapes and dimensions of the resulting craters. Gilbert argued that the volume of each lunar crater cavity should (according to the same assumption he made at Meteor Crater) be less than that of its rim, part of the interior being taken up by the buried impactor—but this was not consistent with lunar observations. However, his projectile experiments showed that the rebound response of the target material during impact could result in variations from this expected behavior, thus mitigating the third obstacle.[40]

Finally, common experience would lead one to expect varying crater shapes, ranging from perfectly circular to elliptical. From geometric arguments, Gilbert derived a distribution of impact angles (assuming impactors arriving uniformly from all directions) showing that the most common impact angle is 45 degrees, with significant numbers expected to strike at shallower angles. The lunar craters, however, are manifestly too uniformly circular to have been produced by such a distribution of meteor strikes. To address this problem, Gilbert introduced his "moonlet theory" of crater formation, which was suggested to him by the rings of Saturn. A similar ring, he posited, might have once orbited Earth, eventually coalescing to form our

---

39. Gilbert, *The Moon's Face*, pp. 258–259. Interestingly, Gilbert recognized that the energy associated with the cosmic velocities of meteors would be large enough to expect melting. He took this argument no further, however, explaining later that only a small portion of the impactor's kinetic energy would go to heating the impactor itself; the rest would go toward heating the target material, forming the crater, and modifying the motions of the Moon.

40. Ibid., pp. 259–260. Gilbert carried out his experiments at the physical laboratory at Columbia College in New York.

Moon. A natural consequence of this formation scenario would be the "scars produced by the collision of those minor aggregations, or moonlets, which last surrendered their individuality."[41] By first gathering the impacting bodies into a ring, Gilbert slowed them from a cosmic velocity to the escape velocity of the Moon, about 1.5 miles per second (2.38 kilometers per second) and allowed them to fall at much steeper impact angles. In this way, the distribution of expected ellipticities could be brought closer into alignment with his analysis of the shapes of craters measured on lunar photographs. Moreover, Gilbert asserted, many disparate observations could be explained by his theory, including the white streaks emanating from some craters and the fine "sculpture lines" radiating from Mare Imbrium, which he interpreted as the product of "the violent dispersion in all directions of a deluge of material—solid, pasty, and liquid."[42]

Toward the end of his exposition, in a section titled "Retrospect," Gilbert reflected on his own thought process in formulating his moonlet theory, invoking the same method of multiple working hypotheses that he would later emphasize in his 1895 presidential address to the Geological Society of America. Initially confronted with two hypotheses, and seeing how neatly the impact theory could explain lunar crater characteristics (in contrast to the volcanic theory, which could not), he was gradually converted "from the attitude of a judge to the attitude of an advocate."[43] Gilbert also noted that meteoric theories may have suffered from an unfair bias, since the plasticity of rocks subjected to extreme pressures lies far outside the realm of common experience.

"The Moon's Face: A Study of the Origin of Its Features" was published in the *Bulletin of the Philosophical Society of Washington* (where Gilbert's address had been given), reprinted in *Scientific American Supplement*, and circulated widely in abstract form. Despite this broad dissemination, however, it was largely ignored, and what response it generated—from astronomers and geologists alike—was overwhelmingly negative. Gilbert had known from the beginning that his investigation of the lunar surface would encounter some

---

41. Ibid., p. 262. To address the concern that a ring of moonlets would not produce a uniform distribution of impacts, he suggested that the impulses imparted to the Moon by the impacts themselves could have modified its rotation axis, thus randomizing the placement of craters (p. 275).
42. Ibid., pp. 261–285.
43. Ibid., p. 287.

pushback from lunar observers, who would disapprove of an outsider treading on their turf, and he was right.[44] The volcanic theory continued to dominate in astronomical circles well into the 20th century, in part because the appeal of a terrestrial analog rendered it something of a default scenario, which astronomers accepted "quite casually and uncritically."[45] As Gilbert's first biographer, William Morris Davis, put it, "The belief in the former volcanic activity of the moon is ingrained in nearly all our standard astronomical literature."[46]

One difficulty that persisted for the meteoric hypothesis in general concerned the size of craters relative to the supposed impactors. To take one example, responding to a paper by R. Parry at the November 1897 annual meeting of the British Astronomical Association, selenographer William Noble found it ridiculous to suppose that objects as large as some lunar craters would be "cruising about in space" only to strike the Moon and leave Earth and other planets unscathed. The minutes report that "[h]e did not think it necessary to discuss this paper seriously."[47] Noble's comments were cited by E. M. Antoniadi in a lively exchange that took place in the pages of *English Mechanic and World of Science* over the next few months. After advocating for the meteoric hypothesis and referencing Gilbert's moonlet theory, commenter "R. P." (very likely the same R. Parry whose paper had so incensed Captain Noble) was quickly outnumbered by his detractors, who argued that Earth's atmosphere would not have protected it from such huge impacts.[48]

If astronomers, for the most part, brushed Gilbert's meteoric theory aside and held on to the volcanic theory, the reaction from his own community of geologists was no better. In his review of *The Moon's Face* for *Science*, Joseph F. James remarked that "he has carried his studies away from things terrestrial and turned his eyes and his attention for a time to things celestial,"[49]

---

44. Gilbert, *The Moon's Face*, p. 242; see also *Scientific American Supplement* 37, no. 938–940 (1893–94); William Graves Hoyt, *Coon Mountain Controversies* (Tucson: University of Arizona Press, 1987), p. 65.

45. Hoyt, *Coon Mountain Controversies*, p. 68.

46. W. M. Davis, "Lunar Craters," *The Nation* 41 (1893): 343.

47. N. E. Green, "Report of the Meeting of the Association Held on November 24, 1897," *Journal of the British Astronomical Association* 8 (1897): 64.

48. E. M. Antoniadi, "Abnormal Planetary Observations," *English Mechanic and World of Science* 67, no. 1740 (29 July 1898): 547; R. P., "The Meteoric Theory of Lunar Formations," *English Mechanic and World of Science* 67, no. 1731 (27 May 1898): 335–336.

49. J. F. James, review of "The Moon's Face," *Science* 21 (1893): 305.

capturing what would become a familiar sentiment. "Theories of great ingenuity and variety," as Gilbert called them, were not generally looked upon with favor by the geologists of the 19th century, and the distant Moon could not help but prove a speculative topic.

Gilbert's work, first at Meteor Crater and subsequently on the lunar craters, serves to illustrate the attitudes among astronomers and geologists concerning the possibility of impact phenomena at the turn of the 20th century. The discovery that there were myriad objects in the solar system far smaller than the known major planets did not automatically translate into acceptance that those bodies might sometimes change Earth's surface. Instead, many scientists would continue to believe well into the 20th century that they played no role in Earth's history at all.

A new chapter for Meteor Crater would open with the new century, and with it, another look at the old problem of the craters on the Moon.

CHAPTER 2

RECOGNIZING IMPACT AS A GEOLOGICAL PROCESS

During the 20th century, geoscientists gradually came to accept that impacts by small solar system bodies, be they comets or asteroids, could alter Earth's surface. Understanding of how an impact event would rearrange the landscape evolved, too. Early thought had focused on impact mechanics, but the development of powerful explosives, their use in the world wars, and then nuclear weapons testing focused researchers on energy instead. Impact craters came to be understood as the result of explosive processes.

The beginnings of space exploration then altered the context of cratering research. Impact craters were found not just on Earth and its Moon, but on Mars and, still later, the moons of the other planets too. Cratering was ubiquitous. The creation of the National Aeronautics and Space Agency, and the parallel foundation of the United States Geological Survey's astrogeology branch, provided institutional homes for this research.

Meteor Crater: An Explosive Event?

Ten years after Gilbert's investigation of the lunar surface, the feature now known as Meteor Crater came to the attention of Philadelphia lawyer and mining engineer Daniel Moreau Barringer, who staked a mining claim at the site after examining the meteoric irons found in the area. Convinced (as Gilbert had been) that the majority of the iron impactor lay buried somewhere beneath the crater, he enlisted the aid of friend and physicist Benjamin C.

Tilghman, and in parallel papers in 1905–06, they laid out their case for an impact origin.[1]

Several lines of evidence introduced by these two papers and subsequent elaborations are recognized today as diagnostic features of the impact process. Tilghman described a huge abundance of fine silica flour made up of "minute fragments of clear transparent quartz with edges and points of extreme sharpness," suggesting an instantaneous blow. In the 1950s, samples of the metamorphosed sandstone would be found to contain phases of silica that can be formed only at the extremely high pressures and temperatures of impact. Drilling shallow shafts around the crater rim, Barringer and Tilghman found fragments of meteoric iron mixed into the layers of rock that stratigraphically must have come from great depth, indicating "absolute synchronism of the two events, namely, the falling of a very great meteor on this particular spot and the formation of this crater." They also noticed that the layers making up the rim formed an inverted stratigraphic sequence, "which is strongly suggestive of the ploughing effect of a projectile."[2] As for "the so-called crucial experiments of Professor Gilbert," Tilghman pointed out that Gilbert's comparison of the cavity and ejecta volumes failed to account for erosion of the rim material, while the null result of the magnetic survey could not discount a distribution of iron bodies, "each having sufficient coercive force of its own to be independent of the earth's inductive action."[3]

Barringer's involvement was largely motivated by what he saw as a potentially lucrative opportunity to mine the iron mass he believed to be beneath the crater, and he formed the company Standard Iron to investigate the matter. Over the next three decades, as they mapped and drilled a series of shafts to explore the area, Barringer worked tirelessly to convince the scientific world of the crater's impact origin (and therefore also of his business venture's

---

1. D. M. Barringer, "Coon Mountain and Its Crater," in *Proceedings of the Academy of Natural Sciences of Philadelphia*, vol. 57 (Philadelphia: Academy of Natural Sciences, 1906), p. 861; Benjamin Chew Tilghman, "Coon Butte, Arizona," in *Proceedings of the Academy of Natural Sciences of Philadelphia*, vol. 57 (Philadelphia: Academy of Natural Sciences, 1905), p. 888.

2. Barringer, "Coon Mountain and Its Crater," p. 875; D. M. Barringer, "Meteor Crater (formerly called Coon Mountain or Coon Butte) in Northern Central Arizona," *National Acad. Sci., Spec. Publ.* (Washington, DC: National Academy of Science, 1909), p. 6.

3. Tilghman, "Coon Butte, Arizona," pp. 888–889.

legitimacy). The first studies received mixed reactions from the scientific community, including a notable silence from the United States Geological Survey. As a mining engineer and businessman, Barringer was a relative outsider to the scientific community, and his attacks on the methodologies of one of the most prominent Survey geologists of the time were not well received. Gilbert himself never publicly reconsidered his earlier stance, a fact that significantly delayed the resolution of this conflict.[4]

Other responses endorsed Barringer's impact origin for the crater, but in ways not entirely to his liking. In 1908, George P. Merrill of the Smithsonian Institution hit upon what would become a severe obstacle to Barringer's impact theory:

> The failure thus far to find a large intact mass within the crater might be further explained on the ground that a considerable portion of it was volatilized by the intense heat generated at the moment of striking the surface.[5]

The possibility that the impactor itself might have been vaporized in the collision would come to form a major theme in subsequent debates over the physics of the impact process in general and Barringer's financial prospects in particular. Among astronomers, the idea was not exactly new, having been mentioned at least obliquely by Proctor in 1878.[6] The account that most likely influenced Merrill's thinking, however, was geologist Nathaniel Shaler's 1903 discussion of the possibility of impact. For lunar impacts in the absence of an atmosphere, he suggested that a meteoric body would impact the surface at high velocity, penetrate to a great depth, and "probably be volatilized by the very high temperature it would attain." Furthermore, the hot, expanding

---

4.   He did privately express to a colleague something to the effect of "he left me with the impression that he considered you and Tilghman had brought forward evidence that entirely changed the conclusions he had drawn regarding the origin of the crater." J. C. Branner to D. M. Barringer, 12 October 1906, Barringer Papers, quoted in Hoyt, *Coon Mountain Controversies*, p. 106.

5.   George P. Merrill, "The Meteor Crater of Canyon Diablo, Arizona: Its History, Origin, and Associated Meteoric Irons," *Journal of Geology* 16 (1908): 496.

6.   "[A]lmost every mass which thus strikes the Moon must be vaporized by the intense heat excited as it impinges upon the Moon's surface." R. A. Proctor, "The Moon's Myriad Small Craters," *Belgravia* 36 (1878): 164, quoted in Hoyt, *Coon Mountain Controversies*, p. 128.

gases of the newly vaporized impactor "would, in effect, explode, the gaseous products being cast forth from the opening made."[7]

Merrill brought Shaler's lunar idea down to Earth, where the atmosphere complicated calculations of the impactor's velocity. Additional estimates concerning the path, dimensions, and cohesiveness of the body that formed Meteor Crater would have to be made to determine with any certainty how much of the original iron body might have survived its entry and concussion.[8] These considerations were eagerly taken up by physicist Elihu Thomson in 1912, who argued that the iron impactor could have been slowed on its passage through the atmosphere and preserved from complete volatilization.[9] Concerned with the suggestion that there might prove to be no iron to mine after all, Barringer also adopted this argument and posited a meteoric swarm or the head of a comet with embedded iron fragments.

In addition to defending Barringer's impactor from volatilization, Thomson's paper also linked Meteor Crater even more strongly to the Moon, mentioning that "in fact if looked down upon from above [it] would appear like a lunar crater transplanted to earth."[10] This comparison, together with A. M. Worthington's 1908 study on splashes captured with high-speed photography, fired Barringer's imagination, leading him to observe the Moon for himself and elaborate his own impact theory for lunar craters in 1914.[11] Two

---

7. Nathaniel Southgate Shaler, *A Comparison of the Features of the Earth and the Moon* (Washington, DC: Smithsonian Institution, 1903), p. 12, quoted in Hoyt, *Coon Mountain Controversies*, p. 128. Shaler goes on to say that the volatilization of impactors should have caused the color of the lunar surface to be completely uniform, and the fact that the maria are still visible as distinctly dark regions argues against the impact hypothesis.

8. The angle of approach would have determined the length of the path taken through the atmosphere, with implications for the amount of drag experienced by the impacting body or bodies and, in turn, the final approach velocity and energy of impact.

9. Elihu Thomson, "The Fall of a Meteorite," in *Proceedings of the American Academy of Arts and Sciences*, vol. 47, no. 19 (1912), p. 728, quoted in Hoyt, *Coon Mountain Controversies*, p. 153 (see also pp. 132–140).

10. Thomson, "The Fall of a Meteorite," p. 730. According to Hoyt, *Coon Mountain Controversies*, p. 155, a 1909 paper was the first to make this direct comparison: F. Meineke, "*Der Meteorkrater von Canyon Diablo in Arizona und seine Bedeutung für Enstehung der Mondkrater,*" *Naturwissenschaftliche Wochenschrift* 8 (1909): 801–810.

11. As he put it, "Anyone who will make a careful study of our Arizona crater, and will then read Worthington's book, studying the diagrams he has made, and will then turn his attention to the lunar craters, cannot escape the conviction that the lunar craters

years later, he went on to tie both Meteor Crater and the craters of the Moon to the Chamberlin-Moulton planetesimal hypothesis, which provided a natural population of impactors that could have been responsible.[12]

These suggestions did little to threaten the volcanic theory's dominance as an explanation for the lunar craters, and in the same period, a barrage of assaults on the impact origin for Meteor Crater appeared from geologists reaffirming Gilbert's steam explosion hypothesis. For example, a 1916 review of Barringer's 1909 Meteor Crater paper in *Nature* emphasized unsolved difficulties with the theory, chief among them "the question of what has become of the vast mass of matter capable of producing the shattering impact."[13]

While the specter of volatilization threw doubt on the presence of a vast buried iron deposit (threatening the business side of operations at Meteor Crater while potentially addressing the mystery of the missing iron), it also offered the possibility of resolving one of the major obstacles to impact theories for the Moon: the circularity of lunar craters. As early as 1909, Nikolai A. Morozov articulated an impact hypothesis in which circular depressions were created by a symmetric explosive force.[14] The explosion theory of impact that

---

are impact craters." (Daniel Moreau Barringer, "Further Notes on Meteor Crater, Arizona," in *Proceedings of the Academy of Natural Sciences of Philadelphia*, vol. 76 (Philadelphia: Academy of Natural Sciences, 1914), p. 562. The reference is to A. M. Worthington, *A Study of Splashes* (New York: Longmans, Green, and Co., 1908).

12. Daniel M. Barringer, "A Possible Partial Explanation of the Visibility and Brilliancy of Comets," in *Proceedings of the Academy of Natural Sciences of Philadelphia*, vol. 68 (Philadelphia: Academy of Natural Sciences, 1916), p. 475; also quoted in Hoyt, *Coon Mountain Controversies*, p. 160.

13. *Nature* (1916): 1697, quoted in Hoyt, *Coon Mountain Controversies*, p. 161; see also N. H. Darton, "Explosion Craters," *Scientific Monthly* 3 (1916): 417–430; C. R. Keyes, "Coon Butte and Meteorite Falls in the Desert," abstract, *Bulletin of the Geological Society of America* 21 (1910): 773–774; C. R. Keyes, "Phenomena of Coon Butte Region, Arizona," abstract, *Science* 34 (1911): 29; "Scientists Study of Meteor Crater," *New York Times* (6 October 1912): 5.

14. N. A. Morozov, "Riddles of the Moon," *Vestnik Znanya* (1909), in Russian, p. 7, quoted in Grzegorz Racki et al., "Ernst Julius Öpik's (1916) Note on the Theory of Explosion Cratering on the Moon's Surface—The Complex Case of a Long-Overlooked Benchmark Paper," *Meteoritics & Planetary Science* 49, no. 10 (1 October 2014): 1851–1874, *https://doi.org/10.1111/maps.12367*. As he put it, the rarity of elliptical craters "indicates apparently that the meteorites at the time of their fall on the Moon exploded from self-heating, and, that is why, discarded the surrounding dust in all directions regardless of their translational motion in the same way as artillery grenades do when falling on the loose earth." This paper and later elaborations by Morozov made an

was only hinted at by discussions of volatilization came into its own during the decades surrounding the First World War (WWI), bringing into focus the role of impact cratering as an agent of geological change on a massive scale.

Wartime innovations and opportunities for cross-disciplinary work directly fueled several early elaborations of impact as an explosive process. In particular, parallel advancements in photography and flight, as well as the development of heavy artillery and trench warfare, provided unprecedented opportunities to view a cratered region of Earth from a perspective similar to that provided by observations and photographs of the Moon. Optical engineer and photography expert Herbert E. Ives, recruited to the U.S. Signal Corps in 1918 to help improve aerial photographic techniques, was struck by the visual similarity of bomb craters viewed from above and the cratered surface of the Moon. In 1919, he published his impact theory for lunar craters based on the crucial idea that the impactor's kinetic energy, abruptly converted into heat, would result in an explosion very much like the T.N.T. bombs developed in the war. The resulting crater would not be elliptical whatever the impact angle because "the shape of the cavity has no reference to the angle at which the bomb strikes, but takes its form from the symmetrical explosive forces."[15]

In France, Meudon Observatory astronomer Jean Bosler published a similar energy conversion argument in 1916, calculating in one example that an impactor might carry two million times more energy than a German "marmite" shell. Like Ives, Bosler included photographs of shell and bomb

---

impression on a young Estonian astronomer named Ernst J. Öpik, who published a brief article in 1916 referencing his impact-as-explosion theory and working out some of its consequences. This paper, later widely cited as the earliest expression of this type of impact theory, was published in Russian in an obscure journal and therefore received little attention. In fact, the key points attributed to Öpik were actually formulated by Morozov. In any case, neither work influenced the development of the explosion analogy, and this case provides a clear illustration of both the language barrier that existed between researchers working on opposite sides of the globe and the fact that the same problem is sometimes solved multiple times throughout history. Also see E. J. Öpik, "*Remarque sur le théorie météorique des cirques lunaires,*" *Bulletin de la Société Russe des Amis de l'Etude de l'Univers* 3, no. 21 (1916): 125–134.

15. Herbert E. Ives, "Some Large-Scale Experiments Imitating the Craters of the Moon," *Astrophysical Journal* 50 (1919): 249.

craters viewed from above and the lunar surface, published side by side for comparison.[16]

German meteorologist Alfred Wegener penned another impact theory for lunar crater formation in 1921, having gained ample experience in aerial observation after carrying out a series of balloon ascents for the Army Weather Service in Riga, Latvia, an area that had seen a barrage of heavy fighting in July and August of 1917. By that time, Wegener (better known for his theory of continental drift) had taken an interest in the origin of the Moon as well as a recent meteorite impact at Treysa. Greene suggests that the similarity of battlefield shell craters as viewed from above would not have been lost on him.[17]

Finally, New Zealand astronomer Algernon Charles Gifford published his "new meteoric hypothesis" in 1924, in which he made the impact-explosion analogy explicit:

> The fact that has not been taken into account hitherto in considering the meteoric hypothesis is that a meteor, on striking the surface of the Moon, is converted, in a very small fraction of a second, into an explosive compared with which dynamite and T.N.T. are mild and harmless.[18]

Another influence on Gifford may have come from the experiences of his colleagues serving in the First World War. While he was exempt from the draft himself, he maintained a brisk correspondence with friends and fellow astronomers overseas, including Charles J. Westland. In a letter dated

---

16. J. Bosler, "Les pierres tombées du ciel et l'évolution du système solaire," *Revue générale des Sciences* 27 (1916): 610–620. After the publication of Ives's paper, which was circulated in abstract form rather widely, Bosler reiterated his argument in another brief article: J. Bosler, "Trous d'obus et cirques lunaires," *L'Astronomie* 34 (1920): 52–56.

17. Mott T. Greene, *Geology in the Nineteenth Century: Changing Views of a Changing World* (Ithaca, NY: Cornell University Press, 1982).

18. A. C. Gifford, "The Mountains of the Moon," *New Zealand Journal of Science and Technology* 7 (1924): p. 135. Gifford had worked closely with A. W. Bickerton on the latter's "partial impact" theory for the formation of the solar system, which involved the collision of two stars, and he was familiar with his colleague's suggestion, expressed at the June 1915 meeting of the British Astronomical Association, that "the normal speed of a meteor in space where the Moon was at the present time would produce an explosive action on impact." Bickerton had relied on a combination of external impact and internal volcanic energy to produce the explosion he suggested, but in his own impact theory, Gifford realized that no volcanic forces were required.

11 April 1917, Westland wrote to Gifford that the shell craters pictured in aerial photographs looked "very like the lunar craters, with rampart and dark shadowed centres."[19]

These physical developments notwithstanding, reactions from astronomers (among whom the volcanic theory of crater formation was still very much entrenched) were overwhelmingly negative. One line of reasoning that appeared over and over in both the WWI-related explosion papers and responses to them was the presence or absence of a terrestrial analog. Ives, Bosler, Wegener, and Gifford all mentioned Meteor Crater as a possible example of an impact crater on Earth in support of their lunar impact theories. By that time, the disparity between the cratered Moon and the relatively blemish-free Earth had become ensconced as a fundamental problem for the impact hypothesis, and a positive identification of even a single terrestrial impact structure significantly influenced its acceptance.

For his part, far from welcoming this connection between Meteor Crater and the impact-explosion analogy, Barringer was incensed. Following the publication of Gifford's theory, suspicions grew that the impactor at Meteor Crater had been completely vaporized, leaving nothing behind. A proponent of the impact hypothesis for lunar craters but not an explosion model of that impact process, Barringer was placed in the ironic position of arguing against supporters won over by Gifford's arguments. In 1929, still unsuccessful in finding an iron deposit and increasingly desperate to show his financial backers that volatilization was not a foregone conclusion, he asked Forest Ray Moulton, an astronomer who had expressed his conviction of the crater's impact origin, to analyze the problem. The initial report that Moulton compiled kicked off an intense debate over the size of the impactor, the influence of the atmosphere in possibly slowing it down, and the partitioning of its kinetic energy.[20] This flurry of private correspondence culminated in a comprehensive second report from Moulton that reaffirmed his earlier conclusion: the impactor would not be found.[21]

---

19. Letter from Charles J. Westland to A. C. Gifford, 11 April 1917, Gifford-Bickerton Papers, MS-Group-1566, Alexander Turnbull Library, Wellington, New Zealand.

20. F. R. Moulton, "The Arizona Meteorite," 24 August 1929, Barringer Family Papers, Folder 1, Box 53, Department of Rare Books and Special Collections, Princeton University Library; hereafter cited as Barringer Papers.

21. F. R. Moulton, "Second Report on the Arizona Meteorite," 20 November 1929, Barringer Papers. Recognizing the conceptual difficulty of imagining such an

Barringer passed away of a heart attack the very next week, his death effectively ending the discussion and most subsequent financial endeavors regarding Meteor Crater. By this time, largely through the efforts of Barringer himself, the crater had come to be accepted by many eminent astronomers and geologists as the remnant of a collision with an extraterrestrial body—a near-Earth object. Reports of another potential impact structure at Odessa, Texas, had already surfaced in 1922 and 1926, and others were soon to follow.[22] At long last, the impact hypothesis had a terrestrial analog, and at least one near-Earth object was accepted as having left its mark on Earth.

The Tunguska Impact Event

Shortly after 7:00 a.m. on 30 June 1908 (00:14 UT), a powerful explosion occurred over the basin of the Podkamennaya Tunguska River in central Siberia. Although contemporary accounts differ somewhat, the event was described as an atmospheric fireball as bright as the Sun. From eyewitness accounts, the bolide likely had an entry angle of about 30 degrees with respect to the horizon and an arrival azimuth angle of about 110 degrees east of north. The terminal height of the explosion was estimated to be 5–12 kilometers, and the shock wave was accompanied by a strong blast of highly heated air.[23] Although not immediately correlated with the Tunguska event, seismic and pressure waves as well as geomagnetic disturbances were recorded in several widespread observatories, and bright nights, or prolongations of twilight, were reported throughout Eurasia.

For more than a decade, no notice of the Tunguska event was taken beyond local newspapers. It was 1922 before Russian geologist Leonid A. Kulik

---

outcome, Moulton wrote: "When we get in the domain of astronomical masses of hundreds of thousands of tons, we are like a child who, after playing a few times with his rattle, suddenly sees and seeks to grasp the moon. If it is explained to him that it is something of quite a different order from the things of his experience, he casts the statements aside as purely theoretical and, relying on his practical experience, reaches again, proves (to himself) the correctness of his views by the assertion that he has always held them and that all other children hold them" (p. 126). See also H. Jay Melosh, *Impact Cratering: A Geologic Process* (New York: Oxford University Press, 1989).

22. Hoyt, *Coon Mountain Controversies*, pp. 233–237.

23. G. Longo, "The Tunguska Event," in *Comet/Asteroid Impacts and Human Society* (Berlin: Springer, 2007), pp. 316–320.

connected the bright nights seen throughout Eurasia with the event and 1925 before the first reported connection was made between the Tunguska event and several anomalous seismic events reported for that day.[24]

Kulik was the first scientist to make an on-site investigation, arriving in 1927. Born in Tartu, Estonia, in 1883, Kulik had spent 1911–12 in a czarist prison for revolutionary activities and would later die of typhus in a World War II Nazi prison camp in April 1942. In early 1918, after the 1917 October Revolution, Kulik went to the Soviet Academy of Sciences in Petrograd (now Saint Petersburg) and began his work on meteorites. In 1921, he was assigned a two-year expedition to gather information on a giant meteor event witnessed in Siberia in June 1908. Working out of a railcar designed for the purpose, Kulik visited many locations in Siberia seeking information on the location of the expected meteorite. However, he soon determined that the most likely location of the event was in the Tunguska area, a region so remote that he was unable to reach the area during this first expedition. He then persuaded the Soviet government to fund another expedition to the Tunguska region, based on the prospect of finding meteoric iron that could be mined to aid Soviet industry.

In 1927, after an arduous two-month journey, Kulik succeeded in visiting the Tunguska blast site and was struck by immense destruction that far exceeded his expectations.[25] A widespread area of the forest had been uprooted and flattened. As a result of this research expedition to the Tunguska blast site and his subsequent returns in 1928, 1929–30, and 1938–39, Kulik attributed a number of boggy depressions to meteorite impacts and made an effort to excavate these depressions to recover the meteorites. Despite these excavations, as well as geodetic and magnetic surveys of the area, Kulik found no meteorites either on the ground or at the bottom of the depressions.[26] However, the on-site investigations and those that followed revealed a curious pattern: the impact blast region outlined a butterfly-shaped pattern of some 80 million felled trees—a great number of them burnt—over an area estimated at 2,150 square kilometers.

24. Ibid.

25. John Baxter and Thomas Atkins, *The Fire Came By: The Riddle of the Great Siberian Explosion* (Garden City, NY: Doubleday & Company, 1976), pp. 62–65.

26. Ursula Marvin, "Leonid Alexyevich Kulik," in *Biographical Encyclopedia of Astronomers*, ed. Thomas Hockey, Virginia Trimble, and Thomas R. Williams (New York: Springer, 2007), pp. 661–662.

The cause of the Tunguska event has been attributed most often to the atmospheric impact of either a comet or an asteroid, though unfortunately the nature of the microparticles found in the soil and tree resins near ground zero do not allow a certain discrimination between those two possibilities. Russian scientists have seemed to favor a cometary explanation, while Western scientists have favored an asteroid impact. The English astronomer F. J. W. Whipple (not to be confused with the American F. L. Whipple) and the American astronomer Harlow Shapley suggested as early as 1930 that the Tunguska event was due to a comet.[27] It would be another two years before the first Earth-crossing asteroid, 1862 Apollo, would be found, so at the time, an asteroid collision may not have been thought possible. The relatively low altitude of the explosion (about 10 kilometers) would favor an asteroidal origin, though, as asteroids are less friable. They also turn out to be far more common in near-Earth space.

Because of the uncertainty of its origin, the Tunguska event has been used repeatedly as evidence for several highly implausible hypotheses, including collisions by antimatter, miniature black holes, unidentified flying objects (UFOs), and even an overly powerful laser signal from intelligent beings on a planet orbiting the star 61 Cygni. Locals, who were reluctant to talk about the event, believed the blast was a visitation by the god Ogdy, who had cursed the area by smashing trees and killing animals.[28]

In an important sense, Tunguska was a non-event. Unnoticed by Western scientists for a couple of decades, it represented an explosive event that did not leave an impact crater.[29] Later generations of scientists would use it to help

------

27. Although he did not explicitly mention a comet, Whipple supposed that the meteor had been traveling about the Sun on a parabolic orbit. Whipple also commented, "It is most remarkable that such an event should occur in our generation and yet be so nearly ignored." F. J. W. Whipple, "The Great Siberian Meteor and the Waves, Seismic and Aerial, Which It Produced," *Quarterly J. Roy. Meteorol. Soc.* 56 (1930): 287–304. Based upon the impact time and the approach direction, Shapley suggested that the event could have been due to a fragment of periodic Comet Pons-Winnecke. Harlow Shapley, *Flight from Chaos. A Survey of Material Systems from Atoms to Galaxies* (New York: McGraw Hill, 1930), p. 58.

28. M. Susan Wilkerson and Simon P. Worden, "On Egregious Theories—The Tunguska Event," *Quarterly Journal of the Royal Astronomical Society* 19 (1978): 282–289.

29. The small bowl-shaped lake Cheko, located some 8 kilometers north-northwest of ground zero, has been suggested as the possible impact site of a small meteoritic fragment from the Tunguska impact, but no fragment has yet been retrieved from the

understand the energy deposited in the atmosphere as small bodies entered—be they comets or asteroids.[30] But it did not immediately influence thinking about cratering.

Impact Cratering as a Geological Process

In the 1930s, distinct themes in cratering research began to emerge, leading over the next few decades to the consensus that collisions were—and are—an ongoing process shaping both the Moon and Earth. These strands of inquiry included the geological investigation of terrestrial impact and cryptovolcanic features, the impact origin of lunar craters, the mechanics of cratering as a physical process, and the identification of unique signatures associated with impact.

While many geologists and astronomers had been convinced by 1929 of Meteor Crater's cosmic origin, the same could not yet be said for lunar craters, as illustrated by the limited success of the Carnegie Committee for the Study of the Surface Features of the Moon. Founded in the mid-1920s and directed by petrologist Frederick E. Wright, the interdisciplinary committee was composed of astronomers, physicists, and geologists and was aimed at definitively settling the question of the origin of lunar craters. While most members of the committee supported the volcanic theory, Wright eschewed the use of terrestrial analogs, focusing instead of determining the nature of lunar materials.[31] Unfortunately, Wright's approach, which focused on polarization

---

lake bottom, and evidence has been raised both for and against this hypothesis. See: G. S. Collins et al., "Evidence That Lake Cheko Is Not an Impact Crater," *Terra Nova* 20 (2008): 165–168; Luca Gasperini, Enrico Bonatti, and Giuseppe Longo, "The Tunguska Mystery," *Scientific American* (June 2008): 80–86.

30. There are still arguments over the magnitude of the Tunguska event. A recent NASA workshop organized by David Morrison of NASA Ames Research Center found an explosive energy of 10–20 megatons of TNT, roughly the same as the Barringer event that formed Meteor Crater. See David Morrison, "Tunguska Workshop: Applying Modern Tools To Understand the 1908 Tunguska Impact," NASA TM-220174, December 2018, *https://ntrs.nasa.gov/archive/nasa/casi.ntrs.nasa.gov/20190002302.pdf* (accessed 18 June 2019).

31. In his words, "Each observer of the moon reasons by analogy from the terrestrial phenomena with which he is acquainted and seeks thus to explain the lunar surface features…. In this process he may, if so inclined, allow his imagination free play and draw almost any conclusion he may fancy at the moment." Wright to Gilbert, 2 March 1925, Box 1, Wright Papers, Huntington Library.

measurements, topographic surveys, and photographic techniques, yielded few results and cut him off from other cratering research going on in the United States at the time.[32] Although the Committee's operations had fizzled by the early 1940s, they represent the first explicit attempt by an institution to marshal expertise spanning multiple disciplines for the study of lunar craters, beginning a tradition that would continue throughout the 20th century.

The 1920s and 1930s also saw developments in both asteroid and meteor research. The number of known asteroids had blossomed from 131 in 1878 to 250 in 1885, and in the 1930s, Mount Wilson Observatory astronomer Walter Baade estimated that about 44,000 would be detectable by the 100-inch reflector. At Yale University and the University of California, Berkeley, in particular, interdisciplinary work on orbital mechanics, including resonances, as well as classification into asteroid families, brought physical, mathematical, and astrophysical methods into play. In his 1938 dissertation, Fletcher Guard Watson, Jr., presented the first estimate of the collision frequency for Earth-crossing asteroids: one in every hundred thousand years.[33] In the meantime, another interdisciplinary project got under way at Harvard University under the direction of Harlow Shapley. Convinced, along with the majority of astronomers, that meteors were primarily of interstellar origin, Shapley put together the Harvard Arizona Meteor Expedition to home in on the average velocity of meteors. The expedition itself, directed by Ernst Öpik and Fred L. Whipple, ran from 1930 to 1932 and used visual methods to track meteor velocities. Öpik found a majority above 42 kilometers per second, indicating hyperbolic (extrasolar) orbits, but this conclusion was later challenged by Whipple's synchronized photographic survey (1934–36), indicating that the overwhelming majority of meteors were indeed solar system objects.[34]

Starting with the Odessa crater in Texas, the number of suspected terrestrial impact structures also skyrocketed over the course of a few short years. I. A. Reinvaldt described the Kaarlijarv crater in Estonia as the result of an explosion caused by a meteorite impact in 1927, just as Kulik's results from his Siberian expeditions were shedding light on the Tunguska explosion and its

32. Ronald E. Doel, *Solar System Astronomy in America, Communities, Patronage, and Interdisciplinary Science, 1920–1960* (1996), pp. 159–161.

33. Ibid., pp. 17, 31.

34. Fred L. Whipple, "The Incentive of a Bold Hypothesis: Hyperbolic Meteors and Comets," *Annals of the New York Academy of Sciences* 198, no. 1 (1972): 219–224, *https://doi.org/10.1111/j.1749-6632.1972.tb12725.x.*

vast area of devastation. In 1931, a collection of craters with surviving meteorites nearby was discovered at Henbury Station in Australia, and a circular lake in Ghana (Ashanti crater) was described as the result of impact. A year later, the double craters at Wabar in Arabia were recognized as meteoritic, and in a 1933 paper, L. J. Spencer argued for the impact origin of the Campo del Cielo craters in Argentina, as well as a half dozen other sites.[35]

In the same paper, Spencer emphasized a new geological marker for terrestrial impact sites: the nickel-iron spherules he found embedded in slaggy glass samples.[36] Condensation from a vapor cloud, he reasoned, indicated temperatures beyond any involved in known volcanic processes. More diagnostic signatures of impact followed, including the local shattering and uplift associated with roughly circular "cryptovolcanic" structures, which John D. Boon and Claude C. Albritton attributed to impact in 1936–37,[37] and the identification of shatter cones as impact-oriented shock features by Robert S. Dietz in 1947.[38]

Alongside these developments in terrestrial impact identification, efforts to understand the Moon's history of bombardment continued to unfold. Dietz

---

35. Ursula B. Marvin, "Impacts from Space: The Implications for Uniformitarian Geology," in Special Publications 150.1 (London: Geological Society of London, 1999), p. 100; Melosh, *Impact Cratering*, p. 7.

36. L. J. Spencer, "Meteoric Iron and Silica-Glass from the Meteorite Craters of Henbury (Central Australia) and Wabar (Arabia)," *Mineralogical Magazine and Journal of the Mineralogical Society* 23, no. 142 (September 1933): 387–404, *https://doi.org/10.1180/minmag.1933.023.142.01*.

37. Ibid., pp. 7–8. These roughly circular, uplifted and severely faulted structures were termed "cryptovolcanic" by Walter H. Bucher. Boon and Albritton applied their impact hypothesis to the Vredefort Dome in South Africa as an explanation for the local shattering and uplift in the absence of volcanic materials. It also provided a testable criterion for impact structures because the deformation should disappear at depth.

38. Robert S. Dietz and Louis W. Butler, "Shatter-Cone Orientation at Sudbury, Canada," *Nature* 204 (1964): 280–281. Bucher had borrowed the term "cryptovolcanic" from W. Branca and E. Fraas, who had found conical, striated structures at Steinheim Basin in Germany in 1905. Studying the Kentland cryptovolcanic structure in Indiana, Dietz argued that these "shatter cones" are shock features that form only in impacts, the nose of each cone generally pointing toward the center of the crater. He later used this property to show that shatter cones at the Sudbury Basin in Ontario pointed to the center when post-impact deformation of the rocks was taken into account, and he discovered shatter cones at the Vredefort Dome as well. Marvin, "Impacts from Space," pp. 100–103; Melosh, *Impact Cratering*, p. 8.

published his impact theory for lunar craters in 1946, pointing out that the morphologies and distribution of craters do not lend themselves to the volcanic theory and that the internal temperatures required for volcanism could not have been supported by the small, cooling Moon.[39] In the same period, Harvard geologist Reginald V. Daly published three papers supporting the formation of Earth from meteoric accretion, describing the early history of the Moon and reinforcing Boon and Albritton's classification of Vredefort Dome as an impact feature.[40] In 1942 and 1943, Ralph B. Baldwin had attempted to have his own meteoric hypothesis published in the *Astrophysical Journal*, but it was rejected and was published in *Popular Astronomy* instead.[41]

These parallel lines of research—terrestrial and lunar—converged as impact was discussed increasingly in terms of energy. Moving in 1946 to work on the proximity fuse at Johns Hopkins's Applied Physics Laboratory, Baldwin put his spare time and access to classified ordinance data to work in developing a quantitative energy scaling for chemical explosions and meteorite impacts, showing that their depth-to-diameter ratios lie on a smooth logarithmic curve and therefore suggesting a common mode of formation. *The Face of the Moon*, published in 1949, brought together lines of reasoning from physics, astronomy, geology, geophysics, and meteorology, and its positive reception—winning over, for example, Harold C. Urey, who was later instrumental in fixing the Moon as a primary target in space exploration—marked a turning point in the acceptance of an impact origin for lunar craters.[42]

If the first decades of the 20th century were characterized by tentative interdisciplinary alliances, the 1950s and 1960s saw a dramatic turn toward institutional backing of cratering research. In Canada, Carlyle S. Beals, director of

39. Robert S. Dietz, "The Meteoritic Impact Origin of the Moon's Surface Features," *Journal of Geology* (1946): 359–375; Marvin, "Impacts from Space," p. 101.

40. Reginald A. Daly, "Meteorites and an Earth-Model," *Geological Society of America Bulletin* 54.3 (1943): 401–456; Reginald A. Daly, "Origin of the Moon and Its Topography," *Proceedings of the American Philosophical Society*, vol. 90.2 (Philadelphia: American Philosophical Society, 1946), pp. 104–119; Reginald A. Daly, "The Vredefort Ring-Structure of South Africa," *Journal of Geology* (1947): 125–145; Doel, *Solar System Astronomy in America*.

41. Ralph B. Baldwin, "The Meteoritic Origin of Lunar Craters," *Popular Astronomy* 50 (1942): 356; Doel, *Solar System Astronomy in America*, pp. 162–163.

42. Doel, *Solar System Astronomy in America*, pp. 164–165; Don E. Wilhelms, *To a Rocky Moon—A Geologist's History of Lunar Exploration* (Tucson, AZ: University of Arizona Press, 1993), p. 19.

Dominion Observatory in Ottawa, initiated an ambitious national program to look for fossil craters in the Canadian Shield and to assess whether Earth's impact history could be distilled from the geologic record. By 1963, using aerial photographic surveys to identify the most promising targets and following up with gravity, seismic, magnetic, and drill core studies, the program quickly identified three new Canadian craters that neatly fell on Baldwin's depth-to-diameter trend. One of these, 13-kilometer-diameter Deep Bay Crater, closed a gap in the size range of known craters, underscoring Baldwin's suggestion that impact cratering is a fundamental geologic process.[43]

In the United States, another institutional shift was occurring, largely thanks to a new impetus: the race to the Moon. Joint U.S. and Soviet participation in the International Geophysical Year (1957–58) precipitated Sputnik's launch and the founding of NASA, and the announcement of the lunar program in 1961 established spaceflight as a major government-funded objective. Yet the notion that American ambitions to land on our satellite would be intimately tied to geological—and more specifically, impact-related—investigations was far from a foregone conclusion. NASA's scientific aspirations for the Moon can be credited in large part to the efforts of a passionate geologist who had already turned his gaze to the sky, noting the striking similarities between lunar and terrestrial craters.

Shoemaker and the Moon

On 27 March 1962, a crowd of raucous USGS geologists in Menlo Park, California, laughed at a skit depicting "Dream Moonshaker," a hapless Survey colleague standing on the Moon and nattering on about impact craters while hot lunar lava burned his feet. The subject of the parody, Eugene M. "Gene" Shoemaker, was not present to witness the spoof, but he was all too familiar with the kind of resistance to lunar and impact studies it represented within the geological community. Shoemaker first became interested in studying lunar craters and understanding the geology of the Moon in the early 1950s. While working for the USGS during a break from his Ph.D. thesis in 1953, he joined the Atomic Energy Commission's Megaton Ice-Contained Explosion (MICE) project, which explored the possibility of creating plutonium by

---

43. Marvin, "Impacts from Space," p. 102; Doel, *Solar System Astronomy in America*, p. 174.

containing the explosion of a nuclear device in a blanket of ice, or—more feasibly—within a relatively pure deposit of salt. This project led him to study both shock wave theory and diatreme volcanism as a means of understanding the ejecta distribution left by underground explosions.[44]

To that end, he also set his sights on Meteor Crater in Arizona. His first glimpse of the crater had been brief and hurried, but "an overwhelming sight" nonetheless, as Carolyn Shoemaker recalled. On an impulsive side trip on the way to Grand Junction, the Shoemakers found that they could not afford the admission fee to see the crater, and they drove down an access road to peer over the rim just as the Sun was setting. After examining a sample of pumiceous silica glass two years later, Shoemaker noted that the temperature required to fuse the quartz far exceeded any achieved by a volcanic agent, and when Project MICE presented the opportunity to study the crater more thoroughly, he eagerly set about mapping its structure.[45]

Continuing his study of underground explosions, Shoemaker followed up his survey of Meteor Crater with a visit to the nuclear bomb test craters Jangle U and Teapot Ess. His comprehensive comparison of the structure of these artificial craters to that of the larger Meteor Crater—including the inverted stratigraphy ("overturned flap") noted by Barringer and Tilghman half a century earlier—convinced him that the one was an "eerily close scaled-up" version of the other. These observations led Shoemaker to develop his own theory of cratering mechanics, which forms the foundation of modern impact physics.[46]

In parallel with these structural revelations, Shoemaker and others were establishing a new mineralogic marker for impact processes. Coesite, a high-pressure phase of silica, had been discovered in 1953 and later found in samples from Meteor Crater by Edward Chao. After mapping the distribution of coesite-bearing rocks in Arizona, Shoemaker and Chao also discovered its presence at the Ries Basin in Germany. A second high-pressure phase of silica,

---

44. David H. Levy, *Shoemaker by Levy—The Man Who Made an Impact* (Princeton: Princeton University Press, 2002).

45. Ibid., pp. 69–80.

46. Ibid.; Melosh, *Impact Cratering*, p. 9; E. M. Shoemaker, "Penetration Mechanics of High Velocity Meteorites, Illustrated by Meteor Crater, Arizona," *Report of the International Geological Congress, XXI Session, Part XVIII* (Norden, 1960), pp. 418–434; Eugene Merle Shoemaker, *Geological Interpretation of Lunar Craters* (Menlo Park, CA: U.S. Geological Survey, 1962).

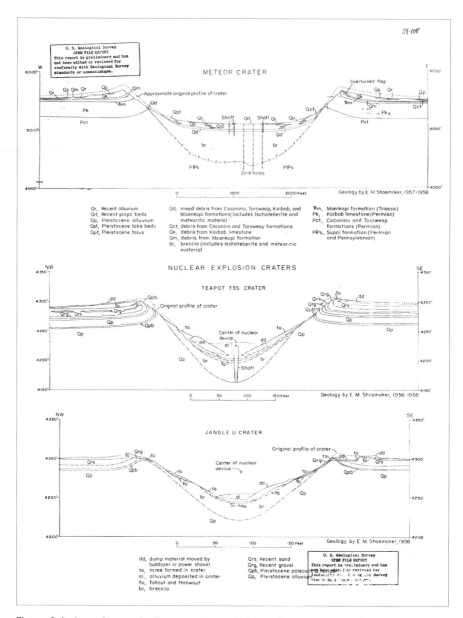

Figure 2-1. Gene Shoemaker's comparison of Meteor Crater to two nuclear test craters in Nevada, Teapot Ess and Jangle U, 1959. (From Eugene M. Shoemaker, "Impact Mechanics at Meteor Crater, Arizona," USGS Open File Report 59-108, 1959, DOI: 10.3133/ofr59108)

stishovite, was identified in 1961 and found at Meteor Crater by Chao a year later. The introduction of a mineralogical diagnostic tool offered the potential to quell debates over ambiguous structures, both on Earth and on the Moon.[47]

Turning again to the sky, Shoemaker applied his new theory of cratering mechanics to the prominent lunar crater Copernicus in 1962, making a strong case for its impact origin. He also noted the significance of overlapping ejecta blankets and secondary craters—smaller craters formed by material thrown out by the initial impact—as a novel means of piecing together a relative history of impact events. The dual advances represented by the recognition of coesite and stishovite as diagnostic markers of impact and the potential of the cratering process itself to record relative ages marked a change in the wind for Shoemaker's impact advocacy. Although his initial efforts to secure support for a lunar mapping endeavor had failed in 1956, they came to fruition four years later in 1960 with the creation of the Branch of Astrogeological Studies within the USGS at Menlo Park. In another two years, while his colleagues there were poking fun at "Dream Moonshaker," Gene Shoemaker was focused on moving his headquarters to Flagstaff, Arizona, preparing for a space program that would invest in and rely on a geologic understanding of the impact process.

As the formation of the Astrogeology Branch attests, the establishment of NASA in 1958 and the announcement of the Apollo project in 1961 both galvanized and supported work on impact cratering in the decades that followed. This was due in part to overestimates of the meteoroid flux by early satellite measurements, perceived to be a major threat to space vehicles. Understanding of cratering mechanics was bolstered by postwar research on high-velocity impact and explosion mechanics, leading to new computational and experimental methods, including the first numerical simulation of the Meteor Crater impact in 1961 and the establishment of the Vertical Gun Range at NASA Ames in 1965.[48] By the end of the decade, astronauts had attested to the dominance of impact cratering on the Moon, the far side of the

---

47. Doel, *Solar System Astronomy in America*; Edward C. T. Chao, Eugene M. Shoemaker, and Beth M. Madsen, "First Natural Occurrence of Coesite," *Science* 132, no. 3421 (1960): 220–222; Eugene M. Shoemaker and Edward C. T. Chao, "New Evidence for the Impact Origin of the Ries Basin, Bavaria, Germany," *J. of Geophys. Res.* 66, no. 10 (1961): 3371–3378; E. C. T. Chao et al., "Stishovite, SiO2, a Very High Pressure New Mineral from Meteor Crater, Arizona," *J. of Geophys. Res.* 67, no. 1 (1962): 419–421.

48. Melosh, *Impact Cratering*, p. 11.

Moon had been imaged, and Mariner IV had sent back photographs revealing a heavily cratered Mars.[49] The impact process—once considered inconceivable—was beginning to be recognized as a fundamental agent of change in the solar system.

49. Robert B. Leighton et al., "Mariner IV Photography of Mars: Initial Results," *Science* 149, no. 3684 (6 August 1965): 627–630.

CHAPTER 3

FINDING AND CHARACTERIZING NEAR-EARTH OBJECTS THROUGH 1990

If the realization that impacts were a fundamental geological process took decades to mature, so did the ability to identify and understand the nature of impactors. Sky surveys designed to discover asteroids and comets were essential to advancing those aspects of science. So too were efforts to understand what small solar system bodies actually were. Early on, astronomers assumed that they were remnants left over from the solar system's formation, but that remained to be proven. And answering that question still left open many others. What were they made of? Were they rocks? Were they balls of dust and ices? The effort to understand the nature of small solar system bodies was known as characterization and employed numerous methods as the 20th century unfolded.

Surveying for Near-Earth Asteroids in the Age of Photography

The first major systematic search for near-Earth asteroids was the Palomar Planet-Crossing Asteroid Survey (PCAS) undertaken by Eleanor "Glo" Helin and Eugene Shoemaker, both of the California Institute of Technology, in 1973 using the Palomar Observatory 0.46-meter (18-inch) Schmidt telescope in southern California. It was funded by NASA's planetary astronomy program, and the stated goal was to discover and obtain precise orbits for a sufficient number of planet-crossing asteroids in order to estimate the population of various classes of these objects.[1] At the time, Shoemaker believed

---

1. Their results were summarized in E. F. Helin and E. M. Shoemaker, "The Palomar Planet-Crossing Asteroid Survey, 1973–1978," *Icarus* 40, no. 3 (1 December 1979): 321–328, *https://doi.org/10.1016/0019-1035(79)90021-6.*

Figure 3-1. Eleanor Helin at the Mount Palomar 18-inch Schmidt telescope.
(Photo courtesy of Caltech Archives and Bruce Helin)

that around two thousand 2-kilometer or larger asteroids might be in Earth-crossing orbits and thus be future Earth impactors. But of course, there would be many more smaller objects as well. Success came slowly, and it took six months before they discovered a highly inclined Apollo asteroid (5496) 1973 NA on 4 July 1973. Helin commented, "There was a lot of sweat and tears for each observing run."[2] The first female observer permitted to use Mount Palomar's facilities regularly, she was initially able to acquire only seven pairs of films per night. On the evening of 7 January 1976, Helin discovered 2062 Aten, the first member and namesake for the orbital class of near-Earth objects whose orbital periods are less than one year, meaning that they orbit, on average, inside Earth's orbit. The PCAS efforts during the first five years netted only about 1–3 per year. But during the period from 1973 through the late 1980s, the PCAS survey was responsible for discovering a large percentage of all near-Earth asteroids.

The PCAS survey imaged 8 to 10 fields of sky twice per night, once with a 20-minute exposure and once with a 10-minute exposure using Kodak IIa-D film with a yellow filter. The PCAS threshold magnitude for the detection of fast-moving objects was initially about 15.5 but improved to about 17.5 in the 1990s. Each 6.5-inch circular film covered an 8.75-degree circular field of view. Upon development of the films, the 20-minute exposures were scanned manually with a binocular microscope for the trailed images that would indicate a relatively fast-moving object on the fixed star background. Any trailed images found would then be compared with the 10-minute exposure for verification. After a trial period of a few years, the PCAS team moved to a more efficient method of detecting NEO images on their films in 1981. Using faster emulsion films, two shorter exposures of approximately 6 minutes were recorded a half hour apart, so any near-Earth asteroid would no longer have obvious trailed images, but these two images would be offset from one another on the two films because of the asteroid's motion between the two exposures. There would be no offsets for the images of the background stars and galaxies that did not move from one film to the next. When the films were viewed through a stereomicroscope, a near-Earth asteroid appeared as one image "floating" above or below the background, depending upon the object's direction of motion. Initially, all films exposed in one night were examined

2.   Quoted in David Levy, *Shoemaker by Levy: The Man Who Made an Impact* (Princeton: Princeton University Press, 2000), p. 167.

before the following night's observations began and each new discovery was followed up, or reobserved, for as many lunations as the diminishing brightness would allow.[3]

During the early years of the PCAS survey, approximate positions of newly discovered objects were obtained during the observing run by plotting their positions on Palomar Sky Survey prints and estimating the right ascension and declination from overlay grids. During each run, these positions were then phoned to the International Astronomical Union's Minor Planet Center (MPC), which was located at the Cincinnati Observatory until 1978, when it was moved to the Smithsonian Astrophysical Observatory in Cambridge, Massachusetts.[4] The Minor Planet Center would then put out a telegram to alert observers worldwide that follow-up observations were needed to prevent the object from being lost. At PCAS, more precise positions were measured at the end of the run using the measuring engine at the Carnegie Institute facility in Pasadena, California, and these revised position measurements were then sent to the MPC and also to James Williams at the Jet Propulsion Laboratory, who provided ephemerides for future observations.[5] Once an interesting object was discovered, Williams would telephone the future ephemeris positions to a group of interested astronomers, who would then schedule telescope observations to provide additional follow-up position measurements. He also phoned ephemerides to observers who could measure the object's spectral and photometric properties. Once a good preliminary orbit was obtained, search ephemerides could be generated for earlier times as well, and searches

3.   David H. Levy, who observed with Shoemaker as well, explained the details of observing in his biography of Gene Shoemaker. See Levy, *Shoemaker by Levy*, pp. 167–174.

4.   For a brief history of the Minor Planet Center, see Brian G. Marsden, "The Minor Planet Center," *Celestial Mechanics* 22 (1980): 63–71.

5.   By the time Kenneth Lawrence began work at PCAS in the late 1980s, a measuring engine was located at Palomar and at JPL so that the positions measured at Palomar could be remeasured upon returning from observing runs. Lawrence interview by Yeomans and Conway, 19 May 2016, transcript in NASA History Division's Historical Reference Collection (HRC). Not only was the scanning of films and photographic plates laborious, but the measurement of celestial positions (right ascension and declination) using the measuring engine was as well. Both asteroid and nearby stars were measured, with the stars providing comparisons with known positions. When CCDs came along, the positions of asteroid and star images on the array of CCD detectors replaced the measuring engine. Digital files of star positions helped automate the determination of celestial coordinates (astrometric data).

for pre-discovery images were conducted using the films taken in earlier survey runs. These pre-discovery images, together with previously unattributed observations in the archives of the Minor Planet Center, would often allow definitive orbits for these objects to be computed.[6]

The PCAS survey personnel and methodology changed slightly in 1980, when they acquired a new stereomicroscope designed and built by McBain Instruments to ease the process of finding asteroid trails on the film. But searching with the new instrument was tiring on the eyes, and Shoemaker's wife, Carolyn, who had been primarily a homemaker until their three children had moved away, began to help out. Carolyn showed a particular talent for using the device to find fast-moving objects. Later, she became a proficient observer as well.

In 1982, Eleanor Helin and the Shoemakers went their separate ways. One reason for the split may have been that Helin felt she was not receiving enough credit for her discoveries and Shoemaker may have felt she was not giving enough credit for discoveries to the students who assisted her. Shoemaker, who controlled the project funds, transferred them to JPL, and arrangements were made for Helin to be moved there as well.[7] Mom and Pop Shoemaker, as they jokingly referred to themselves, moved back to the U.S. Geological Survey in Flagstaff, Arizona, and Gene set up a new survey program called the Palomar Asteroid and Comet Survey (PACS). With part-time help from Henry Holt and David Levy, the Shoemakers continued their photographic search program until the mid-1990s.[8] Both groups continued to use the Palomar 0.46-meter Schmidt telescope for their respective surveys. Eleanor Helin's PCAS program continued until the program morphed into JPL's Near-Earth Asteroid Tracking (NEAT) program in the mid-1990s.[9]

---

6. E. F. Helin and R. S. Dunbar, "Search Techniques for Near Earth Asteroids," *Vistas in Astronomy* 33, no. 1 (1990): 21; J. Williams, interview by Yeomans, 9 June 2016, transcript in NASA History Division HRC.

7. Levy, *Shoemaker by Levy*, pp. 172–173.

8. Carolyn Shoemaker, interview by Rosenburg, 10 February 2017, transcript in NASA History Division HRC; Lawrence interview, 19 May 2016. In July 1997, Gene Shoemaker was killed in a tragic auto accident in Australia.

9. E. F. Helin and E. M. Shoemaker, "The Palomar Planet-Crossing Asteroid Survey, 1973–1978," *Icarus* 40 (1979): 321–328; Donald K. Yeomans, *Near-Earth Objects: Finding Them Before They Find Us* (Princeton: Princeton University Press, 2013), chapter 5.

Figure 3-2. Gene and Carolyn Shoemaker.

As early as 1978, the PCAS and PACS efforts sometimes used the 1.2-meter Schmidt telescope at Palomar to search for near-Earth asteroids in an effort to extend the survey to fainter objects, including distant comets. A technique called exposure gating was sometimes used, whereby a single-plate exposure was interrupted for a short interval so that fast-moving objects appeared with two unequal streaks that could be used to determine their apparent speed and direction without the need for a second plate. The 14-inch square plates of the 1.2-meter telescope would not fit into the stereomicroscope for scanning, so inspection of the plates was carried out using a low-power microscope or hand magnifier.[10]

---

10. Helin and Dunbar, "Search Techniques for Near Earth Asteroids," pp. 21–37.

The photographic and time-intensive Helin and Shoemaker surveys of the 1970s, 1980s, and through the mid-1990s, with their modest numbers of new discoveries, would soon give way to the era of electronic charged-coupled devices (CCDs) that would revolutionize the survey efforts. Although the early CCDs had slow readout times and limited fields of view, they would soon completely dominate the NEO search efforts. As soon as microcomputers were put in play to handle the vast amount of imaging data that modern CCDs generate, the advantages of CCD detectors over photographic films were apparent. CCDs are far more sensitive (faster) than films; they are more linear in their response to light levels; and they have a far wider spectral bandwidth than films. The CCD revolution for near-Earth object discovery surveys began with Spacewatch.

Spacewatch

Tom Gehrels, an immigrant from the Netherlands who had joined Gerard Kuiper's Lunar and Planetary Laboratory in 1961, had long been interested in asteroids, in particular looking for near-Earth asteroids. He pioneered the photometric study of asteroids during the 1950s. In 1960, he joined with astronomers Cornelis Johannes van Houten and Ingrid van Houten-Groeneveld of the Leiden Observatory to conduct the Palomar-Leiden Survey of dim minor planets.[11] Gehrels left the search for minor planets for a little over a decade to become the Principal Investigator (PI) of the Imaging Photopolarimeter instruments on the Pioneer 10 and 11 missions to Jupiter and Saturn, respectively, but returned to the search again in the 1970s, extending the original Palomar-Leiden Survey with surveys to identify Trojan asteroids.[12]

These surveys were done photographically, with Gehrels exposing plates on the 1.2-meter (48-inch) Schmidt telescope on Mount Palomar and then shipping them to his collaborators at Leiden to be studied with a blink comparator. The resulting discoveries were then sent to the Minor Planet Center to have their orbits computed and be published in the *Minor Planet Circular.*

11. C. J. van Houten, I. van Houten-Groeneveld, P. Herget, and T. Gehrels, "The Palomar-Leiden Survey of Faint Minor Planets," *Astronomy and Astrophysics Supplement Series* 2 (1970): 339–448.

12. I. van Houten-Groeneveld, C. J. van Houten, M. Wisse-Schouten, C. Bardwell, and T. Gehrels, "The 1977 Palomar-Leiden Trojan Survey," *Astronomy and Astrophysics* 224 (1 October 1989): 299–302.

Figure 3-3. Tom Gehrels in his office in August 1997.
(Photo courtesy of Robert S. McMillan)

This was a slow process. The 130 plates Gehrels exposed at Palomar in the fall of 1960 did not see their full results published until 1970; the 68 plates he exposed in 1977 were finally fully published in 1989.

Gehrels saw promise in the new technology of charge-coupled detectors, or CCDs, for improving the speed with which new minor planets, and especially those passing close to Earth, could be discovered. CCDs had been invented by Bell Labs in 1970; within NASA, both JPL and Goddard Space Flight Center launched partnerships with Texas Instruments to develop CCDs for spaceborne use. At JPL, they were first embedded in a planned outer planets mission that gradually evolved into the Galileo mission to Jupiter. Several astronomers at Caltech built instruments to use CCDs made available by JPL; the first reported use on Mount Palomar was in May 1976.[13] In 1977, James Westphal of Caltech proposed a CCD-based camera for NASA's Large Space Telescope; his winning proposal became the Hubble Space Telescope's Wide Field and Planetary Camera.

---

13. J. E. Gunn, E. B. Emory, F. H. Harris, and J. B. Oke, "The Palomar Observatory CCD Camera," *Publications of the Astronomical Society of the Pacific* 99, no. 616 (1987): 518–534; James E. Gunn and James A. Westphal, "Care, Feeding, and Use of Charge-Coupled Device Imagers at Palomar Observatory," *SPIE Solid State Imagers for Astronomy* 290 (1981): 16–23.

Gehrels wanted to use the new technology for asteroid hunting. In 1977, he had proposed a dedicated "Spacewatch Telescope," similar to the 1.2-meter Schmidt he had used on Palomar, but larger and dedicated exclusively to the hunt for near-Earth objects.[14] That proposal went nowhere. In early 1980, he tried again, teaming up with Gene Shoemaker, Robert McMillan, and several other colleagues, this time proposing a dedicated 1.8-meter CCD-based Spacewatch camera and data-processing system that would be installed in an existing dome on Kitt Peak, operated by the University of Arizona's Steward Observatories. This proposal was not funded either; at a June 1980 meeting with William E. Brunk and Geoffrey Briggs, the astronomy program officials at NASA Headquarters, he was told that NASA did not see the construction of ground telescopes as part of its mission, and that they were concerned about "the possibility of unforeseen technical problems that would cause delays and cost overruns." The "agonizing delays and overruns" of the new Infrared Telescope Facility, built on Maunakea and completed in 1979, were fresh on their minds.[15]

But Brunk and Briggs supported the scientific goals of the proposal, which included finding Earth-approaching asteroids that could be visited by future Space Shuttle–based missions and improving understanding of what Gehrels had called the "hazard aspect" of near-Earth objects. NEOs were a hazard due to the "low but finite" probability of a catastrophic impact, but the actual risk could not be quantified adequately. "The problem would no longer be statistical in nature if all the Earth-crossing asteroids down to about 150 meters in diameter were detected and their orbits accurately determined. *Any collision that might occur in the near term (next few decades) could be forecast,*" Gehrels had contended.[16] So Brunk was willing to fund a design study for the facility—in NASA's jargon, this was to be a Phase A study.

---

14. Tom Gehrels and Richard P. Binzel, "The Spacewatch Camera," *Minor Planet Bulletin* 11, no. 1 (1984): 1–2.

15. T. Gehrels notes on a meeting at NASA Headquarters, 20 June 1980, and undated typewritten attachment, Spacewatch Historical Files, Lunar and Planetary Laboratory (LPL), University of Arizona, courtesy of Robert McMillan; S. J. Bus, J. T. Rayner, A. T. Tokunaga, and E. V. Tollestrup, "The NASA Infrared Telescope Facility," *arXiv Preprint arXiv:0911.0132,* 2009, *http://www.ifa.hawaii.edu/-tokunaga/Plan_Sci_Decadal_files/IRTF%20white%20paper.pdf.*

16. Emphasis in original. Tom Gehrels (PI) and E. M. Shoemaker (Co-PI), "A Scanning Facility for Earth Crossing Asteroids," Steward Observatory and Lunar and Planetary

The Spacewatch Camera Phase A study did not lead to Gehrels's dedicated 1.8-meter telescope—at least not directly. The 1980 election put Ronald Reagan into the White House, and his administration levied significant cuts across most federal agencies, including NASA. So while the phase A study resulted in a completed conceptual design for the facility, what NASA ultimately funded was just the CCD-based camera and its data-processing system. The camera was designed to be mounted on an existing 0.9-meter Newtonian telescope at Kitt Peak. The Newtonian design had two light paths for two instruments, and Gehrels had to share the telescope with a radial velocity instrument intended to search for planets around other stars developed by Krzysztof Serkowski, later taken over by Robert S. McMillan when Serkowski passed away in 1981. This arrangement was less problematic than it might seem, as the radial velocity experiment needed bright stars and could observe during the part of the month astronomers call "bright time," the two weeks surrounding the full Moon, while the Spacewatch camera needed the "dark time" surrounding the new Moon to hunt for much dimmer asteroids.[17]

The first Spacewatch camera was designed and built by Jack Frecker, who had also designed Gehrels's balloon-based precursor to his polarimeters on Pioneers 10 and 11, the Polariscope. It used an RCA 320- by 512-pixel CCD as its detector with a coffee can loaded with dry ice to keep the detector cold. It was built quickly, seeing "first light" in May 1983. In his newsletter, *The Spacewatch Report*, Gehrels reported that the camera's first targets were parts of the sky with known asteroids, as a means of demonstrating that the complex new system actually worked. "The asteroids were promptly found in the computer processing of the data; it was probably the first time in history that an asteroid was found with a CCD system."[18]

James V. "Jim" Scotti, hired as an undergraduate to write the data-processing software, later described how the system worked. Initially, the idea had been to have the telescope quickly image the target areas of the sky and do this several times each night. That proved beyond the capabilities of both the

Laboratory of the University of Arizona, 7 November 1980, courtesy of Robert S. McMillan.

17. Robert McMillan interview with Conway, 17 August 2016, transcript in NASA History Division HRC.

18. T. Gehrels and M. S. Matthews, eds., *Spacewatch Report*, no. 2 (1 July 1983), folder 14, box 22, Spacewatch Camera Fundraising 1985, Gehrels papers (MS 541), University of Arizona Special Collections.

old 0.9-meter telescope they were using and the computers' data-processing capacities. So they instead developed a technique called "drift scanning," in which they shut off the telescope's drive system and matched the CCD's readout time to Earth's rotation. Readout was done by a Data General Nova computer, with the data written to tape. Each scan could be up to 29 minutes long, and they imaged the same part of the sky three times in an observing night—twice in quick succession and a third time later in the night. Fast-moving, and therefore probably closer to Earth, asteroids would appear to have moved between the first two images, while slower-moving asteroids would appear to move between the first and third. The tapes from a night's observing were brought down from Kitt Peak each morning for analysis by a Perkin-Elmer mainframe computer that had been supplied by NASA.[19]

The pacing element in the Spacewatch program was Scotti's software. With a working camera to provide data, he initially developed software that just subtracted one image from its partner. "I would actually take two images, register them and then subtract them, and all the stars would be these black-and-white smudgy areas and all the asteroids would be a black-and-white pair."[20] Then he had to manually identify the asteroids. He did not implement usable software capable of automatically identifying asteroids until the second half of 1984, and even then it was not reliable until early 1985.[21]

Scotti also reflected that the initial RCA CCD had not been very useful for the discovery of new near-Earth objects. For its first several years of existence, Spacewatch focused on recovering known asteroids and comets, though by

---

19. James Scotti, interview by Conway, 2 August 2016, transcript in NEO History Project Collection; T. Gehrels, "CCD Scanning," in *Asteroids, Comets, Meteors II; Proceedings of the International Meeting, Uppsala, Sweden, June 3–6, 1985* (A87-11901 02-90) (Uppsala, Sweden: Astronomiska Observatoriet, 1986), *http://adsabs.harvard.edu/abs/1986acm..proc...19G*; Tom Gehrels and Richard P. Binzel, "The Spacewatch Camera," *Minor Planet Bulletin* 11 (1 March 1984): 1–2; T. Gehrels and M. S. Matthews, eds., *Spacewatch Report*, no. 3 (20 April 1984), folder 14, box 22, Spacewatch Camera Fundraising 1985, Gehrels papers (MS 541), University of Arizona Special Collections.

20. Scotti interview.

21. Scotti interview; T. Gehrels and M. S. Matthews, eds., *The Spacewatch Report*, no. 3 (20 April 1984), folder 14, box 22, Spacewatch Camera Fundraising 1985, Gehrels papers (MS 541), University of Arizona Special Collections.

the fall of 1986, Gehrels's team had found 69 new asteroids, too.[22] According to Jim Scotti, the original RCA chip let them

> learn how to find asteroids, how to measure them where they're at on the sky, how to find them the next night, how to find them the next month, how to recover lost ones, how to do all that kind of stuff that's very important in the process of observing asteroids. Most people only think of the discovery of asteroids when they think about finding the dangerous ones, but there's a huge industry behind that in order to not lose those asteroids. If you can't follow up an asteroid and find out what its orbit really is, you might as well have not found it in the first place.[23]

Follow-up observations of asteroids and comets were crucial to refining knowledge of their orbits, even though doing it did not gain one much scientific credit. It also was a good training ground for learning how best to exploit the new digital technologies.

Gehrels had launched a newsletter for his project as part of an effort to raise funds for the project. The NASA grant paid for the first Spacewatch camera and some of the data-processing equipment, but not all of it; it also did not pay for improvements to the telescope, dome, or supporting equipment. For these expenses, Gehrels estimated needing another $200,000 in 1982; he raised more than half from a single donor, Bernard M. Oliver, in the form of a matching grant. Known as Barney, Oliver had been the founder of Hewlett Packard Laboratories and directed it until he retired in 1981. His retirement lasted only a couple of years, after which he became director of NASA Ames Research Center's Search for Extraterrestrial Intelligence (SETI) office.[24] About 55 individual and institutional donors had matched Oliver's grant by 1985.

Gehrels also sought Department of Defense (DOD) funding for Spacewatch. In 1983, he approached Hans Mark, then Deputy Administrator of NASA, about helping to arrange DOD funding. Mark, who had been

---

22. R. S. McMillan, T. Gehrels, J. V. Scotti, and B. G. Marsden, "Astrometry with the Spacewatch Scanning CCD," *Bulletin of the American Astronomical Society* 18 (September 1986): 1013, *http://adsabs.harvard.edu/abs/1986BAAS...18Q1013M.*
23. Scotti interview.
24. Steven J. Dick and James E. Strick, *The Living Universe: NASA and the Development of Astrobiology* (New Brunswick, NJ: Rutgers University Press, 2004), p. 134.

Director of Ames Research Center during the Pioneer 10 and 11 missions and Secretary of the Air Force from 1979 to 1981, was not immediately able to help. He wrote back that he had "been completely unsuccessful in raising Bob Cooper's interest level in this business. There is no doubt that what you are talking about is extremely important for them but I don't think they understand this yet. I will keep trying."[25] Cooper was the Director of the Defense Advanced Research Projects Agency (DARPA) between 1981 and 1985; by "this business," Mark was likely referring not just to the use of automated CCD-based imaging for asteroid discovery, but the more general problem of space surveillance. Gehrels tried again in 1984, and in 1985, he finally gained DARPA funding for Spacewatch.[26] Among other things, these funds enabled Spacewatch to upgrade the 0.9-meter (36-inch) telescope with a Tektronix 2,048- by 2,048-pixel CCD of much greater sensitivity.

David Rabinowitz joined the Spacewatch team just in time to write the second version of the automated moving-object detection software for a new, more powerful computer. He and Jim Scotti made their first near-Earth asteroid discovery, eventually named 1989 UP, on 27 October 1989. It was not detected by the new software—Scotti remembers that Rabinowitz saw it on their computer screen and woke him up. "It wasn't really fast, but it was trailed enough that you could see that it was an interesting object," he commented years later.[27] They put the discovery out on the International Astronomical Union's Astronomical Telegrams service, and it was rapidly followed up by Robert McNaught at the Siding Spring Observatory in Australia and by JPL's Eleanor "Glo" Helin using the Mount Palomar 0.5-meter (18-inch) Schmidt telescope. Combined, there were seven observations of the new find by 30 October.[28]

Rabinowitz had the new detection software (called, unexcitingly, MODP for Moving Object Detection Program) working by the following year. Asteroid

---

25. Hans Mark to Tom Gehrels, 3 October 1983, folder 22, box 5, Gehrels papers (MS 541), University of Arizona Special Collections.

26. Gehrels to Mark, 1 June 1984 and 5 July 1985, both folder 22, box 5, Gehrels papers (MS 541), University of Arizona Special Collections.

27. Scotti interview, 2 August 2016.

28. "IAUC 4887: 1989 UP," available at the Central Bureau for Astronomical Telegrams, *http://www.cbat.eps.harvard.edu/iauc/04800/04887.html* (accessed 29 November 2016); D. L. Rabinowitz, "Detection of Earth-Approaching Asteroids in Near Real Time," *Astronomical Journal* 101 (1991): 1518–1529.

1990 SS, an "Apollo"-class asteroid, was their first fully automated near-Earth object discovery, reported by Scotti in October. That first year of automated operations, Spacewatch averaged two Earth-crossing asteroid discoveries per month while also detecting about two thousand other asteroids monthly.[29]

Characterization of Near-Earth Objects

Discovering near-Earth objects was one activity pursued via optical astronomy; another was characterization. Asteroids were understood to be not just points of light, but three-dimensional bodies with shapes, spins, densities, probably colors, and certainly compositions. Astronomers deployed various other tools to try to understand these qualities.

It was a full century after the discovery of the first asteroid, Ceres, that the Austrian astronomer, Egon von Oppolzer, noted periodic variations in an asteroid's brightness. Taking advantage of the Earth approach by Eros in late 1900 (to within 0.32 au), Oppolzer observed it at the German Potsdam Observatory and was surprised to find that the asteroid faded by a factor of 4 (an increase of 1.5 magnitudes) in 79 minutes and then returned to its original brightness over the next few hours, only to wane yet again.[30] The entire brightness period took 5 hours and 16 minutes, a value in agreement with more modern rotation period determinations. It was recognized that this was an effect due to the rotation of Eros and one of the following possibilities: the body had light and dark sides; two co-orbiting bodies were eclipsing one another; or the body of Eros was irregularly shaped, so that its apparent brightness depended upon which of its sides was being viewed.

By observing an object's apparent brightness over long intervals of time, one can determine the object's rotation period from a plot of the asteroid's brightness as a function of time. In this process, the photometer, which measures incident light levels and was attached to the telescope, was first guided

---

29. J. F. Scotti, D. L. Rabinowitz, and T. Gehrels. "Automated Detection of Asteroids in Real-Time with the Spacewatch Telescope" (paper presented at the International Conference on Asteroids, Comets, Meteors, Flagstaff, AZ, 24–28 June 1991), p. 191, *http://adsabs.harvard.edu/abs/1991LPICo.765..191S*; "IAUC 5117: 1990ad; 1990 SS," *http://www.cbat.eps.harvard.edu/iauc/05100/05117.html* (accessed 29 November 2016).

30. Fletcher G. Watson, *Between the Planets*, 2nd rev. ed. (Cambridge, MA: Harvard University Press, 1956), pp. 33–34; Egon von Oppolzer, "Notiz. betr. Planet (433) Eros," *Astronomische Nachrichten* 154 (1901): 297.

to the appropriate asteroid to determine its brightness, followed by calibration measurements for a neighboring standard star. With enough of these data, the light curve can be inverted to determine a rough model of the asteroid's shape.[31] Light curve observation and analysis was a relatively slow process requiring hours at the telescope until the 1990s, when automation would make this a far less tedious process.

If light curves were useful in understanding asteroid rotations and shape, another tool, spectral observations, offered the possibility of linking asteroid color, or spectra, to composition. In a perfect world, telescopic spectral observations of asteroids could uniquely determine their compositions. Alas, it is not a perfect world.

In 1929, N. T. Bobrovnikoff conducted a pioneering work on asteroid colors, and he was the first to consider that asteroids might not be just colorless grey reflectors of sunlight. Using the Lick Observatory 36-inch (0.9-meter) refractor, a light prism spectrograph, and a 12-inch camera, he took spectra for 12 bright main-belt asteroids, noted the differences between them, and, for Vesta, even determined its rotation period from the color variations alone.[32] However, Vesta was the only observed object for which this was possible. Bobrovnikoff noted no bright lines or bands and concluded that their light is wholly due to reflected sunlight. He compared the similar spectra of Comet Halley with the spectrum of asteroid 9 Metis and commented that "it

---

31. Alan W. Harris (of JPL) points out that a paper written by Henry Norris Russell in 1906 inadvertently set back light curve inversion work 75 years by making the statement, "It is quite impossible to determine the shape of the asteroid (from its light curve)." A careful reading of his short article reveals that he meant that it is only impossible to reconstruct an asteroid's albedo features (not its shape) from a light curve. Henry Norris Russell, "On the Light-Variations of Asteroids and Satellites," *Astrophysical Journal* 24 (1906): 1–18.

Against all odds, there are two asteroid astronomers named Alan William Harris. The somewhat younger one is English-born and working at the German Aerospace Center, while the other works out of his home in southern California. To avoid confusion, they refer to themselves as "Harris the younger" and "Harris the elder."

In a 1984 work, Steve Ostro and colleagues showed how a convex-profile shape model can be determined from a light curve.

32. N. T. Bobrovnikoff, "The Spectra of Minor Planets," *Lick Observatory Bulletin* 407 (1929): 18–27. The modern rotation period for Vesta is 5 hours and 20.5 minutes.

is becoming increasingly clear that there is no essential difference between comets and asteroids."[33]

The first standardized photometric system was the Johnson-Morgan system, or UBV system, introduced in 1953 for determining stellar colors.[34] UBV photometry of asteroids was introduced in the mid-1950s, and it soon became possible to obtain precision photometry with more narrow filters and for a more extended range of wavelengths.[35] Tom McCord and colleagues published the first narrow-band reflectance spectrum of an asteroid in 1970.[36] Using the 60-inch (1.5-meter) telescope at Cerro Tololo (Chile) and the Mount Wilson 60-inch (1.5-meter) and 100-inch (2.6-meter) telescopes near Pasadena, California, McCord noted a strong absorption feature near 0.9 microns for Vesta, which was attributed to the silicate mineral pyroxene. A comparison of this Vesta spectral band with the laboratory measurements of meteorites and Apollo 11 lunar samples indicated that the surface of Vesta had a very similar composition to that of certain basaltic achondrite meteorites. This was a significant step toward linking meteorite compositions with a specific asteroid. However, nearly a half century later, in 2016, after an enormous amount of effort spent on asteroid classifications, Schelte "Bobby" Bus, a recognized authority on asteroid spectral classifications, commented that this was a rare link between the composition of a meteorite type and an asteroid spectral class. Bus explains:

> It's very easy to look at an asteroid reflected spectrum and say it's got this wiggle and I'm going to stick it into this class, but to say what that class is, is

---

33. Ibid. The albedos of Vesta and Ceres determined by the Dawn spacecraft mission are 0.42 and 0.09 respectively.

34. These so-called ultraviolet (U), blue (B), and visual (V) broadband filters were centered on 365 nanometers, 445 nanometers, and 551 nanometers respectively, with bandwidths of several tens of nanometers. H. L. Johnson and W. W. Morgan, "Fundamental Stellar Photometry for Standards of Spectral Type on the Revised System of the Yerkes Spectral Atlas," *Astrophysical Journal* 117 (1953): 313–352. A nanometer is one-billionth of a meter. An Angstrom unit equals 0.1 nanometers, and a micron equals 1,000 nanometers.

35. Clark R. Chapman and Michael J. Gaffey, "Reflectance Spectra for 277 Asteroids," in *Asteroids*, ed. Tom Gehrels (Tucson, AZ: University of Arizona Press, 1982), pp. 655–687.

36. T. B. McCord, J. B. Adams, and T. V. Johnson, "Asteroid Vesta: Spectral Reflectivity and Compositional Implications," *Science* 168 (1970): 1445–1447.

very difficult. There's only one class of meteorites and spectra that I feel fairly confident about and that is the objects that are tied to Vesta. The heavily basaltic asteroids have a unique spectral signature, and if you take meteorites that you know are basaltic achondrites, and you grind them up, you get basically the same spectrum.[37]

While the goal of linking known meteorite compositions with asteroid spectral types is still a work in progress, there are ongoing efforts to use color photometry to classify asteroids. In 1975, Clark Chapman, of the Planetary Science Institute; David Morrison, then of the Institute for Astronomy, University of Hawai'i; and Ben Zellner, of the University of Arizona's Lunar and Planetary Laboratory, developed an asteroid classification system that utilized spectra and albedos to define two main groups. The C-types were low-albedo objects with relatively flat, featureless spectra in the 0.3- to 1.1-micron wavelength range, while the S-types often had spectral features and were more pronounced at the red end of the spectrum and with higher albedos.[38] A catchall type U included objects for which C and S did not apply. The C and S types were identified as "carbonaceous" and "silicaceous" and while often these descriptors might apply, subsequent research has shown that these types cannot in general be interpreted in terms of an asteroid's composition, structure, or origin.[39]

In the 1970s and 1980s, both the number of filters used and the wavelength range increased with a concomitant increase in the number of asteroid types suggested. Edward Bowell and colleagues expanded the C and S taxonomy

---

37. Schelte Bus, interview by Yeomans and Conway, 25 January 2016, transcript in NASA History Division HRC. Basaltic achondrites are stony meteorites, without chondrules, formed by igneous processes. Most of these meteorites are thought to come from Vesta (the so-called howardites, eucrites, diogenites).

38. The polarization of light from a diffusely reflecting surface is intimately connected with its albedo. A plot of an object's percentage polarization as a function of the observation phase angle (Earth-asteroid-Sun angle) can be used to estimate the asteroid's albedo.

39. C. Chapman, D. Morrison, and B. Zellner, "Surface Properties of Asteroids: A Synthesis of Polarimetry, Radiometry and Spectrophotometry," *Icarus* 25 (1975): 104–130. Massachusetts Institute of Technology (MIT) Professor Richard Binzel, an authority on asteroid spectral classification (taxonomy), emphasizes that asteroid taxonomy is not the same thing as asteroid mineralogy. Richard Binzel, interview by Yeomans and Conway, 17 October 2016, transcript in NASA History Division HRC.

in 1978 to include the classes M, R, and E.[40] The M class is often interpreted as "metallic," and some M types do represent iron-nickel objects, but others do not. In 1984, David Tholen, then a graduate student at the University of Arizona, proposed a revision and expansion of the existing asteroid classification system. It employed eight filters over the 0.31- to 1.06-micron visible wavelength region and put asteroids into 14 different types.[41] Lebofsky's 1978 identification of a 3-micron infrared spectral feature on a spectrum of Ceres was attributed to hydrated minerals, thus beginning the ongoing search for water resources on asteroids (see chapter 9). Lebofsky's observations were made with the 0.7-meter (28-inch) Mount Lemmon telescope near Tucson, Arizona.[42]

In large part, the eagerly sought links between the known composition of meteorite types and corresponding spectral classification of asteroids have not fully materialized. Even so, by the 1980s, a few general conclusions relating to asteroid spectral types were becoming evident. In general, asteroids in the outer main belt beyond 3.8 au and in the two Jupiter Trojan regions were systematically redder and darker than asteroids closer to the Sun, and they formed a new D-type classification. As early as 1971, A. F. Cook of the Smithsonian Astrophysical Observatory suggested that the D-type Trojan asteroid 634 Hektor was a binary object based upon its asymmetric light curve that changed with time. He postulated that its binary nature may have been formed as a result of its rapid rotation period of 6.9 hours, which produced stresses upon the object that may well have exceeded the crushing strength of meteoritic stone, thus forming a binary system.[43] The formation

40. Edward Bowell, Clark R. Chapman, Jonathan C. Gradie, David Morrison, and Benjamin Zellner, "Taxonomy of Asteroids," *Icarus* 35, no. 3 (1978): 313–335.

41. D. Tholen, "Asteroid Taxonomy from Cluster Analysis of Photometry" (Ph.D. thesis, University of Arizona, Tucson, 1984.

42. L. A. Lebofsky, "Asteroid 1 Ceres: Evidence for Water of Hydration," *Monthly Notices of the Royal Astronomical Society* 182 (1978): 17–21. In 1979, NASA's Infrared Telescope Facility (IRTF) achieved "first light" at an altitude of 13,600 feet near the summit of Maunakea in Hawai'i. This telescope is optimized for infrared observations and can provide spectroscopic data over the wavelength range 0.8 to 5.4 microns. Although its initial purpose was to provide observations of the outer planets in support of the Voyager spacecraft that were launched in 1977, it has provided numerous infrared (IR) observations of near-Earth asteroids.

43. A. F. Cook, "624 Hektor: A Binary Asteroid?" in *Physical Studies of Minor Planets*, ed. T. Gehrels (Washington, DC: NASA SP-267, 1971), pp. 155–163. In July 2006,

of asteroid binaries as a result of rapid spin rates would become an often-mentioned mechanism in the coming years as more and more near-Earth asteroids were discovered.

Early Radar Efforts

A superb tool used to characterize small bodies is radar. The basic technique, bouncing radio waves off various objects, had been used as early as the 1920s for ionospheric research. It became famous during World War II for its military utility, and it was again used for scientific research at the war's end. In September 1945, the U.S. Army Signal Corps' Evans laboratory in New Jersey started trying to bounce radio waves off the Moon, though they were not successful in detecting the returned signals until January 1946. Similar experiments were conducted almost immediately in Britain and Hungary; war surplus radars were also quickly turned to the study of meteors.[44]

One of the fundamental weaknesses of radar as a tool for solar system research is that the returned signal strength declines with the inverse fourth power of the distance. So detecting returned signals from celestial objects requires powerful transmitters, extremely sensitive receivers, and large antennas. The MIT Lincoln Lab's Millstone Hill radar, built as a prototype ballistic missile early warning radar, was used in 1958 to try to receive echoes from Venus but was probably not successful. (They attempted to verify apparently positive results with a second set of experiments in 1959 but were unsuccessful.) In March 1961, using a "bistatic radar" consisting of separate transmitting and receiving antennas built originally for communications purposes, JPL's Goldstone Deep Space Information Facility near Barstow, California, was unequivocally successful.[45]

---

observations using the Keck 10-meter telescope on Maunakea, Hawai'i, confirmed a 12-kilometer satellite of Hektor. Hektor is the only asteroid at the L4 position (leading Jupiter) named after a Trojan hero. The remaining members at L4 all have Greek names. At the L5 position (trailing Jupiter), which is reserved for asteroids with Trojan names, asteroid (617) Patroclus is the sole exception. So there is a single "spy" in each of the Trojan and Greek camps.

44. Andrew J. Butrica, *To See the Unseen: A History of Planetary Radar Astronomy* (Washington, DC: NASA SP-4218, 1996), pp. 1–26.

45. Ibid., pp. 27–41.

The first radar detections of an asteroid occurred in June 1968, when the 37-meter Haystack antenna in Westford, Massachusetts, and the Goldstone 64-meter antenna successfully observed near-Earth asteroid 1566 Icarus as it passed within 0.3 au of Earth in June 1968. With the exception of some limited observations of near-Earth asteroids 4179 Toutatis in 1996 and 367943 Duende in 2013, this was the only time that the Haystack radar was used to observe near-Earth objects. Using estimates of the rotation period and spin axis direction provided by Arizona astronomer Tom Gehrels, JPL scientist Richard Goldstein, at Goldstone, determined that the diameter of Icarus was greater than or equal to 980 meters and that its radar reflectivity was less than or equal to 0.13.[46] MIT scientist Gordon Pettengill and colleagues analyzed their Haystack radar data of Icarus and deduced a radar cross section of about 1 square kilometer, a radar reflectivity of about 0.05, and a diameter of about 2 kilometers.[47] Astrometric radar data were first used to refine the orbit of an asteroid with the radar observations of Icarus in 1968. Irwin Shapiro and colleagues considered radar data in their orbital analysis of Icarus to compare (and confirm) its perihelion advance with the predictions from general relativity.[48]

There were successful Goldstone radar observations of 1685 Toro in 1972 and of 433 Eros three years later.[49] For the latter observations, two frequencies were broadcast and received in two opposite circular polarizations. Since a circularly polarized signal would reverse sense upon reflection from a smooth object, information on the object's surface roughness was obtained by noting to what extent a particular circular polarization had been reversed when received. The surface of Eros was determined to be rougher than the lunar surface, with total dimensions 37.2 and 15.8 kilometers—not too different from the actual dimensions of 34 by 11 by 11 kilometers determined by the

46. T. Gehrels, E. Roemer, R. C. Taylor, and B. H. Zellner, "Minor Planets and Related Objects. IV. Asteroid (1566) Icarus," *Astronomical Journal* 75, no. 2 (1970): 186–195; R. M. Goldstein, "Radar Observations of Icarus," *Icarus* 10 (1969): 430–431.

47. G. H. Pettengill, I. I. Shapiro, M. E. Ash, R. P. Ingalls, L. P. Rainville, W. B. Smith, and M. L. Stone, "Radar Observations of Icarus," *Icarus* 10 (1969): 432–435.

48. I. I. Shapiro, M. E. Ash, W. B. Smith, and S. Herrick, "General Relativity and the Orbit of Icarus," *Astronomical Journal* 76 (1971): 588–606.

49. R. M. Goldstein, "Minor Planets and Related Objects. XII. Radar Observations of (1685) Toro," *Astronomical Journal* 78, no. 6 (1973): 508–509; Raymond F. Jurgens and Richard M. Goldstein, "Radar Observations at 3.5 and 12.6 cm Wavelength of Asteroid 433 Eros," *Icarus* 28 (1976): 1–15.

Near Earth Asteroid Rendezvous (NEAR) space mission some 25 years later. In a paper published in 1982, Ray Jurgens at JPL outlined a method for determining the dimensions of a triaxial ellipsoid asteroid model using an extensive radar dataset that included the apparent radar cross section, the center frequency, and the effective bandwidth as a function of time.[50] This effort marked the beginning of asteroid shape modeling using radar data—a technique that would develop to provide remarkable radar "images" in the years to come.

The largest radio telescope built by the United States, the 305-meter Arecibo radio observatory in Puerto Rico, began its life as a concept for ionospheric observation in 1958. Planetary radar capability was added very quickly to make the very expensive facility more attractive to its funder, the Advanced Research Projects Agency (ARPA, the same agency as DARPA; its name has changed back and forth over time). The observatory began operations in 1963. Despite its original funding source, Arecibo was initially operated by Cornell University exclusively for scientific research. A 1969 agreement transferred the facility to the National Science Foundation, although Cornell remained the facility's manager.[51] NASA agreed to fund upgrades to the facility in 1971 to raise the frequency at which it operated. In 1975, Arecibo scientist Don Campbell and MIT's Gordon Pettengill made the first Arecibo asteroid radar observations, including ranging data, when they successfully observed Eros with the soon-to-be-replaced 70-centimeter transmitter (430 MHz).[52] They also noted that the Eros surface was rough compared with the terrestrial planets and the Moon. The next year, operating with the new S-band system (13-centimeter rather than 70-centimeter wavelength), they successfully observed 1580 Betulia. Steve Ostro, then a graduate student but soon to become a leader in the field of planetary radar, did the analysis and determined a lower limit on the diameter equal to 5.8 ± 0.4 kilometers.[53]

Planetary radars are not useful for discovering near-Earth objects because their fields of view, about 2 arc minutes, are too narrow. However, once an

50. Raymond F. Jurgens, "Radar Backscattering from a Rough Rotating Triaxial Ellipsoid with Applications to the Geodesy of Small Asteroids," *Icarus* 49 (1982): 97–108.

51. Butrica, *To See the Unseen*, pp. 88–103.

52. D. Campbell, G. Pettengill, and I. I. Shapiro, "70-cm Radar Observations of 433 Eros," *Icarus* 28 (1976): 17–20.

53. G. H. Pettengill, S. Ostro, I. I. Shapiro, B. G. Marsden, and D. Campbell, "Radar Observations of Asteroid 1580 Betulia," *Icarus* 40 (1979): 350–354.

accurate ephemeris is available for a target body, modern planetary radar observations are invaluable in that they can be used to model the shape and size of a near-Earth object with a resolution of a few meters—vastly superior to the ground-based optical image resolution for the same object. Only close-up spacecraft optical images can improve upon radar "images" of near-Earth objects. Modern radar observations can help determine an object's size, shape, rotation rate, and surface roughness, as well as whether it has a satellite. Moreover, radar line-of-sight velocity (Doppler) and range measurements, when used with optical plane-of-sky angle measurements, can dramatically improve the accuracy of the orbit and long-term ephemerides for recently discovered objects.[54] Almost all planetary radar observations have been carried out using the 64-meter (upgraded to 70 meters in May 1988) movable antenna at NASA's Deep Space Network (DSN) Goldstone facility in southern California or the fixed 305-meter antenna at Arecibo. While the Goldstone and Arecibo facilities dominate the radar observations of near-Earth objects, other facilities have been involved as well. For example, a Russian-German collaboration between a transmitter at the Evpatoria 70-meter antenna in Crimea and the 100-meter receiving antenna at Effelsberg, Germany (near Bonn), was undertaken to observe 4179 Toutatis in November 1992, and some other antennas, including the Green Bank Observatory facility in West Virginia, the Very Large Array (VLA) in New Mexico, and the widespread Very Long Baseline Array (VLBA) have been used as receiving stations for the Goldstone or Arecibo transmitters.[55]

By 1990, 24 near-Earth asteroids had been observed with radars; by mid-2000, this value had climbed to 58, and by the end of March 2018, it had reached 751. By contrast, radar observations for only 20 long- and short-period comets had been made.[56] Radar observations of comets are far less numerous

---

54. D. K. Yeomans, P. W. Chodas, M. S. Keesey, S. J. Ostro, J. F. Chandler, and I. I. Shapiro, "Asteroid and Comet Orbits Using Radar Data," *Astronomical Journal* 103, no. 1 (1992): 303–317; S. J. Ostro and J. D. Giorgini, "The Role of Radar in Predicting and Preventing Asteroid and Comet Collisions with Earth," *Mitigation of Hazardous Comets and Asteroids*, ed. M. J. S. Belton, T. H. Morgan, N. Samarasinha, and D. K. Yeomans (Cambridge: Cambridge University Press, 2004), pp. 38–65. The one-way light time delay is a measure of the distance or range to the asteroid, and the Doppler shift in frequency can be used to measure the rate of change of this distance.

55. A comprehensive list of all small-body radar observations is available at *http://echo.jpl.nasa.gov/asteroids/PDS.asteroid.radar.history.html* (accessed 16 April 2017).

56. Ibid.

because there are far fewer of them in Earth's neighborhood. The 2003 NASA Near-Earth Object Science Definition Team report noted that the number of Earth close approaches by long-period comets is less than 1 percent of comparably sized near-Earth asteroids.[57] Nevertheless, there were radar observations of comets, beginning with Arecibo radar observations of periodic Comet Encke on 2–8 November 1980, that were carried out from a distance of just over 0.3 au. Paul Kamoun of MIT and colleagues estimated that the nucleus diameter of Comet Encke was 3 kilometers (with large uncertainties of +4.6 and –2 kilometers) using the observed limb-to-limb bandwidth and values for the comet's rotation period and rotation axis direction determined from optical data.[58]

Kamoun made successful Arecibo radar observations of periodic Comet Grigg-Skjellerup in late May and early June 1982, and the very narrow bandwidth of less than 1 Hertz suggested either a very slow nucleus rotation or that the radar was looking nearly along the rotation axis of the comet's nucleus.[59] Successful Arecibo and Goldstone radar observations of long-period Comet IRAS-Araki-Alcock were carried out during this comet's close Earth approach (0.03 au) on 11 May 1983.[60] These were the first successful comet radar observations from Goldstone; the rapid motion of the comet on the sky, due to its relatively high 73-degree inclination, highlighted an advantage of the steerable Goldstone antenna. Whereas the larger, but fixed, Arecibo antenna is more sensitive than the Goldstone antenna, the Goldstone antenna could track the target and hence observe it over larger regions of sky and integrate the signal over longer intervals of time. The relatively large nucleus was estimated to be in the range of 6–12 kilometers, with a rotational period of 1–2 days. One month after the IRAS-Araki-Alcock radar observations, both Arecibo and Goldstone attempted radar observations on long-period Comet Sugano-Saigusa-Fujikawa when it passed 0.06 au from Earth on 12 June 1983. Oddly enough, these two comet passages, one month apart in 1983, were the only

57. NASA Near-Earth Object Science Definition Team Report, *Study to Determine the Feasibility of Extending the Search for Near-Earth Objects to Smaller Limiting Diameters* (Washington, DC: NASA, 22 August 2003).

58. P. G. Kamoun, D. B. Campbell, S. J. Ostro, and G. H. Pettengill, "Comet Encke: Radar Detection of Nucleus," *Science* 216 (1982): 293–295.

59. P. G. Kamoun, "Comet P/Grigg-Skjellerup," NAIC QR, Q2 (1982), pp. 7–8.

60. R. M. Goldstein, R. F. Jurgens, and Z. Sekanina, "A Radar Study of Comet IRAS-Araki-Alcock 1983 d," *Astronomical Journal* 89, no. 11 (1984): 1745–1754.

known comets to pass less than 0.1 au from Earth during the entire 20th century. Goldstone observations of Comet Sugano-Saigusa-Fujikawa were unsuccessful, and Arecibo received only weak signals. This comet was farther away, and apparently smaller, than Comet IRAS-Araki-Alcock.[61]

The predicted arrival of Comet Halley in late 1985 provided another opportunity to study a comet with radar. Halley's inbound trajectory carried it within Arecibo's view during late November 1985, though only marginally; its outbound trajectory in April 1986 brought it closer to Earth, but not within Arecibo's limited declination coverage. An observed large broadband feature with a high radar cross section and large Doppler bandwidth was interpreted as resulting from particles larger than 4 centimeters being ejected from the nucleus. Observer Don Campbell remarked that "Halley is the first comet to give a stronger echo from particles than from the nucleus itself."[62] Attempts to view Halley with the Goldstone radar during this apparition were unsuccessful. The Arecibo radar observations of Comet Halley would be the last successful radar observations of a comet for 10 years, until the Goldstone observations of Comet Hyakutake in March 1996.[63]

The First Comet Visitors

For the first two decades of its existence, NASA paid little attention to comets and asteroids. Advocacy by many American space scientists, including one of us (Yeomans), failed to get a U.S. spacecraft built as part of an international flotilla of spacecraft intended to visit Comet Halley in 1986. While American scientists had endorsed various mission concepts for a Halley spacecraft, including solar sail and ion drive rendezvous concepts, ultimately a U.S. Halley mission was done in by a shortage of resources and a less-than-complete endorsement from NASA's advisory groups.[64]

But Robert Farquhar, then of Goddard Space Flight Center, was able to convince NASA managers to repurpose the International Sun-Earth Explorer

61. Butrica, *To See the Unseen*, p. 220.

62. D. Campbell, J. Harmon, and I. I. Shapiro, "Radar Observations of Comet Halley," *Astrophysical Journal* 338 (1989): 1094–1105; Butrica, *To See the Unseen*, p. 221.

63. An outline of key radar characterization results for more recent times is included in chapter 7.

64. J. M. Logsdon, "Missing Halley's Comet: The Politics of Big Science," *Isis* 80, no. 2 (1989): 254–280.

(ISEE)–3 heliophysics spacecraft after the end of its primary mission into a comet visitor. It became the International Comet Explorer (ICE), and, after some celestial gymnastics, it was maneuvered to make passage through the tail of periodic Comet Giacobini-Zinner in September 1985. The ICE spacecraft did not carry a camera, but its particles and fields instruments were designed to monitor the solar wind of charged particles that would collide with Earth's ionosphere. These same instruments detected the Sun's magnetic field lines wrapped around the comet's ion atmosphere.[65]

In March 1986, the international armada of five spacecraft flew rapidly past Comet Halley, reaching distances as close as 596 kilometers to more than 7 million kilometers. The Halley Armada enabled the first view of an active comet nucleus. When a comet enters the inner solar system, it becomes active, and the resulting dust and gas effectively hide the nucleus from view by ground-based sensors. Thus, the science results from the Halley spacecraft flybys included the first close-up of a cometary nucleus.

In the first half of the 20th century, an accepted model for the cometary nucleus envisioned a bound or unbound collection of separate particles flying in formation about the Sun. The English astronomer Raymond Lyttleton, in the mid-20th century, maintained that comets were flying clouds of dust captured from interstellar dust clouds. At about the same time, the Soviet astronomer Sergey Vsekhsvyatskij argued that comets formed from volcanic eruptions from the outer planets or their satellites. However, most of the scientific community accepted the solid, icy conglomerate model for the cometary nucleus put forward by Fred Whipple in 1950.[66]

Whipple's comet model was extremely influential since it could explain the large cometary gas production rates, the jetlike structures in coma, erratic activity, and especially the so-called nongravitational motions of comets that were thought to be due to the rocket-like thrusting of the outgassing nucleus. However, Whipple's model was not actually confirmed until the Giotto imaging results and, to a lesser extent, the Soviet Union's Vega mission imaging

65. Wing-Huen Ip, "Global Solar Wind Interaction and Ionospheric Dynamics," in *Comets II*, ed. M. C. Festou, H. U. Keller, and H. A. Weaver (Tucson, AZ: University of Arizona Press, 2004), pp. 605–629.

66. F. L. Whipple, "A Comet Model I. The Acceleration of Comet Encke," *Astrophysical Journal* 111 (1950): 375–394. Whipple's, Lyttleton's, and Vsekhsvyatskij's ideas, along with the earlier history of comets, are outlined in Donald K. Yeomans, *A Chronological History of Observation, Science, Myth and Folklore* (New York: Wiley, 1991).

results. Prior to the Halley missions, a widely held view of its nucleus considered that it would be solid, nearly spherical, with a diameter of 4–5 kilometers, a bulk density of 1.0–1.7 g/cm^3, and a fairly bright albedo of 26 percent. It was expected to be outgassing over its entire sunward surface with a gas-to-dust mass ratio of one or more.[67] None of these predictions turned out to be correct.

Expecting a relatively bright, icy nucleus, engineers had designed the Giotto camera system to follow the brightest region in its field of view. That turned out to be a gas and dust jet near one end of the dark nucleus, and the nucleus itself was both darker and larger than expected, with an albedo of only 4 percent and triaxial dimensions of 7.2, 7.2, and 15.3 kilometers. After extensive image processing, the Giotto images revealed more than a dozen narrow jets, some of which were surprisingly well collimated. Surface features were noted, but there were no obvious impact craters—at least none that could be identified at Giotto's best image resolution.[68] Most of the gas (with entrained dust) from the comet came from the vaporization of water ice at day-side active regions, and organic particles rich in carbon, hydrogen, oxygen, and nitrogen (CHON particles) were evident in the mass spectrometer data. It was estimated that Comet Halley loses about 0.5 percent of its mass at each return to perihelion.[69] The mass of Comet Halley's nucleus was estimated from its orbital acceleration due to outgassing rocket-like effects, and dividing this mass by its estimated volume (computed from the observed dimensions) provided a nucleus bulk density of only 0.28 grams per cubic

---

67. M. J. S. Belton, "P/Halley: The Quintessential Comet," *Science* 230 (1985): 1229–1236. JPL researcher Neil Divine, in 1981, put forward an expected, preflight model of Comet Halley as a solid, spherical nucleus, 6 kilometers in diameter, having a bulk density of 1 g/cm^3 and with a ratio of dust production to gas production of 0.5. N. Divine, "Numerical Models for Halley Dust Environments," in *The Comet Halley Dust & Gas Environment: Proceedings of a Joint NASA/ESA Working Group Meeting* (held in Heidelberg, Germany) (Washington, DC: NASA SP-174, 1981), pp. 25–30.

68. The name Giotto arose from an early-14th-century painting by the Italian artist Giotto di Bondone that showed a realistic comet as the star of Bethlehem.

69. H. U. Keller et al., "In Situ Observations of Cometary Nuclei," in *Comets II*, ed. M. C. Festou, H. U. Keller, and H. A. Weaver (Tucson, AZ: University of Arizona Press, 2004), pp. 211–222; B. C. Clark, L. W. Mason, and J. Kissel, "Systematics of the CHON and Other Light Element Particle Populations in Comet P/Halley," *Astronomy and Astrophysics* 187 (1987): 779–784.

centimeter.[70] This low density implied a very porous nucleus rather than a solid monolith. The dust-to-gas ratio was expected to be about 0.5 but turned out to be greater than 1, suggesting that the nucleus was mostly dust, rather than ices. Whipple's icy conglomerate model—affectionately termed a "dirty snowball" model prior to the Halley armada—was apparently more like an "icy dirtball." The nucleus of Comet Halley was dark, under-dense, and porous, with icy active areas on (or just below) the surface that generated gas and dust jets. Once initiated, the nucleus outgassing activity and mass loss then continuously altered the nucleus surface features, erasing any impact craters. These basic comet nucleus characteristics would be upheld by several subsequent comet missions.[71]

Not many near-Earth objects had been identified by 1990, and while visits to comets had somewhat illuminated their nature, near-Earth asteroids remained relative mysteries. The decades of radar, light curve, and spectral research had raised many questions, but quite a few remained unanswered. Chemical composition would remain a mystery until sample-return missions in the 2000s began to shed some light on the subject.

By the time that happened, the science/policy landscape had dramatically shifted, in part driven by a set of arguments advanced in the 1980s and largely accepted by the early 1990s, that cosmic impacts could not only leave marks on Earth's surface like Meteor Crater but could affect the course of evolution. Truly enormous impacts might even transport life between the major planets. The following chapter will discuss this transformation in understanding.

---

70. H. Rickman, "The Nucleus of Comet Halley: Surface Structure, Mean Density, Gas and Dust Production," *Advances in Space Research* 9 (1989): 359–371.

71. Keller et al., "In Situ Observations of Cometary Nuclei."

CHAPTER 4

COSMIC IMPACTS AND
LIFE ON EARTH

Introduction

In 1980, near-Earth object research was transformed by a far-reaching argument: not only did asteroids and comets strike Earth occasionally, but such
impacts could redirect—and had in the past altered—the course of life itself.
Known as the Alvarez hypothesis, after the father-son pair of lead authors,
physicist Luis W. Alvarez and geologist Walter Alvarez, it posited that a
roughly 10-kilometer asteroid had impacted somewhere on Earth about 65
million years ago, at the boundary between the Cretaceous and Paleogene
periods of Earth's history.[1] The enormous amount of pulverized rock and
dust thrown into the atmosphere would have produced a years-long darkness,
explaining the mass extinction already known to have occurred.[2] Their argument was based on discovery of a thin clay layer enriched in iridium, a metal
that is rare on Earth's surface but common in meteorites. While many other
scientists had proposed that large impacts could produce mass extinctions,

---

1. The Paleogene period (66 mega annum [Ma] to 23 Ma) was formerly part of the Tertiary
   period (66 Ma to 2.5 Ma), and the extinction event referred to as the Cretaceous–
   Tertiary extinction, often shorthanded as "K–T." The International Commission on
   Stratigraphy abandoned the Tertiary nomenclature in 2008, and the current term for
   the timing of the extinction event is "K–Pg."
2. Luis W. Alvarez et al., "Extraterrestrial Cause for the Cretaceous-Tertiary Extinction,"
   Science 208, no. 4448 (6 June 1980): 1095–1108.

the Alvarez team was the first to produce geochemical evidence of such a co-occurrence.[3]

During the 1980s, the Alvarez hypothesis spawned a number of meetings and workshops on the subject of large impacts and their influence on life on Earth. Some scientists embarked on a search for the crater; while a few large craters were known by 1980, none were of the correct age. Over the next 15 years, most scientists involved with the search for the impactor's crater came to accept that a roughly 200-kilometer circular magnetic and gravitational anomaly under the Yucatán Peninsula that had been discovered in the 1950s was, in fact, the crater remnant.

Other scientists sought mechanisms for the extinction. Not everyone believed that the dust-induced climatic cooling that the Alvarez team had hypothesized was adequate to explain the mass extinction apparent in the fossil record. Indeed, many paleontologists were initially skeptical that the impact could be responsible for the mass extinction at all. In their view, the disappearance of the non-avian dinosaurs was gradual, not sudden, and thus the impact could not be the principal cause.

By the early 1990s, though, most researchers involved in the controversy had also accepted a set of mechanisms that together could explain the end-Cretaceous mass extinction. Along the way, the effort to understand the effects of this impact raised questions about the effects of the still larger impacts that would be expected to have occurred earlier in Earth's history.

The Cretaceous–Paleogene Impactor and Extinction Mechanisms

The Alvarez paper was not the first suggestion that cosmic impacts might have influenced the course of life on Earth. As early as 1694, in an address to the Royal Society, Edmond Halley had noted that the biblical deluge could have been caused by the shock of a comet impact, followed by the seas rushing violently toward the impacted area. The Caspian Sea and other large lakes in the world could have been created by cometary impacts, and finally, in a very prescient comment, Halley noted that widespread extinction of life, before the

---

3.   Geochemist Harold Urey, for example, argued that comets might cause mass extinctions in a 1973 letter to *Nature*. Harold C. Urey, "Cometary Collisions and Geological Periods," *Nature* 242 (2 March 1973): 32–33.

creation of man, would be the likely consequence of a comet striking Earth.[4] In 1903, arguing against the possibility of impacts, N. S. Shaler contended that "the fall of a bolide of even ten miles in diameter would have been sufficient to destroy organic life of the earth."[5] As early as 1916, and continuing sporadically through his career, the Estonian astronomer Ernst Öpik explored impact energies and their effects through modeling. In a 1958 abstract, he even argued that hot rock gas from large impacts could completely annihilate life on land.[6] In his 1969 Presidential Address to the Paleontological Society, Digby McLaren of the Geological Survey of Canada argued for catastrophic extinction at the Frasnian–Famennian boundary, about 372 million years ago.[7] And in a 1973 *Nature* communication, famed geochemist Harold Urey suggested that comet impacts might have been responsible for the Cretaceous–Paleogene extinction, among other mass extinctions.[8] None of these arguments had much effect on the prevailing scientific opinion of their respective times about the role of impacts in the evolution of Earth or its biosphere.

In the early 1970s, Walter Alvarez started out trying to explain the origin of a specific limestone formation in Italy known as the Scaglia Rossa. This sequence of layers had been deposited during the Cretaceous–Paleogene boundary, and he and some colleagues were working through it, mapping its record of magnetic field reversals. It was during this work that he noticed a pattern that had already been reported: a sudden change in the type and abundance of foraminifera—the microfossils that were widely used to perform relative dating of rock units—in the layers of rock. "With a hand lens," he wrote, "you could spot the near extinction of the forams, which are abundant and

---

4. Edmond Halley, "Some Considerations About the Cause of the Universal Deluge, Laid Before the Royal Society, on the 12th of December 1694. By Dr. Edmond Halley, R. S. S.," *Philosophical Transactions of the Royal Society of London Series I* 33 (1724): 118–123.

5. N. S. Shaler, *A Comparison of the Features of the Earth and the Moon*, Smithsonian Contributions to Knowledge, vol. XXXIV (Washington, DC: Smithsonian Institution, 1903), p. 14. He argued that since life persisted on Earth, then it could be safely assumed that such an impact had never occurred.

6. E. J. Öpik, "On the Catastrophic Effect of Collisions with Celestial Bodies," *Irish Astronomical Journal* 5 (1 March 1958): 34.

7. D. J. McLaren, "Time, Life, and Boundaries," *Journal of Paleontology* 44, no. 5 (1970): 801–815.

8. Harold C. Urey, "Cometary Collisions and Geological Periods," *Nature* 242 (2 March 1973): 32–33.

as big as sand grains in the top beds of the Cretaceous, but with only the very smallest ones surviving into the first beds of the Tertiary."[9] In addition, a thin clay layer that was devoid of fossils separated the limestones of the two periods. This delicate layer would become a flash point of contention for a number of years.

Walter Alvarez wondered about the extinction event and its relationship to this clay. He asked his father, Luis W. Alvarez, a nuclear physicist who had been involved with the Manhattan Project and received the 1968 Nobel Prize, what might be measurable to help pin down the length of time represented by the clay layer. Walter Alvarez's interpretation of the surrounding magnetic field changes ruled out a span of time greater than a half million years. But had the clay layer been deposited over 10 years, or thousands? His father's first suggestion, to measure the amount of radioactive beryllium-10, did not work out. The half-life of beryllium-10 is too short to have left a measurable amount after tens of millions of years. But later, Luis suggested iridium, which is very rare on Earth's surface and crust because it has a high affinity for iron—most of

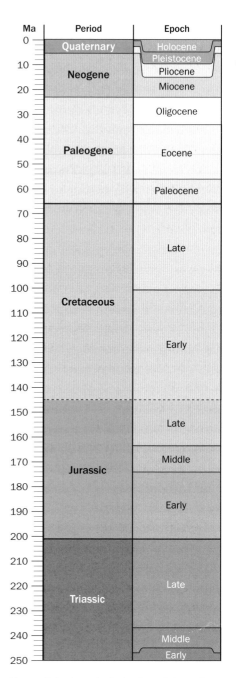

Ma	Period	Epoch
0	Quaternary	Holocene
		Pleistocene
10	Neogene	Pliocene
20		Miocene
30		Oligocene
40	Paleogene	Eocene
50		
60		Paleocene
70		
80		Late
90		
100	Cretaceous	
110		
120		Early
130		
140		
150		Late
160	Jurassic	
170		Middle
180		Early
190		
200		
210		
220	Triassic	Late
230		
240		Middle
250		Early

Figure 4-1. A geologist's view of time since the beginning of the Mesozoic Era, 252 million years ago. Column "Ma" represents millions of years.

9. Walter Alvarez, *T. Rex and the Crater of Doom* (Princeton, NJ: Princeton University Press, 1997), p. 40.

which was carried to the center of our planet early in Earth's history to form the core. Asteroids, though, can bring iridium to Earth's surface, and at a relatively constant rate. Thus, measuring the amount of iridium in the clay layer would provide a proxy measurement of the layer's deposition period. At least, that is what the Alvarez team thought initially.

They had samples of the layer tested by Frank Asaro, but the result did not answer their question at all. If it were true that iridium came from space at a fixed rate, then the amount of iridium in this thin layer of clay was far too high—20 times as high as it should have been, given the expected rate of inflow. To clarify the situation, they went looking for another site with continuous deposition across the Cretaceous–Paleogene boundary (not many were known at the time) and found one in Denmark. It also showed a highly elevated level of iridium.[10] There were not many possible causes of such a huge spike in iridium.

One possible cause for such an elevated level had been proposed back in 1971: a nearby supernova. Such an event would have been devastating to life on Earth and would have deposited lots of iridium. But a supernova would have left other smoking guns as well, notably plutonium-244 (Pu-244). With an 80-million-year half-life, any plutonium-244 that was present when Earth formed more than 4 billion years ago has long since decayed to undetectable levels. So any Pu-244 detected around the Cretaceous–Paleogene boundary could only be the product of a nearby stellar detonation—conclusive evidence. But no such signature was found in the mysterious clay layer, ruling out the supernova hypothesis.[11] That left a giant meteor strike as the likely source of the iridium enrichment they had discovered.

Walter attributed to his father the realization that the atmospheric effects of an impact might provide a mechanism for extinction. A large impactor would literally blow away part of the atmosphere and flood the rest with enormous amounts of rock vapor and dust. Based on their iridium measurements and impact frequency estimates, they calculated that such an impactor would likely have been around 10 kilometers in diameter or larger. Some fraction of that large amount of material would remain in the stratosphere for years. "This dust," they wrote in their 1980 paper, "effectively prevented sunlight from reaching the surface for a period of several years, until the dust settled to

---

10. Ibid., pp. 69–71.
11. Ibid., pp. 73–74.

Earth. Loss of sunlight suppressed photosynthesis, and as a result, most food chains collapsed and the extinctions resulted."[12] The Alvarez team did not postulate a crater location, though they did comment that since the impactor had a two-thirds chance of impacting an ocean, there was a good chance that the resulting crater had already been subducted and had thus been erased.

The Alvarezes found the iridium component of their work rapidly corroborated by others. Walter presented first results at a September 1979 meeting in Copenhagen and met Jan Smit, who had already found a similar anomaly at an outcrop in Spain. (He had not yet published it because he had been ill.) But others followed, too. By the time their paper was published, elevated iridium had been found in a Cretaceous–Paleogene–age outcrop in New Zealand as well.[13]

Three main points of controversy followed their 1980 publication. One is familiar already: the time-honored argument that volcanoes were really responsible—either for the iridium layer or for the extinction—whenever impact is invoked. The Deccan Traps volcanism, which spanned the same period but was much longer in duration, formed the touchstone of this argument.[14] The second point involved extinction mechanisms: the impactor would not just have thrown enormous amounts of dust into the atmosphere. Scientists conceived, and sought evidence for, other killing mechanisms initiated by such an impact. The third point of controversy was simply the location of the crater, which many scientists wanted to find.

Among those who immediately dove into analysis of the Alvarez claim was James Pollack's research group at NASA Ames Research Center. A specialist in planetary atmospheres, Pollack had focused on understanding the atmospheres of Mars and Venus in the 1970s, assembling a group of atmospheric modelers with diverse interests and developing models for studying atmospheric chemistry, as well as radiative transfer. A good deal of their effort in the late 1970s concerned the effects of large volcanic eruptions on Earth's climate; they compiled and reviewed the available evidence from historic

---

12. Alvarez et al., "Extraterrestrial Cause for the Cretaceous-Tertiary Extinction," p. 1105.
13. Ibid.
14. See James Lawrence Powell, *Night Comes to the Cretaceous: Dinosaur Extinction and the Transformation of Modern Geology* (New York: W. H. Freeman and Co., 1998), pp. 85–95, for this line of argumentation.

eruptions just when a fortuitous series of eruptions occurred that could be studied directly, including the 1980 eruption of Mount St Helens.[15]

Thomas P. Ackerman, the most junior member of Pollack's group, was primarily interested in the effects of aerosols on climate. He had been involved in studying aerosols regionally—having worked on Los Angeles's smog problem for a while—and he recalled that their initial impression was that the Alvarez group had misunderstood what actually produced the climatic cooling in volcanic eruptions. The Alvarez analysis used the 1883 explosion of Krakatoa in Indonesia as an analogy for their asteroid impact. Krakatoa had thrown about 18 cubic kilometers of material into the atmosphere, reducing the amount of sunlight reaching the surface. This had produced a notable global cooling lasting about 2.5 years.[16] But large volcanoes do not produce dust exclusively. They also often inject sulfate aerosols into the stratosphere, and work done by Pollack and O. Brian Toon, also from NASA Ames, in the 1970s had suggested that the sulfate aerosols were in fact the principal cause of the cooling.[17]

Applying their models to the much larger amounts of dust that the impactor would have injected into the atmosphere, they found that the majority of the dust would settle out in three to six months, and sunlight levels would be too low for photosynthesis for between two months and a year.[18] This would, they speculated, cause the collapse of food chains and generate some extinctions, but not the 70 percent extinction that the paleontological record seemed to show.

Brian Toon presented the results of their model study in September 1981 at a special Geological Society of America meeting held to consider the implications of large impacts on Earth. The meeting was held at the Snowbird

15. Lawrence Badash, *A Nuclear Winter's Tale: Science and Politics in the 1980s* (Cambridge, MA: MIT Press, 2009), pp. 34–35; Adarsh Deepak, ed., "Atmospheric Effects and Potential Climatic Impact of the 1980 Eruptions of Mount St. Helens," in NASA Conference Publication 2240 (Hampton, VA: NASA, 1982); Alvarez et al., "Extraterrestrial Cause," p. 1105.

16. Alvarez et al., "Extraterrestrial Cause," p. 1105.

17. Badash, *Nuclear Winter's Tale*, p. 28.

18. O. B. Toon, J. B. Pollack, T. P. Ackerman, R. P. Turco, C. P. McKay, and M. S. Liu, "Evolution of an Impact-Generated Dust Cloud and Its Effects on the Atmosphere," in *Geological Implications of Impacts of Large Asteroids and Comets on the Earth*, ed. Leon T. Silver and Peter H. Schultz, Geological Society of America Special Papers, no. 190 (Boulder, CO: Geological Society of America, 1 January 1982), pp. 187–200, doi:10.1130/SPE190-p187.

ski resort in Utah and is often referred to as Snowbird I; to further confuse things, another meeting related to near-Earth objects was held the same summer at the Snowmass resort in Colorado—it was a busy year for scientists interested in impacts. Toon took the opportunity to comment on another potential mechanism that had been suggested by Emiliani, Kraus, and Shoemaker: that heat, not cold, had been the killer. An asteroid impact in the ocean (they hypothesized a North Pacific impact site) would have injected enormous amounts of water vapor into the atmosphere. Since water vapor is a greenhouse gas, it might have led to a rapid, but short, warming event that caused the extinction.[19] Toon pointed out that the dust and aerosols, which would be produced by an impactor this size regardless of impact site, would initially prevent much sunlight from reaching the surface, so there would be no immediate warming. But there might be a warming pulse after the dust had settled. This pulse was hard to quantify because some of the water vapor would have been removed with the dust.[20]

At the Snowbird meeting, John O'Keefe and Thomas Ahrens of the California Institute of Technology (Caltech) Seismological Laboratory examined the impact from the standpoint of energy. An impactor of the size expected by the Alvarez team would not be slowed by atmospheric entry (nor would it be by seawater), and thus the energy it supplied to the atmosphere and/or ocean would come from the shock waves and ejecta produced by its impact on the solid surface of Earth. The impact would produce a "jet" of superheated material—molten and vaporized rock as well as dust—that would reach space. A small portion of this material would reach escape velocity and leave Earth entirely; the rest would reenter, with the smallest particulate sizes remaining in the stratosphere and above for weeks and possibly months. The ejecta blanket would be global (the Alvarez clay layer was the global expression of the finest ejecta sizes) and roughly sorted so that the closer one got to the impact site, the thicker the blanket would be.

O'Keefe and Ahrens also addressed the question of heat. Their calculations suggested that a little more than a third of the impact energy would be transferred to the atmosphere by the ejecta via several different mechanisms, while

---

19. Cesare Emiliani, Eric B. Kraus, and Eugene M. Shoemaker, "Sudden Death at the End of the Mesozoic," *Earth and Planetary Science Letters* 55, no. 3 (1 November 1981): 317–334, doi:10.1016/0012-821X(81)90161-8.

20. Toon et al., "Evolution of an Impact-Generated Dust Cloud," pp. 187–200.

only around 5 percent would be transferred directly by the shock waves. They estimated that additional energy might result in "a global average temperature increase of at least 15°C," though they did not think this was likely. "We believe that a localized heat pulse would be more probable and that global heating via the ejecta interaction mechanism would be difficult."[21] The shock waves and heating produced by the impact would also generate nitric oxide, an ozone scavenger and a component of acid rain, providing still more stressors to surviving post-impact ecosystems.[22]

After his talk at the Snowbird meeting, Toon was approached by someone from the Defense Department—probably from the Defense Nuclear Agency—and asked if his team had applied their models to nuclear weapons detonations. They had not, of course, but that would soon change.[23] In the early 1980s, the Defense Department was planning to shift nuclear policy away from primarily targeting cities (which could be destroyed most effectively with air bursts) toward targeting Soviet missile installations, for which ground bursts would likely be more effective. That would throw a lot of dust into the air, just as volcanoes and asteroid impacts would, making the Pollack group's work very relevant to nuclear policy. They eventually received a formal letter from the Defense Nuclear Agency asking them to study the atmospheric effects of nuclear warfare.[24]

Initially, Ackerman did not think that nuclear weapons would have a major effect on the climate. A detonation would loft orders of magnitude less material into the atmosphere than the Cretaceous–Paleogene asteroid, and it would not produce a great deal of sulfate aerosol, either. So nuclear war did not seem capable of causing a climate catastrophe. But one of his collaborators, Richard Turco, who worked at a private research company, send him a draft of a paper in March 1982 that completely changed that interpretation.[25]

---

21. John D. O'Keefe and Thomas J. Ahrens, "The Interaction of the Cretaceous/Tertiary Extinction Bolide with the Atmosphere, Ocean, and Solid Earth," in *Geological Implications of Impacts of Large Asteroids and Comets on the Earth*, ed. Leon T. Silver and Peter H. Schultz, Geological Society of America Special Papers, no. 190 (Boulder, CO: Geological Society of America, 1982), p. 119.

22. Ibid.; Toon et al., "Evolution of an Impact-Generated Dust Cloud," pp. 187–200.

23. Thomas P. Ackerman, interview with Conway, 12 July 2016, transcript in NEO History Project collection.

24. Ibid.

25. Ackerman interview, 12 July 2016.

Written by Paul Crutzen and John Birks, it was part of a special issue of the Swedish journal *AMBIO* focused on the effects of nuclear weapons. Titled "The Atmosphere After Nuclear War: Twilight at Noon," the paper drew an analogy between large forest fires and nuclear war. Burning forests produce soot in addition to dust, and soot is black—it is carbon, after all. Black soot absorbs sunlight, heating itself and the air around it. That heated air, in turn, helps keep the soot aloft, creating a feedback loop ensuring that soot does not fall back out of the atmosphere as quickly as dust.

Worse, forests would not be the only thing to catch fire under nuclear attack. Cities are full of petrochemicals and their derivatives, plastics, which also burn and produce soot. And their calculations suggested that the petrochemical contribution of soot would be several times that of the forest fires. Crutzen and Birks concluded that

> the average sunlight penetration to the ground will be reduced by a factor between 2 and 150 at noontime in the summer. This would imply that much of the Northern Hemisphere would be darkened in the daytime for an extended period of time following a nuclear exchange.[26]

This was the first presentation of nuclear winter.

The Ames group's summary article was published in December 1983, filling 10 pages of *Science*. They presented results for several nuclear warfare scenarios, concluding: "[W]e find that a global nuclear war could have a major impact on climate—manifested by significant surface darkening over many weeks, subfreezing land temperatures persisting for up to several months, large perturbations in global circulation patterns, and dramatic changes in local weather and precipitation rates—a harsh 'nuclear winter' in any season."[27]

The soot finding from nuclear winter studies quickly influenced the "impact winter" research. Wendy Wolbach, then of the University of Chicago, led a team that studied the clay layer deposited in the handful of Cretaceous–Paleogene layers that had been sampled in the early 1980s, but not originally from a post-impact climate standpoint. They were interested in the question

---

26. Quoted from Badash, *A Nuclear Winter's Tale*, p. 52.
27. R. P. Turco, O. B. Toon, T. P. Ackerman, J. B. Pollack, and Carl Sagan, "Nuclear Winter: Global Consequences of Multiple Nuclear Explosions," *Science* 222, no. 4630 (23 December 1983): 1283–1292, doi:10.1126/science.222.4630.1283. The authorship was arranged so that the first initial of the five authors spelled out "TTAPS."

of whether impacts could supply organic material to Earth, or more precisely, what sizes of impacts might be efficient suppliers. "It has been clear all along that only bodies in the meteoritic size range could survive atmospheric entry," they wrote in 1985, since larger impactors would largely be vaporized on impact.[28] But they were surprised to find that the destruction of the impactor was far more complete than they had anticipated: no more than 4 percent survived vaporization. Thus, comets and asteroids would not have been efficient suppliers of cosmic organic material.

They also found that the clay layer contained soot particles, which they argued were similar enough to those produced by wildfires to make post-impact fires the likely source. They contended that the amount represented about 10 percent of the total biomass carbon on Earth (in 1985), meaning that "the scale of the putative wildfires must have been enormous." In addition to the climate impact of the soot, the fires would have produced pyrotoxins that "would harm most land life. Carbon monoxide alone, if produced in the same amount as soot, would reach 50 ppm in the atmosphere, a distinctly toxic level."[29] These fires would have occurred even if the impactor had struck an ocean site and not land, because the enormous, expanding cloud of vaporized rock would reach continents from any impact site and ignite vegetation. Later, others would point out that debris blasted out into space by the impact would reenter and provide still another potential mechanism for igniting fires and spreading them globally.[30]

By the time a second "Global Catastrophes in Earth History" meeting was held at Snowbird in 1988, numerous possible extinction mechanisms from the impact had appeared in the scientific literature. Extreme heat, prolonged cold, darkness shutting down photosynthesis, global wildfires, and various toxic gases were under discussion. Enormously destructive tsunamis would have ravaged coastal populations in the event of an oceanic impact; while they would not produce global extinctions, they would leave traces in the geologic record useful for identifying an impact location. But not all scientists accepted

---

28. Wendy S. Wolbach, Roy S. Lewis, and Edward Anders, "Cretaceous Extinctions: Evidence for Wildfires and Search for Meteoritic Material," *Science* 230, no. 4722 (11 October 1985): 167–170, doi:10.1126/science.230.4722.167.

29. Ibid.

30. H. J. Melosh, N. M. Schneider, K. J. Zahnle, and D. Latham, "Ignition of Global Wildfires at the Cretaceous/Tertiary Boundary," *Nature* 343, no. 6255 (January 1990): 251–254, doi:10.1038/343251a0.

that the impact had caused the contemporaneous mass extinction despite the many potential mechanisms.

A 1984 survey published in *Geology* revealed that while most American geophysicists already accepted the impact extinction hypothesis, most paleontologists accepted only that an impact had occurred, but not that it had been the cause of the extinction event.[31] Perhaps the most skeptical during that decade were dinosaur specialists. Many believed that the dinosaurs were already in decline by the end of the Cretaceous, based on their apparent disappearance 3 meters below the Cretaceous–Paleogene boundary at the richest site for dinosaur collecting, in Hells Creek, Montana. This 3-meter "gap" suggested that the impact might have been only a kind of coup de grâce. Other fossil types, most significantly planktons, did seem to disappear abruptly at the boundary.[32] At the 1981 Snowbird meeting, Phillipp Signor and Jere Lipps of the University of California, Davis, had argued that apparent gradual declines in the fossil record could well be the result of sampling bias and not representative of actual population declines. Thus, the fossil record could not be taken at face value, adding another layer of debate to the extinction controversy—which, it seemed, would not be solved through paleontology alone.[33]

Finding the Crater

By the time the Alvarez paper was published in 1980, only about a hundred impact craters had been identified on Earth. Only three were of approximately the right size to be considered candidates for the Cretaceous–Paleogene

31. Antoni Hoffman and Matthew H. Nitecki, "Reception of the Asteroid Hypothesis of Terminal Cretaceous Extinctions," *Geology* 13, no. 12 (1 December 1985): 884–887, doi:10.1130/0091-7613(1985)13<884:ROTAHO>2.0.CO;2; see also Powell, *Night Comes to the Cretaceous*, pp. 162–163.

32. Hans R. Thierstein, "Terminal Cretaceous Plankton Extinctions: A Critical Assessment," in *Geological Implications of Impacts of Large Asteroids and Comets on the Earth*, ed. Leon T. Silver and Peter H. Schultz, Geological Society of America Special Papers, no. 190 (Boulder, CO: Geological Society of America, 1982), pp. 385–400.

33. Philip W. Signor and Jere H. Lipps, "Sampling Bias, Gradual Extinction Patterns and Catastrophes in the Fossil Record," in *Geological Implications of Impacts of Large Asteroids and Comets on the Earth*, ed. Leon T. Silver and Peter H. Schultz, Geological Society of America Special Papers, no. 190 (Boulder, CO: Geological Society of America, 1982), pp. 291–296.

impact, and none of these were of the right age.[34] Unlike those on the Moon, craters on Earth are gradually destroyed by a suite of mechanisms. If the impactor had crashed into one of the ocean basins, the resulting crater might already have been subducted into the mantle over the ensuing 66 million years. If it had fallen into Antarctica, it would be buried in ice. And if it had struck continental shelf, it might be buried in sediments. Even in the "best case" for geologists interested in looking for signs of the crater—an impact on land—the structure would be eroded, tectonically deformed, and potentially very hard to identify.

If there had been a big impact, there would also be ejecta, and lots of it—somewhere. Some of this ejecta would have come from the asteroid, but most of it would be from the "target rock" that had been struck. Since the ejecta would be scattered the world over, the composition of that target rock would provide an important clue as to the location of the crater. Oceanic crust and continental crust rocks have different compositions, so analysis of the clay layers at the known exposures of the Cretaceous–Paleogene transition would (at least) indicate whether the impact had occurred in land or water. But in 1980, there were not many known complete transitional sequences for that era. And compositional results turned out to be at best ambiguous, or at worst, misleading. Some of Walter Alvarez's collaborators got very solid results indicating an oceanic impact, while others, drawing samples from possible sites in the North American west, found shocked varieties of quartz, a mineral typically found in continental, not oceanic, crust.[35]

The answer to this little mystery took until 1990 to emerge, and the story is complex. No single person deserves credit for the crater's discovery, although one person, Alan Hildebrand, is most responsible for getting the geological community to accept it as the impact site.[36] His story provides an interesting parallel to the controversy surrounding Meteor Crater because the structure that turned out to be the crater had already been identified but had been widely dismissed as a potential impact crater.

Hildebrand started graduate school at the University of Arizona in 1984 after a short career in the mining industry in northwestern Canada. His

---

34. Alvarez, *T. Rex and the Crater of Doom*, p. 89.
35. Ibid., pp. 92–93, 96–97.
36. See Alan Hildebrand, "The Cretaceous/Tertiary Boundary Impact (or The Dinosaurs Didn't Have a Chance)," *J. Roy. Astron. Soc. Can.* 87, no. 2 (1993): 77–118.

advisor, William Boynton, ran a neutron-activation lab for isotopic analysis. During his first year, Hildebrand heard a talk by Walter Alvarez and had also started reading the literature on the impact mystery as part of his classwork. He was intrigued by the boundary layer clay composition, which he felt, along with many others, held the clues to the impact site. One recent analysis of neodymium isotopes in the clay had further supported the idea of an oceanic impact, he remembered, but the composition differed enough from oceanic crust to suggest that some amount of continental crust might have been present in the target area, too. In other words, the impact could have occurred someplace where oceanic crust was overlain by continental sediments.

Hildebrand thought that an impact at an oceanic site would produce an enormous tsunami, and he found papers describing the kinds of evidence that a tsunami would have left in the geologic record. A tsunami would have churned up the sea floor, and that disturbed layer would show up in sediment cores near the impact site. It would also have carried quartz grains from the impact site and dropped them on the shores bounding whichever ocean the impactor had hit. He already knew that there were deposits of quartz grains at various sites in the U.S. Gulf Coast states, although they had been interpreted as the result of low-standing seas present at the end of the Cretaceous, not the result of impact-generated giant waves.[37] Perhaps that interpretation was wrong.

The most studied end-Cretaceous deposit was in Texas, known as the Brazos River deposit. The outcrop was not far from College Station, so paleontologists Stephan Gartner and Ming-Jung Jiang at Texas A&M University had sent samples to be tested for the iridium anomaly, which was subsequently identified. Hildebrand visited the section with them in 1987 and came away convinced that it actually represented wave deposits. The impact, then, had to have occurred in the Caribbean. Hildebrand was not alone in interpreting the Brazos River deposit as wave-related. Walter Alvarez's collaborator Jan Smit had suggested it in 1984; Joanne Bourgeouis and Thor Hansen had again in 1987.[38]

The evidence notwithstanding, Hildebrand visited other sites in Missouri, Arkansas, and even Illinois, looking for more corroboration. In April 1988, he

---

37. Alan Hildebrand interview with Conway, 20 September 2016, transcript in NEO History Project collection.

38. Alvarez, *T. Rex and the Crater of Doom*, pp. 108–109; Joanne Bourgeois et al., "A Tsunami Deposit at the Cretaceous-Tertiary Boundary in Texas," *Science* 241, no. 4865 (29 July 1988): 567–569.

went through the records of the Deep Sea Drilling Project, looking for sediment cores that contained disturbed layers around the Cretaceous–Paleogene boundary—and he found them, in cores from the Gulf of Mexico and Columbia Basin. "That was literally the light-bulb moment, the impact must be between North and South America."[39]

The last important piece of evidence he needed was ejecta; to date, nobody had found an ejecta blanket. Or, as he soon learned, it had been found but had not been interpreted as such. A Haitian-American geologist named Florentin Maurrasse had discovered a thick Cretaceous–Paleogene–era deposit on the southern peninsula of Haiti. It was known as the Beloc formation after a nearby town, and he had interpreted it as volcanic in origin. Hildebrand visited Maurrasse's lab in June 1989 and decided that Maurrasse was wrong; it was actually a piece of the ejecta blanket he was looking for. And when he collected samples at the outcrop himself, he discovered that it contained the largest shocked-quartz grains and tektites yet found. Since gravity ensures that grain size declines with distance from the impact site, the impact site had to be relatively nearby.[40]

Hildebrand presented some of his findings at the 1988 Snowbird meeting, published abstracts in the meeting's abstracts volume with Boynton, and got a fuller discussion published in May 1990 in *Science*.[41] In their discussion, they dismissed a few candidate sites before discussing two that could not be rejected easily: one in the Columbia Basin underlying a large area of the Caribbean Sea, and the second an approximately 200-kilometer-diameter circular structure, which had been reported in 1981, on (or rather, mostly under) the Yucatán Peninsula.

The Yucatán Peninsula site was a late addition to the paper because Hildebrand had only just heard of it, from a *Houston Chronicle* reporter named Carlos Byers, shortly before publication. Byers knew of it from an abstract published in 1981 by two scientists who had worked for Mexico's state oil company, PEMEX. The abstract's authors, Antonio Camargo Zanoguera and Glen Penfield, had spent the mid-1970s resurveying the northern Yucatán area,

---

39. Hildebrand interview, 20 September 2016.

40. Ibid.; Powell, *Night Comes to the Cretaceous*, p. 102.

41. Alan R. Hildebrand and William V. Boynton, "Proximal Cretaceous-Tertiary Boundary Impact Deposits in the Caribbean," *Science* 248, no. 4957 (18 May 1990): 843–847, doi:10.1126/science.248.4957.843.

revisiting a circular magnetic anomaly that had been discovered by PEMEX in the 1950s. The 1950s survey had concluded that the structure was likely to have been volcanic in origin. Penfield and Camargo were not convinced of that conclusion, and they had wondered whether it was the remnant of an impact crater. But since they had never published more than the abstract, the idea had vanished until Hildebrand's encounter with Carlos Byers.

Hildebrand got another lucky break that quickly presented him with new evidence: a job interview with the Canadian Geological Survey in Ottawa. At the time, they were involved in a major project to produce volumes of geophysical data for all of North America. Hildebrand recalled years later seeing a gravity map of the Yucatán, which surprised him because PEMEX, like other oil producers, did not typically release their data. "So I happened to look at the Yucatán Peninsula, so I could see this big circular negative anomaly where there had been the suggestion of a buried crater, which was very interesting, because, (a), there was something there, the data showed it; and, (b), data were available."[42]

Hildebrand shifted his efforts from the Columbia Basin site, which had been the focus of his 1990 paper, to the Yucatán site, which was named Chicxulub for a village near the anomaly's center, and started to work with Penfield to find other kinds of data. Penfield had done magnetic survey work in the area during the 1970s, although those data were not yet available from PEMEX. And PEMEX had drilled a series

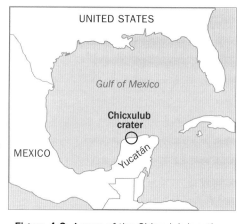

Figure 4-2. A map of the Chicxulub location.

of six cores across the peninsula in the same decade, though the resulting core data were not available either. But within a few months, Penfield and Hildebrand had located some surviving samples of a core called Yucatán 6, in the possession of Alan Weidie at the University of New Orleans. The core had been drilled from within what Penfield and Hildebrand thought was likely to be the crater, and Hildebrand found shocked quartz in these samples, too.

---

42. Hildebrand interview, 21 September 2016.

Hildebrand told Weidie of the discovery, and then apparently Weidie told Kevin Pope, a consultant in La Cañada, California. Pope called Hildebrand and told him that he intended to publish a paper identifying the crater via still another means, Landsat imagery. He, Adriana Ocampo at JPL, and Charles Duller at NASA Ames Research Laboratory in Sunnyvale, California, had noted in the imagery that what the inhabitants of the Yucatán Peninsula called "cenotes" were arranged in a semicircle about 170 kilometers in diameter. The cenotes were sinkholes, formed by the collapse of the limestone subsurface. Mayan civilization had been built around these cenotes, as they provided a ready, and mostly reliable, source of freshwater. The Landsat imagery, taken by itself, was suggestive of a buried crater but was hardly definitive. Pope, Ocampo, and Duller had submitted a paper on this to *Science*, which was rejected after a lengthy review—Ocampo remembered it taking about 10 months—and then they had sent it to *Nature* as a scientific correspondence in early 1991. It was published in May.[43] *Nature* rejected Hildebrand's submission, though, and he ultimately published a more substantive paper in *Geology*, with Penfield and others, in September 1991.[44]

Identification of the putative impact crater's location did not quite end the story. Other geologists continued looking for more evidence. A team led by Walter Alvarez surveyed northern Mexico looking for further evidence of an impact, still trying to prove beyond a doubt that the buried structure was not volcanic in origin. In 1992, they found extensive evidence of an impact-generated tsunami there, most prominently in a formation known as Lajilla.[45] Pope and Ocampo, with the aid of the Planetary Society in Pasadena, organized a series of expeditions to Belize to look for outcrops close to where the crater rim was supposed to have been. They had been unable to acquire samples from PEMEX's drilling, and that drove them to make their own search. In a quarry on Albion Island in the Rio Honde river, they found a unique exposure of what they believed was the closest ejecta surviving from the impact. Fifteen meters thick, it was composed of dolomite fragments. Pope and Ocampo invited Walter Alvarez and some of his collaborators along

43. K. O. Pope, A. C. Ocampo, and C. E. Duller, "Mexican Site for K/T Impact Crater?" *Nature* 351 (1 May 1991): 105, doi:10.1038/351105a0.

44. Alan R. Hildebrand et al., "Chicxulub Crater: A Possible Cretaceous/Tertiary Boundary Impact Crater on the Yucatán Peninsula, Mexico," *Geology* 19, no. 9 (1991): 867–871.

45. Alvarez, *T. Rex and the Crater of Doom*, pp. 126–127.

to see it on a 1995 trip; Alvarez was skeptical that the formation was really an impact blanket and had needed to see it for himself.[46]

In 1994, a third Snowbird conference on catastrophes in Earth's history was held, this time in Houston. Robert Ginsburg of the University of Miami organized a field trip to one of the giant wave deposits in northeast Mexico, hoping to settle the controversy over whether these deposits had been laid down nearly instantly or over a substantial span of time. His participants included both sedimentologists who supported the impact hypothesis and those who were skeptical of it. *Science* magazine journalist Richard Kerr reported that while the field trip did not change many minds, it did settle one question: whether the formation was the result of thousands of years of deposition or at most a few months. There was no evidence of reprocessing of the sand by burrowing animals, meaning that the entire several-meter-thick formation was laid down in a geological instant. One sedimentologist from Conoco, Inc., commented to Kerr that "if you don't have [an impact], you'd be hard put to come up with an alternative."[47]

At the third Snowbird conference in Houston, Ginsburg also presented the results of a "blind test" of the Cretaceous–Paleogene impact extinction hypothesis, hoping to settle the issue. He and two collaborators had sent coded but otherwise unidentified samples of the El Kef formation in Tunisia to four specialists in foraminifera, asking them to identify the species present in the samples and their numbers, and to return the data to Ginsburg. The idea was to determine whether species went extinct gradually or nearly instantly. As one might expect, this exercise did not end the debate. Richard Kerr reported that a majority of the attendees seemed to interpret the results as supporting a rapid extinction, but one outspoken critic attacked Kerr for this conclusion and also attacked one of the participants in the test.[48] The gradualist idea did not go into extinction easily.

In 1991, a group from the University of Rhode Island's School of Oceanography analyzed impact glasses from the Beloc formation in Haiti for their composition. They found two main types: black glass that was high in silica, deriving from continental crust, and high-calcium yellow glasses

---

46. Ibid., pp. 135–136.

47. Richard A. Kerr, "Searching for the Tracks of Impact in Mexican Sand," *Science* (11 March 1994): 1372.

48. Richard A. Kerr, "Testing an Ancient Impact's Punch," *Science* (11 March 1994): 5152.

that were rich in sulfur. The yellow glasses derived from sulfur-rich sediments, which are relatively rare on Earth's surface—but not on the Yucatán Peninsula, which consisted of continental crust with a few kilometers of alternating carbonate and anhydrite (anhydrous calcium sulfate) layers on top of it. In other words, the chemistries were similar enough to link the glasses to the Yucatán impact site.

But the main point of the paper focused on the implications of this mineralogical association. The impact energy would have vaporized an essentially unimaginable amount of rock, trillions of tons of it, and released billions of tons of sulfates from the anhydrite layers into the atmosphere. The Cretaceous–Paleogene impact would have injected this mass far higher into the atmosphere than a volcano, too, prolonging the sulfate aerosols' residence time to perhaps a decade.[49] This also suggested that the asteroid-induced winter would last not months, but years. Suppression of photosynthesis for a decade would collapse nearly all of Earth's ecosystems. The unusual mineralogy of the Chicxulub site, then, may have made the extinction worse than it would have been if the impactor had hit somewhere else on Earth.

These discoveries did not quite end all debate about the linkages between the impact and the mass extinction. Gerta Keller of Princeton University, for example, argued in a 2003 review that then-current evidence suggested not a single impact but three, over a period of about 400,000 years.[50] And she did not accept that the Chicxulub impactor (which was second in her chronology) was the "coup de grâce," as she believed it to be 300,000 years too young. Rather, a third, even larger impactor whose crater had not been identified might have been the culprit.[51] But it might not have been. It bothered her enormously that no other mass extinction event could be tied to a

---

49. H. Sigurdsson, S. D'Hondt, and S. Carey, "The Impact of the Cretaceous/Tertiary Bolide on Evaporite Terrane and Generation of Major Sulfuric Acid Aerosol," *Earth and Planetary Science Letters* 109, nos. 3–4 (April 1992): 543–559, doi:10.1016/0012-821X(92)90113-A.

50. G. Keller et al., "Multiple Impacts Across the Cretaceous–Tertiary Boundary," *Earth-Science Reviews* 62, no. 3 (1 September 2003): 327–363, doi:10.1016/S0012-8252(02)00162-9. By the time this book was published, four large impact craters had been identified within a few hundred thousand years of the K–Pg boundary, though Chicxulub was by far the largest.

51. G. Keller, "Impacts, Volcanism and Mass Extinction: Random Coincidence or Cause and Effect?" *Australian Journal of Earth Sciences* 52, nos. 4–5 (1 September 2005): 725–757, doi:10.1080/08120090500170393.

large impact by the early 2000s; instead, most coincided with very-large-scale volcanism and associated climatic and sea-level change events.

Keller was not alone in arguing that asteroids were not likely to be the cause of most mass extinctions. A 2007 review found that while a number of large flood basalt emplacements and associated climatic and oceanic changes coincided with Phanerozoic extinction events, only the Chicxulub impactor coincided in time closely enough to be the cause of its simultaneous extinction.[52] A decade later, another review found five cases of large impact events (totaling six craters) that coincided with extinction events during the latter half of Phanerozoic time.[53] The same researchers also found that seven extinction events were coincident with flood basalt eruptions like those that formed the Deccan Traps. Thus "these statistical relationships argue that most mass extinction events are related to climatic catastrophes produced by the largest impacts and large-volume continental flood-basalt eruptions."[54]

The Chicxulub impact occurred during the Deccan Traps volcanism in what is now India, a series of massive eruptions that emplaced over 200,000 cubic kilometers of basalt. That scale of volcanism would have imposed dramatic climatic changes globally without the need of an impact. And indeed, some scientists, including Walter Alvarez, later argued that the Chicxulub impactor could have triggered some of the later Deccan eruptions.[55] With generally widespread acceptance that both large-scale volcanism and large impacts could trigger extinction events, much of the remaining controversy swirled around geochronology, a notoriously inexact science.

52. Simon Kelley, "The Geochronology of Large Igneous Provinces, Terrestrial Impact Craters, and Their Relationship to Mass Extinctions on Earth," *Journal of the Geological Society* 164, no. 5 (1 September 2007): 923–936, doi:10.1144/0016-76492007-026.

53. Michael R. Rampino and Ken Caldeira, "Correlation of the Largest Craters, Stratigraphic Impact Signatures, and Extinction Events over the Past 250 Myr," *Geoscience Frontiers* 8, no. 6 (1 November 2017): 1241–1245, doi:10.1016/j.gsf.2017.03.002.

54. Richard B. Stothers, "The Period Dichotomy in Terrestrial Impact Crater Ages," *Monthly Notices of the Royal Astronomical Society* 365, no. 1 (1 January 2006): 178–180, doi:10.1111/j.1365-2966.2005.09720.x.

55. Mark A. Richards, Walter Alvarez, Stephen Self, Leif Karlstrom, Paul R. Renne, Michael Manga, Courtney J. Sprain, Jan Smit, Loyc Vanderkluysen, and Sally A. Gibson, "Triggering of the Largest Deccan Eruptions by the Chicxulub Impact," *Geological Society of America Bulletin* 127, nos. 11–12 (November 2015): 1507–1520, doi:10.1130/B31167.1.

Impacts and the History of Life on Earth

Beyond the specifics of the Cretaceous–Paleogene extinction event, the discussion of cosmic impacts launched in 1980 opened two other important lines of research and argumentation: periodicity in extinction events and the effects of even larger impact events than Chicxulub represented. The periodicity argument triggered efforts to find a cause, or causes, that could explain a cyclic recurrence of impacts on Earth, while the likelihood of even larger impacts had implications for astrobiology.

The periodicity argument largely derived from the work of David Raup and J. John Sepkoski, Jr., of the University of Chicago, though in his own memoir, Raup credits a 1977 analysis as the progenitor of the idea. Raup had rejected it as lacking in statistical rigor when it had been published.[56] His work with Sepkoski came about after a series of workshops sponsored by NASA at its Ames Research Center on the evolution of life. Sepkoski had developed a database of marine fossil ranges covering the past 250 million years, to which the two had then applied a series of statistical tests. They found that mass extinctions recurred about every 26 million years during that 250-million-year period. If that were true, they argued,

> the implications are broad and fundamental. A first question is whether we are seeing the effects of a purely biological phenomenon or whether the periodic extinctions result from recurrent events or cycles in the physical environment. Does this reflect an earthbound process or something in space? If the latter, are the extraterrestrial influences solar, solar system, or galactic?[57]

They favored extraterrestrial explanations because it seemed "incredible" that a purely earthbound process could occur on such a fixed schedule over tens of millions of years.

Before their paper had even been published, other scientists had already proposed similar extraterrestrial extinction mechanisms. Sepkoski had given a presentation on the subject at a meeting in Flagstaff, Arizona, and George

---

56. David M. Raup, *The Nemesis Affair: A Story of the Death of Dinosaurs and the Ways of Science*, 2nd ed. (New York: W. W. Norton, 1999), p. 107.

57. D. M. Raup and J. J. Sepkoski, "Periodicity of Extinctions in the Geologic Past," *Proceedings of the National Academy of Sciences* 81, no. 3 (1 February 1984): 801–805, quoted from p. 805.

Alexander of the *Los Angeles Times* published a story on it. That resulted in other scientists asking Raup and Sepkoski for preprints of the paper, so by the time it was actually published, *Nature* had received a number of papers proposing cosmic mechanisms.[58] The journal chose to publish them in a single issue: 19 April 1984. Gene Shoemaker had pointed out to Raup in 1983 that passage of the solar system through the "arms" of our galaxy might disturb the Oort cloud at the far edge of our solar system enough to unleash a comet bombardment on the inner solar system, but that would occur on a 100-million-year time scale, not a 10-million-year one.[59]

One of the *Nature* papers published in April 1984 advanced the hypothesis that an "unseen companion to the Sun, travelling in a moderately eccentric orbit," periodically passes through the Oort cloud, triggering the release of a billion comets into the inner solar system on the 26-million-year cycle. A few of these, they proposed, had hit Earth over the ensuing million years. The Sun's mysterious companion might have remained hidden, being small, dim, and relatively slow-moving against the background of more distant stars. They also understood some professional risk in hypothesizing an invisible star: "We worry that if the companion is not found, this paper will be our nemesis."[60] Their proposal is sometimes known as the "Nemesis hypothesis"—or the "Siva hypothesis," after Stephen J. Gould argued in favor of the Hindu god of destruction instead.[61]

In January 1985, still another possible extraterrestrial agent for periodic bombardment was identified—"Planet X." Daniel Whitmire and John Matese of the University of Southwestern Louisiana proposed the existence of a 10th planet as the solution to a basic problem of orbital mechanics: that well-known discrepancies in the orbits of the outer planets required the presence of some other large body between 1 and 5 times the mass of Earth as an explanation. This undiscovered planet could be the source of periodic comet bombardment, though the comets would not come from the distant Oort cloud

58. Raup, *The Nemesis Affair*, pp. 135–136.

59. E. M. Shoemaker, "Large Body Impacts Through Geologic Time," in *Patterns of Change in Earth Evolution*, ed. H. D. Holland and A. F. Trendall, Dahlem Workshop Reports Physical, Chemical, and Earth Sciences Research Reports (Berlin, Heidelberg: Springer, 1984), pp. 15–40, *https://doi.org/10.1007/978-3-642-69317-5_3*.

60. Marc Davis, Piet Hut, and Richard A. Muller, "Extinction of Species by Periodic Comet Showers," *Nature* 308, no. 5961 (April 1984): 715–717, doi:10.1038/308715a0.

61. Stephen Jay Gould, "The Cosmic Dance of Siva," *Natural History* (August 1984): 14.

but from a nearer "disk" of comets beginning about 30 au from the Sun that itself had been hypothesized but not yet seen.[62] More recently, physicist Lisa Randall has argued that the solar system's passage through or near regions of "dark matter" might be the trigger for periodic comet bombardments.[63]

The recognition of the Chicxulub impactor's enormous effects on Earth life also raised more general questions about the evolution of life on Earth and on other planets. The pockmarked face of the Moon provided clear evidence that early in the solar system's existence, much larger collisions than Chicxulub's must have happened. Mars displays such evidence, too. Analyses of the Chicxulub impact suggested that the thermal radiation produced by the ejecta would subject Earth to around 10 kilowatts of power on every square meter of its surface (with variations from place to place), igniting fires as well as causing "widespread mortality and desiccation of plant life," which would then itself be vulnerable to ignition by lightning.[64] Larger impacts would create even larger heat pulses at the surface. At some point, the ocean surface would boil, producing a global steam bath that would serve as an effective sterilizing agent (and, of course, cook the dwellers of the ocean shallows). What was that level? How big would an impactor have to be to completely boil away the oceans?

In 1989, a team led by Norman Sleep of Stanford University raised these questions in the context of establishing when the last impact large enough to sterilize the early Earth could have happened. To researchers interested in the evolution of life, that was a key question. Knowing the answer would effectively bound life's timeline. They found that an impactor about the size of main-belt asteroids Vesta and Pallas, around 440 kilometers in diameter, would be sufficient to completely boil away Earth's oceans. The world would be enveloped by rock vapor and molten rock that would take several months to condense out. Once it had, the "runaway greenhouse effect" that currently keeps Venus superheated would prevent heat from radiating out into space quickly. The steam bath would last one to two thousand years in the case of

62. Daniel P. Whitmire and John J. Matese, "Periodic Comet Showers and Planet X," *Nature* 313, no. 5997 (January 1985): 36–38, doi:10.1038/313036a0.

63. Lisa Randall, *Dark Matter and the Dinosaurs: The Astounding Interconnectedness of the Universe* (New York: Ecco, 2015).

64. H. J. Melosh et al., "Ignition of Global Wildfires at the Cretaceous/Tertiary Boundary," *Nature* 343, no. 6255 (January 1990): 251–254, doi:10.1038/343251a0.

the "minimal ocean-vaporizing impact," as they phrased it, and even longer for a larger impact.[65] Nothing would likely survive this.

The impact of an object 190 kilometers or so in diameter would be sufficient to boil the top 200 meters of the oceans away (the photic zone, the maximum depth to which enough sunlight reaches for photosynthesis to take place). Returning to "normal" global temperatures would take merely 300 years, which would not be a significantly better outcome for surface life. Statistically, the smaller impact could have happened more recently than the larger "minimum ocean evaporating" impact by a few hundred million years, raising the possibility that life evolved very early, was extinguished by a more recent impact, and restarted. In effect, giant impacts could serve as reset buttons. The last reset for Earth could have been as recent as 3.8 billion years ago.[66] In the case of the smaller impactor, while surface life might be annihilated, the ocean floors might have served as a kind of refuge for the small number of surviving species, if early life had existed in that environment. But in 1989, there was no evidence of deep ocean life at 3.8 billion years ago.[67]

In a 1998 paper, Sleep and Kevin Zahnle of NASA Ames Research Center suggested that giant impacts might also serve as a means to transfer life between planets. Huge impacts on Earth would inevitably blast rocks containing microscopic life into space, some of which would just as inevitably wind up crashing onto Venus, Mars, Earth's Moon, etc.—and some of these rocks would eventually return to Earth. Rocks returning to Earth could serve as a space-based "refugia" for life and reseed Earth after a sterilizing impact. Conversely, many astrobiologists believed Mars was habitable before Earth had cooled enough to become habitable, meaning that life on Earth might have originated on Mars and then been seeded here by a giant impact. They suggested that this possibility might make it difficult to distinguish Martian life (should there be any, perhaps in subsurface refuges) from terrestrial life, as we might actually have the same common ancestor.[68]

---

65. Norman H. Sleep, Kevin J. Zahnle, James F. Kasting, and Harold J. Morowitz, "Annihilation of Ecosystems by Large Asteroid Impacts on the Early Earth," *Nature* 342, no. 6246 (November 1989): 139–142, doi:10.1038/342139a0.

66. Ibid.

67. Ibid.

68. Norman H. Sleep and Kevin Zahnle, "Refugia from Asteroid Impacts on Early Mars and the Early Earth," *Journal of Geophysical Research: Planets* 103, no. E12 (25 November 1998): 28529–28544, doi:10.1029/98JE01809.

This was not idle speculation, nor entirely the province of modeling. In 1982, scientists at NASA Johnson Space Center had identified a meteorite found in Antarctica as likely originating from Mars; in 1983, others had more conclusively shown a lunar origin for another Antarctic meteorite.[69] In 1996, a team of scientists had even published a claim (accompanied by a widely viewed press conference at NASA Headquarters) that they had found traces of ancient life in a Mars meteorite collected from Antarctica's Allen Hills.[70] That claim was under fierce dispute in 1998, but by then it was no longer controversial to believe that bits of Mars had reached Earth, and possibly vice versa. The question that remained as of this writing was whether living organisms had been able to make the trip, too.

The larger theme of this research on impact periodicity and effects was that cosmic impacts had repeatedly influenced the evolution of life on Earth. They should also have seeded life to Venus and Mars (or from Mars to Venus and Earth). Impacts were important not only to the geology and geophysics of the solar system, but to its biology as well.

The two decades following the Alvarez team's discovery of iridium in the Cretaceous–Paleogene boundary clay witnessed a sweeping reevaluation of the evolution of Earth and its place within the solar system. Earth was no longer seen by scientists as an entity unto itself, alone in the cold darkness of space. It had been—and still is—directly affected by events outside its atmosphere and must be studied in that larger context.[71]

Many more impact craters, or astroblemes, were identified on Earth in the wake of the Alvarez hypothesis, in part due to the efforts to find the Cretaceous–Paleogene impact site. In 2018, the Earth Impact Database at the University of New Brunswick, Canada, listed 190 confirmed impact structures on Earth. The largest accepted crater, the 160-kilometer-diameter

69. Steven J. Dick and James E. Strick, *The Living Universe: NASA and the Development of Astrobiology* (New Brunswick, NJ: Rutgers University Press, 2004).

70. This story is well told in Kathy Sawyer, *The Rock from Mars: A Detective Story on Two Planets* (NY: Random House, 2006).

71. Also see Valerie A. Olson, "Political Ecology in the Extreme: Asteroid Activism and the Making of an Environmental Solar System," *Anthropological Quarterly* 85, no. 4 (2012): 1027–1044.

Vredefort Structure in South Africa, is also among the oldest, at about 2 billion years, and it is slightly larger than the Chicxulub structure.[72]

Neither Nemesis/Siva nor Planet X had been located as of this writing, and while a group at Caltech was busily hunting for a "Planet 9" even larger than the putative Planet X was supposed to be,[73] they were not arguing that it might be the cause of a periodic comet bombardment. The periodicity argument was also in considerable dispute. The idea that impacts showed periodicity was accepted by some within the scientific community and rejected by others, and the actual period was also contested (a 2006 review put it at 35 million years).[74]

The recognition that impacts could affect the course of life on Earth would also gradually filter into the public consciousness and into public policy circles, and this growing recognition began to demand more policy response.

---

72. "Earth Impact Database," *http://www.passc.net/EarthImpactDatabase/index.html* (accessed 6 October 2016).

73. Michael E. Brown, "The Planet Nine Hypothesis," *Physics Today* 72, no. 3 (2019): 70–71.

74. Richard B. Stothers, "The Period Dichotomy in Terrestrial Impact Crater Ages," *Monthly Notices of the Royal Astronomical Society* 365, no. 1 (1 January 2006): 178–180, doi:10.1111/j.1365-2966.2005.09720.x.

CHAPTER 5
RECOGNIZING COSMIC HAZARD

Introduction

If the research efforts spawned by the 1980s effort produced a radical shift in scientific thought about the development of life on Earth, they had a somewhat more delayed effect on public policy. Scientists initially found it difficult to quantify and relay to the public and policy-makers the credible risks of near-Earth objects. In 1980, there were only 51 known near-Earth asteroids, and the discovery rate was only 2–3 per year.

In the last two decades of the 20th century, an accelerating pace of NEO discoveries, together with the resulting increase in predicted close-Earth approaches and the spectacular impact of Comet Shoemaker-Levy 9 on Jupiter in 1994 brought about the gradual recognition in the policy community of the existence of cosmic risks. Incorrect predictions for a possible Earth impact by Comet Swift-Tuttle in 2126 and by asteroid 1997 XF11 in 2028 made it even more clear to scientists that an organized effort would be necessary to discover, follow up, accurately track, and physically characterize the population of near-Earth objects. In the late 20th century, complex computer simulations were used to compute and quantify the mechanics and energies of near-Earth object impacts.[1] These simulations were used to model the nearly forgotten 1908 Tunguska event in Russia, as well as the remarkably energetic impacts on

---

1. A simplified, interactive program to compute asteroid-Earth impact energies, crater sizes, and effects has been made available by Gareth Collins, Jay Melosh, and Robert Marcus. See *Impact Earth!* at *https://www.purdue.edu/impactearth/* (accessed 16 April 2017).

Jupiter by fragments of Comet Shoemaker-Levy 9 that were measured as part of a worldwide observing campaign in 1994 that included extensive coverage by the Hubble Space Telescope.

In the final decades of the 20th century and the first decade of the 21st, several NASA and international NEO conferences and working groups were convened to outline the issues relating to NEO risks, including the recommended discovery goals and possible Earth impact mitigation options. In the United States, NASA and the U.S. Air Force were the two most involved organizations, and by the end of the 20th century, NASA had taken an international leadership role for near-Earth object issues.

Early Studies of the Asteroid Hazard

As we witnessed earlier in the book, the idea that impacts could have major geophysical and biological effects on Earth was controversial within the scientific community during the first half of the 20th century but gradually gave way to acceptance. That large-scale community view had not prevented some tentative, but influential, studies of the asteroid threat and its mitigation from being made. One early effort was an MIT student study of threat mitigation in 1967.

In early 1967, MIT professor Paul Sandorff gave his systems engineering graduate students a novel assignment based on the expected close pass of Apollo asteroid 1566 Icarus the following year.[2] Discovered in 1949, this Earth-crossing object has an orbital period that forms an integer ratio with that of Earth, 19 to 17, resulting in a close approach every 19 years. Near the time of its discovery, Icarus passed within 9 million miles of Earth, about 38 times the distance to the Moon. It would pass within half that distance on 14 June 1968, just 15 months from the time Sandorff announced the assignment. The exercise was intended to give the students experience in improvising

---

2. Luis A. Kleiman, ed., "Project Icarus," Massachusetts Institute of Technology Report, interdepartmental student project in systems engineering at the Massachusetts Institute of Technology (MIT), spring term, 1967 (Cambridge, MA: MIT Press, 1967); Dwayne A. Day, "Giant Bombs on Giant Rockets: Project Icarus," *The Space Review,* http://www.thespacereview.com/article/175/1; MIT Student Project in Systems Engineering, *Project Icarus* (Cambridge, MA: MIT Press, 1979), *https://mitpress.mit.edu/books/project-icarus-systems-engineering* (accessed 5 March 2021).

under pressure, and they were charged with using this period to devise a plan to stop Icarus—by whatever means necessary.

With such a short time window, Sandorff's engineering students quickly realized that a fast intercept was the only feasible option, and nuclear weapons were integral to the plan from the outset. The caliber of the nuclear bomb required to fragment a solid rock the size of the asteroid (about 1,000 megatons) was out of reach, and it was too late to launch a mission to reach it at aphelion—the slowest point in its orbit and hence the best opportunity to efficiently change its velocity. They would have to reach Icarus on its inward trajectory as it barreled toward Earth. At that point, the asteroid would be traveling too fast for a gentle rendezvous, so a spacecraft would need to be launched quickly, intercept Icarus, and detonate its payload on arrival.

In the end, the students found that a single such explosion would not be sufficient, and they settled on a multi-stage mission concept. In the spring of 1967, six Saturn V rockets were in production, and the first test was not scheduled until November. The Icarus Project plan called for all six rockets (plus three more for test flights, requiring an increase in the speed of production) to be modified to carry a Payload Module containing a 100-megaton nuclear bomb. The payload would itself have been a challenge to produce, given that the largest nuclear weapon then in American hands topped out at 25 megatons. These would be detonated sequentially, gradually nudging the asteroid out of Earth's path. The first launch would need to take place by April 1968, with the others following every two weeks. Additional probes called Intercept Monitoring Satellites would be launched ahead of and alongside the primary mission to observe the process, since next to nothing was known about the true effects of detonating a nuclear weapon in space near the surface of an asteroid, and the 1963 Nuclear Test Ban Treaty effectively banned any experimentation to find out.

Project Icarus was only a fictional exercise, and 1566 Icarus's close pass on 14 June 1968 came and went without incident. Nevertheless, the project and resulting reports of the students' plan left a legacy that would influence subsequent deflection studies, not least because they directly inspired the movie *Meteor!* that debuted in 1979, bringing public attention to the possibility of Earth impact and potential defense strategies. As the final report states, confronting the complexities encountered in the exercise "resulted in what the team felt to be much more than a pure academic study; it resulted in a solution to a problem perhaps more imminent than anyone realizes, and the goal of that solution is the most rewarding of all goals, the saving of human lives."[3]

Besides the glaring unknowns inherent in the use of nuclear devices in space, the plan conceived during Project Icarus suffered from another important flaw. Little was known in 1967 about the physical characteristics of 1566 Icarus, and that problem was general to all near-Earth objects that might be considered targets of deflection attempts. By the late 1970s, studies considering collisions in the asteroid belt and the presence of collisional families of objects spawned the term "rubble pile" to describe a structure composed of disrupted fragments held together rather loosely by gravity. Future asteroid deflection and mitigation studies would have to wrestle with the additional problem that a rubble-pile asteroid might simply split into pieces after a mitigation attempt, converting a single, massive impact into a shotgun blast of smaller, but not much less destructive, impacts.

Perhaps the first effort by asteroid scientists to gain policy salience for impacts was a 1981 study chaired by Gene Shoemaker. It was the result of a NASA Advisory Council effort to help define NASA's future programs. These outside consultants and technologists proposed several future research themes and one subpanel, which included Barney Oliver (vice president of Hewlett-Packard), suggested investigating the potential hazard of Earth-colliding comets and asteroids.[4]

Participants in this four-day workshop at Snowmass, near Aspen, Colorado, included Barney Oliver, Tom Gehrels, George Wetherill, Clark Chapman, General Theodore Taylor, astrodynamicist Alan Friedlander, geoscientist David Roddy, and NASA program chief for planetary astronomy William Brunk.[5] Perhaps because of Gene Shoemaker's busy schedule, the resulting Snowmass report, consisting of about 100 draft pages, was never published, though a typescript was widely distributed.[6] Normally, an unpublished report would have little influence on the field, but some of the ideas and concepts in this report would be carried forward in a 1989 popular book,

---

3.   MIT Student Project in Systems Engineering, *Project Icarus*, pp. 3–4.

4.   Clark Chapman, "History of the Asteroid/Comet Impact Hazard," online essay at *http://www.boulder.swri.edu/clark/ncarhist.html* (accessed 16 April 2017).

5.   Ibid. Oliver would later contribute support for Gehrels's efforts to establish the Spacewatch telescope.

6.   Snowmass report (unpublished, draft dated July 1981), pp. 84–86, courtesy of Donald K. Yeomans, copy in NEO History Project collection.

Box 5-1. Key concepts from the unpublished 1981 Shoemaker report.

The 1981 Snowmass report was forward-thinking in a number of areas, including observations that

- early detection of threatening objects is vital (Spacewatch should go forward);

- orbiting IR telescopes should be looked at for detecting near-Earth objects;

- accurate orbit determination requires astrometric data—especially for the small objects that cannot be frequently viewed;

- asteroid deflection via means of explosive devices appears to be the best approach in terms of payload delivery weight, cost, and fast action time;

- asteroid exploration, apart from the scientific value, will be needed to gather engineering data, and such missions would also serve any future endeavors involving asteroid resource utilization; and

- NASA should cooperate with the military to acquaint them with the atmospheric signature of impacting bodies to differentiate these natural atmospheric explosions from nuclear blasts; likewise, the military should be asked to participate with NASA in accumulating and analyzing small-body information.

Cosmic Catastrophes, by Clark Chapman and David Morrison, as well as the Spaceguard Survey report of 1992 (chaired by David Morrison).[7]

The Snowmass report introduced a number of important concepts (see box 5-1). Perhaps most significant was its recognition that early detection of potentially threatening objects is vital. That is because numerous observations are necessary to determine orbits accurately, and if an object turns out to have a very high impact likelihood, doing anything about it will require many years of warning.[8] Thus the report supported development of Tom Gehrels's Spacewatch camera while also arguing that better survey technology would

7. Clark R. Chapman and David Morrison, *Cosmic Catastrophes* (New York: Plenum Press, 1989); David Morrison, "The Spaceguard Survey: Report of the NASA International Near-Earth-Object Detection Workshop" (Moffett Field, CA: NASA Ames Research Center, 1992), *http://ntrs.nasa.gov/search.jsp?R=19920025001* (accessed 8 March 2021).

8. Snowmass Report. See, for example, p. 85.

be necessary for finding smaller (and therefore more common) objects than it would be able to discover.

The Snowmass report also argued that the long-term rates of mortality due to asteroid impacts are comparable to other, far more frequent, disasters that the public takes very seriously and for which significant resources are allocated (e.g., airline accidents). "On the other hand, we are dealing with a qualitatively different kind of threat, in which the risk is the product of an extremely small probability of occurrence multiplied by huge apocalyptic hazard."[9] Shoemaker introduced an analysis showing the frequency of asteroid impacts versus the impact energy of the impactor. This analysis used population estimates for various asteroid sizes (size-frequency distribution) based upon the lunar cratering record. The Snowmass report emphasized the large uncertainties in this analysis due to a lack of knowledge of asteroid population numbers for various

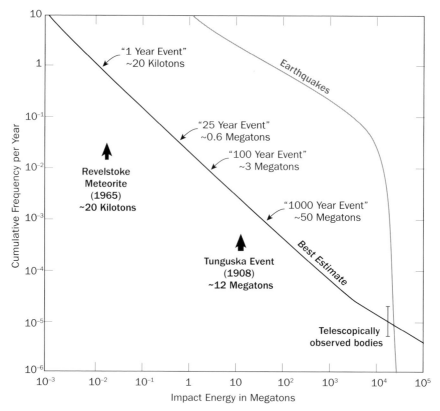

Figure 5-1. Size-frequency-consequence chart from the unpublished 1981 Shoemaker report. (Image courtesy of Donald K. Yeomans)

---

9.   Snowmass report, p. 55.

sizes as well as uncertainties as to their compositions. Only a few dozen near-Earth asteroids had been discovered by that time. The population, sizes, and compositions of long-period comets were called out as particularly uncertain, so it was unclear if they represented a significant hazard.

Shoemaker's unpublished 1981 study did not ultimately result in a new NASA initiative to study NEOs during that decade, and there would not be a specific NEO program launched until the late 1990s. NASA's funds and energies went into its Space Shuttle and space station programs during the 1980s.[10] Planetary science had a near-death experience, with an explicit threat to close it down in 1981, while astronomy suffered poor funding until the Hubble Space Telescope was finally launched in 1990.[11] While the 1980s were an exciting decade for the American geosciences, that excitement was largely outside NASA's purview.

Establishing the Spaceguard Goal

NASA's gradual movement toward an organized program of near-Earth object research started with an asteroid known as 1989 FC. On 19 April 1989, the Agency issued a press release making note of the asteroid, which was claimed to be 0.8 kilometers or more in diameter, that had zipped closely by Earth on 23 March, some eight days before its discovery by Northern Arizona University astronomer Henry Holt. The NASA release went on to say that this close pass—about twice the distance between Earth and the Moon—had been the closest known Earth approach by an asteroid since the similarly close approach of asteroid 69230 Hermes in 1937.[12] Had the asteroid impacted

---

10. E.g., Ray A. Williamson, "Developing the Space Shuttle," in *Exploring the Unknown*, vol. 4, *Accessing Space*, ed. John M. Logsdon with Ray A. Williamson, Roger D. Launius, Russell J. Acker, Stephen J. Garber, and Jonathan L. Friedman (Washington, DC: NASA SP-4407, 1999), pp. 161–193; Howard McCurdy, *The Space Station Decision: Incremental Politics and Technical Choice* (Baltimore, MD: Johns Hopkins University Press, 1991).

11. Amy Paige Snyder, "NASA and Planetary Exploration," in *Exploring the Unknown*, vol. 5, *Exploring the Cosmos*, ed. John M. Logsdon with Amy Paige Snyder, Roger D. Launius, Stephen J. Garber, and Regan Anne Newport (Washington, DC: NASA SP-2001-4407, 2001), pp. 263–300.

12. 1989 FC made a close Earth approach, to within 0.0046 au, late on 22 March 1989. Hermes made an Earth approach (0.0050 au) on 30 October 1937 and an even closer Earth approach (0.0042) on 26 April 1942.

Earth, "the impact would have been equivalent to the explosion of 20,000 hydrogen bombs…enough to destroy a good-sized city."[13] Subsequent contemporary reports noted (incorrectly) that Earth had been at the same position in space as the asteroid only 6 hours earlier, lending credence to the idea that an Earth collision might have taken place.[14] As Clark Chapman, an asteroid scientist at the Southwest Research Institute, later recalled, this close call made the front page of the *New York Times*.[15] But by 2018 (when this paragraph was written), this close Earth approach would not have raised an eyebrow, given that at least 10 other known Earth approaches at this distance or less by objects of comparable size have been recorded in the 20th century.[16] However, in 1989, before the discovery of thousands more near-Earth asteroids, this was big news and set in motion a string of events that ultimately led to the articulation of a NEO discovery policy.

The 1989 FC close approach prompted the American Institute of Aeronautics and Astronautics (AIAA) Headquarters to ask its Space Systems Technical Committee to determine if the threat of asteroid impacts was real.[17] It fell to Edward Tagliaferri to examine the issue. Decades later, he remembered, "What really had people concerned was that it hadn't been detected until three weeks after it passed by the Earth. Nobody saw it coming. It was coming at us out of the Sun, so nobody saw it. And then by the time it got

---

13. "NASA Astronomer Discovers 'Near-Miss' Asteroid That Passed Earth," NASA Press Release 89-52, 19 April 1989. The equivalent energy quoted here is likely a significant over-estimate since the diameter of 1989 FC, now (4581) Asclepius, is only about 250 meters.

14. The Minimum Orbital Intersection Distance (MOID) between the orbits of Earth and 1989 FC is about 0.0036 au, far too large to allow an Earth collision.

15. Clark Chapman comment in review, personal communication, March 2018; Warren E. Leary, "Big Asteroid Passes Near Earth Unseen in a Rare Close Call," *New York Times* (20 April 1989): sec. U.S., *https://www.nytimes.com/1989/04/20/us/big-asteroid-passes-near-earth-unseen-in-a-rare-close-call.html* (accessed 9 March 2021).

16. The closest known comet approach to Earth was 1770 I Lexell, approaching to within 0.015 au on 1 July 1770. See Donald K. Yeomans, *Comets: A Chronological History of Observation, Science, Myth and Folklore* (New York: Wiley, 1991), p. 157.

17. Edward Tagliaferri, "Dealing with the Threat of an Asteroid Striking the Earth: An AIAA Position Paper," April 1990, 3 pp, copy in NEO History Project collection; Edward Tagliaferri, "The History of AIAA's Interest in Planetary Defense" (paper AIAA 96-4381, 1996 AIAA Space Programs and Technologies Conference, 24–26 September 1996, Huntsville, AL); Edward Tagliaferri, interview by Donald K. Yeomans, 14 December 2016.

into opposition to the solar, the Sun was illuminating it, it had already been three weeks since it passed."[18] Tagliaferri consulted the few known authorities in the field—Gene and Carolyn Shoemaker, Tom Gehrels, Eleanor Helin, and a few others—and spent six months drafting a white paper for the AIAA. His conclusions were dramatic: the threat is indeed real, we are almost totally ignorant of where these things are in space, and no one had really done a systematic study of how we would cope with an impending impact event.[19]

Tagliaferri also recalled that he had not considered the "giggle factor" when he sent the white paper into review. The AIAA's board of directors had to approve the paper, and that was where he ran into trouble. "There was an enormous giggle factor about rocks falling from the sky. Chicken Little. 'This is nonsense. What are you guys talking about? No one's ever been killed by an asteroid. Why are we concerned about this thing? The last time we ever saw one was in 1908, Tunguska.'" A friend of his on the board told him later that there had been a "knock-down, drag-out fight" over the white paper, and it passed by only a single vote.[20]

Once it had been published, a copy of the white paper was delivered to every House Representative and Senator, and a key briefing was given to Congressman George E. Brown and Dr. Terry Dawson, then a senior staffer on the House Space Subcommittee on Science, Space and Technology.[21] At Congressman Brown's request, Dawson inserted wording into the committee report that accompanied the 1991 NASA Authorization bill directing NASA to sponsor two workshops, one on NEO search programs and another on systems that could deflect or destroy threatening near-Earth objects.[22] Jurgen Rahe, from NASA's Office of Space Science and Applications, then established an international conference on Earth-approaching bodies, a workshop on NEO search programs, and a second workshop on the deflection

---

18. Tagliaferri interview, 14 December 2016.

19. Summarized from Tagliaferri, "The History of AIAA's Interest in Planetary Defense" (AIAA Paper 96-4381, AIAA Space Programs and Technologies Conference, Huntsville, AL, 24–26 September 1996), p. 2.

20. Tagliaferri interview. Also see Tagliaferri, "The History of AIAA's Interest in Planetary Defense."

21. Dawson, in turn, invited David Morrison to talk to other staffers about the threat. Morrison interview, 17 May 2016.

22. Tagliaferri interview; see also Tagliaferri, "The History of AIAA's Interest in Planetary Defense."

of threatening near-Earth objects. The international conference was held in San Juan Capistrano, California, in July 1991 and was chaired by Clark Chapman. Three search program workshops, held in 1991, were chaired by NASA Ames Research Center's David Morrison, and the Near-Earth Object Interception Workshop was held at Los Alamos, New Mexico, the next year and was chaired by John Rather from NASA's Office of Aeronautics and Space Technology.

The report summarizing Morrison's search program workshops, known as the 1992 Spaceguard Survey report, defined the spectrum of NEO threats and discussed the most efficient means for "dramatically increasing the detection rate of Earth crossing objects."[23] The Workshop participants were international in scope, and their report was the first to note that discovering the largest objects should take priority. Based upon the research of NASA Ames Research Center's O. Brian Toon, it was noted that an impact by an NEO larger than about 1 kilometer in diameter, while rare, would have global consequences.[24] This group recommended a goal of finding 90 percent of Earth-crossing asteroids larger than 1 kilometer in diameter and proposed a global network of six search telescopes, each 2.5 meters in aperture, to accomplish this task. The effort would cost about $50 million in capital expenditures and be expected to reduce the risk of an unforeseen cosmic impact by more than 75 percent over the course of 25 years, with annual operations costs of about $10 million. The proposed survey would be expected to discover fully 90 percent of the NEOs larger than 1 kilometer in diameter.[25]

Nothing much happened immediately after the Spaceguard workshops. Fundamentally, neither NASA nor the Air Force was interested in the task of finding near-Earth objects. Tagliaferri's contacts in NASA told him, "…there's really not a lot of interest in this. We don't have any budget for it…" and "It's not our job. That's the military's job." But his Air Force contacts told him much the same thing. "We only defend against terrestrial enemies. We

---

23. David Morrison, ed., "The Spaceguard Survey: Report of the NASA International Near-Earth-Object Detection Workshop," 25 January 1992, p. 3, as quoted from the NASA Multiyear Authorization Act of 1990, 26 September 1990.

24. David Morrison, interview with Yeomans and Conway, 17 May 2016, transcript in NASA History Division HRC.

25. Morrison, "Spaceguard Survey," p. vi.

don't worry about things coming from outer space."[26] Asteroids were someone else's problem.

The second ("Interception") workshop initiated after Congressman Brown's legislative support, which dealt with near-Earth object interception, took place over two and one-half days (14–16 January 1992) at the U.S. Department of Energy's (DOE's) Los Alamos National Laboratory (LANL). It was tasked with defining "systems and technologies to alter the orbits of such asteroids or to destroy them if they should pose a danger to life on Earth."[27] The meeting was chaired by John Rather, with Jurgen Rahe as cochair and Gregory Canavan of LANL as official host. The 94 participants arrived from varied backgrounds, including academic NEO scientists as well as several weapons specialists from U.S. national laboratories that included LANL, Lawrence Livermore National Laboratory (LLNL), Sandia National Laboratory, and the Strategic Defense Initiative Organization (SDIO), which had been set up in 1984 by the U.S. Department of Defense to develop a United States defense system against incoming nuclear missiles. NEO scientists and weapons physicists, coming from such different cultures, were likely to interact in a contentious fashion—and they did.

The SDIO had developed a concept called Brilliant Pebbles, whereby nonexplosive U.S. defensive missiles about the size of watermelons would be deployed to ram incoming offensive missiles, destroying them. Lowell Wood (LLNL), a protégé of Edward Teller (LLNL), had circulated a paper noting that small, few-meter-sized asteroids ("the stuff between a truck and a house in scale") represented a significant threat and that Brilliant Pebble–type interceptors could be counted upon to knock them off course, or destroy them, before they had a chance to impact Earth's atmosphere. Perhaps to explore the tradespace without a size limit imposed, Wood's analysis failed to properly consider the mitigating effect of Earth's atmosphere on these small asteroids; most would not reach Earth's surface anyway. To counter large impact threats, some of the weapons physicists were eager to test nuclear deflection techniques in space, and others suggested that missiles be placed at the ready on Earth's surface or in Earth orbit to fend off any incoming asteroids that might prove a threat. Teller, who had been the architect of the U.S. hydrogen bomb program, noted that a new super-bomb might have to be developed to

---

26. Tagliaferri interview.
27. Morrison, "Spaceguard Survey," p. 3.

deal with the enormous energies of a large threatening asteroid.[28] Teller celebrated his 84th birthday at the meeting, and in honor of his birthday, Eleanor Helin announced the naming of an asteroid after Teller.[29]

Three types of nuclear deflection approaches emerged in the case of long lead time before impact: ablation of material irradiated by a nearby nuclear blast, the impulse from ejecta created in a near-surface explosion, and a buried charge intended to fragment the asteroid.[30] These scenarios were considered in detail by Tom Ahrens of Caltech and Alan Harris of JPL, first in a paper presented at the 1992 meeting and later in a review article published in *Nature* that December. In the first case, detonation of a radiatively efficient nuclear explosive near the object (a "stand-off explosion") would spray the surface with high-energy neutrons, irradiating an outer shell on one side of the asteroid. This layer—which Ahrens and Harris estimated to be about 20 centimeters in depth—would then be removed as material is imparted with velocity in excess of the escape velocity. This blow-off would in turn impart an impulse to the remainder of the asteroid in the opposite direction, slightly changing its course. With a lead time of about a decade, the deflection velocity required to divert an asteroid was found to be about 1 centimeter per second. To achieve this value for a 100-meter impactor, this approach would call for the detonation of an approximately 0.01- to 0.1-kiloton nuclear explosive. For an asteroid 10 times larger, this energy would increase by 1,000 to approximately 0.01 to 0.1 megatons. Likewise, a 10-kilometer object would require an approximately 10- to 100-megaton explosion. This method had two advantages: first, it depended somewhat less on detailed knowledge of the physical characteristics of the asteroid; and second, "the development of the nuclear weapons required to deflect large Earth-crossing objects seems to be feasible technologically."[31]

---

28. At one point in the discussions on the use of nuclear devices to deflect or destroy threatening asteroids, Lowell Wood shouted out, "We're all here, we're after the same thing. Nukes forever," which many attending NEO scientists found rather chilling. Morrison interview, 17 May 2016.

29. 5006 Teller, discovered by Helin at Palomar Observatory on 5 April 1989. See *https://minorplanetcenter.net/db_search/show_object?object_id=5006* (accessed 7 May 2017).

30. G. J. Canavan, J. C. Solem, and J. D. G. Rather, "Proceedings of the Near-Earth-Object Interception Workshop" (Los Alamos, NM: Los Alamos National Laboratory, February 1993), p. 89; T. J. Ahrens and A. W. Harris, "Deflection and Fragmentation of Near-Earth Asteroids," *Nature* 360 (6403): 429–433.

31. Ahrens and Harris, "Deflection and Fragmentation of Near-Earth Asteroids," p. 432.

Another nuclear option would involve orchestrating a surface explosion to produce a crater on the surface of the asteroid. Material ejected by the explosion, itself too dispersed to pose a threat to Earth, would induce a velocity change in the object. According to calculations by Ahrens and Harris, this method could be as efficient in deflecting an asteroid as the nuclear ablation method, but it comes with a major drawback: the asteroid might be inadvertently disrupted, producing large fragments that could themselves threaten Earth. This was more likely in the case of a "rubble pile" asteroid, so one needed decent knowledge of the asteroid's composition before execution.

The third strategy, fragmentation via a buried explosive charge, also carried the threat of producing large and dangerous pieces of the original asteroid. To mitigate this risk, the object would need to be disrupted one or more orbits prior to its arrival at Earth to allow the debris to spread out along and transverse to its orbit. In situ drilling would also be required to optimally bury the nuclear charges, presenting another difficulty, particularly in low-gravity environments. Thus, fragmentation was not considered to be a particularly safe choice except perhaps in the case of a small body or a very long lead time of decades. In the opposite situation (a large impactor discovered on close approach), fragmentation was assessed to be unavoidable.[32]

Ahrens and Harris considered two non-nuclear approaches as well: the kinetic impactor and the mass driver. In the former strategy, the impact of a spacecraft-borne mass could directly transfer enough kinetic energy to deflect a small asteroid, on the order of 100 meters in size. For larger sizes approaching 1 kilometer, however, they found that the thousandfold increase in asteroid mass, together with the decrease in cratering efficiency due to the body's stronger gravitation, would make direct-impact deflection impractical. In the mass driver scenario, a spacecraft would be sent to the asteroid to excavate and electromagnetically accelerate material from a particular area, ejecting it from the body. The mass driver would likely have to operate for many years to achieve the required deflection velocity. While the study authors deemed such a system technically possible, they nevertheless felt that "nuclear energy offers a much more efficient solution."[33] Both the kinetic impactor and mass driver techniques, as well as several other novel ideas, would continue to be studied in the decades to come, but the Interception Workshop's assessment of current

---

32. Ibid.
33. Ibid., p. 431.

and future technologies concluded that "chemical or nuclear rockets with nuclear explosives are the only present or near-term technology options available that have significant probability of success without significant research and development activities."[34]

The Interception Workshop report issued in February 1993 (hereafter "Interception Workshop Report") acknowledged that the use of nuclear devices for deflection purposes would "require appropriate international agreements and protocols."[35] Moreover, Ahrens and Harris advocated holding off on any actual engineering designs for the approaches discussed in their study "because of the low probability of impact of hazardous asteroids, the high cost in the face of a low risk factor, and the rapid changes that are expected in defence systems technology."[36] Even so, the eagerness of some participants to resort to nuclear weapons alarmed many NEO scientists present at the Interception Workshop and generated a divide between the two communities. At a meeting hosted by Tom Gehrels at the University of Arizona in Tucson in January 1993, the AIAA's Tagliaferri recalled a long conversation he had with Carl Sagan, a vocal opponent of developing deflection technologies. Sagan was convinced that the ability to divert an asteroid from its course would inevitably fall into the wrong hands and be turned to ill purposes.[37] Tagliaferri understood the concern but felt the impact hazard warranted the risk because the argument against nuclear deflection could not hold up against calculations like those of Ahrens and Harris. "The numbers don't lie. If you're going to move these things, you're going to need that kind of energy density, and nothing else will do that."[38]

Many of the NEO scientists at the meeting found these suggestions both unnecessary and dangerous, and some felt that the weapons scientists were

34. Canavan, Solem, and Rather, "Proceedings of the Near-Earth-Object Interception Workshop," p. 233.
35. Ibid., p. 9.
36. Ahrens and Harris, "Deflection and Fragmentation of Near-Earth Asteroids."
37. Carl Sagan, *Pale Blue Dot: A Vision of the Human Future in Space* (NY: Ballantine Books, 1994), pp. 255–256.
38. Tagliaferri interview; Remo offers a recent study of this dilemma: "The peaceful and critically effective use of nuclear energy to prevent a civilization-threatening collision is at odds with its potential for inducing a catastrophic thermonuclear war on Earth." J. L. Remo, "The Dilemma of Nuclear Energy in Space," *Bulletin of the Atomic Scientists* 71, no. 3 (2015): 38–45.

simply looking for a justification to continue the SDIO program, which was drawing to a close.[39] Two years later, Carl Sagan and radar astronomer Steve Ostro would call out the dangers involved with the premature deployment of nuclear weapons for asteroid deflection, noting that a deflected asteroid could conceivably be used as a horrific offensive weapon.[40] Clark Chapman was so upset at the Interception Workshop's misplaced emphasis upon small impactors, space-based nuclear tests, a standby nuclear defense option, and some ridiculously futuristic deflection ideas (e.g., antimatter devices), that he demanded his name be removed from the final report.[41] Chapman was not the only offended scientist. In his "Star Warriors on Sky Patrol," published in the *New York Times,* University of Maryland physicist Robert L. Park commented, "As calls for more and bigger bombs continued, Lowell Wood, Dr. Teller's protégé at the Lawrence Livermore National Laboratory, could not contain his excitement; from the back of the auditorium he shouted, 'Nukes forever!'" And Park continued, "In defending Earth against this minuscule threat, the Star Warriors would create a vastly greater hazard of nuclear missiles at the ready. Who will protect us from the 'nukes forever' mentality?"[42]

In an effort to foster better relations between the U.S. defense community and the international near-Earth object scientists, Colonel Simon "Pete" Worden organized a retreat in Erice, Sicily, in May 1993. As David Morrison later recalled, "He really felt that the two sides were so fractured…that it was really creating a huge public relations problem, and that the answer was to get us all in the same place and literally not let us out of the room until we'd agreed on a common statement."[43] The international community, including

---

39. In 1993, the SDIO program was recast as the Ballistic Missile Defense Organization (BMDO), including a shift in emphasis from global to regional defensive systems.

40. Carl Sagan and Steve Ostro, "Dangers of Asteroid Deflection," *Nature* 369 (1994): 501; see also Sagan, *Pale Blue Dot*, pp. 254–255.

41. Chapman did have his name associated with the meeting proceedings, in which his numerous objections were noted.

42. Robert L. Park, "Star Warriors on Sky Patrol," *New York Times* (25 March 1992): 23, *https://www.nytimes.com/1992/03/25/opinion/star-warriors-on-sky-patrol.html? searchResultPosition=1* (accessed 29 May 2019). In the *Wall Street Journal*, Bob Davis wrote a very skeptical article entitled "Never Mind the Peace Dividend, the Killer Asteroids Are Coming." This article is reprinted in P. R. Weissman, "The Comet and Asteroid Impact Hazard in Perspective," in *Hazards Due to Comets and Asteroids*, ed. T. Gehrels (Tucson, AZ: University of Arizona Press, 1994), pp. 1191–1212.

43. Morrison interview.

members of the space policy community and media, were invited, and for the first time since the unpublished Snowmass report in 1981, there was a discussion of impact-generated tsunamis. Participants also discussed the use of nuclear explosions to deflect sub-kilometer near-Earth objects for which there would be little warning. Worden's goal was to forge an agreement between the astronomical community and Edward Teller and his colleagues—and he succeeded. After lengthy negotiations, an agreed-upon joint statement was issued that encouraged additional surveys and studies of NEOs, expressed a concern that an atmospheric explosion of a bolide could be mistaken for a nuclear attack, and noted that the study of mitigation options should be continued. Although a unanimous agreement could not be reached on the use of standby mitigation systems (e.g., nuclear weapons), the meeting summary statement ended with "Many of us believe that unless a specific and imminent threat becomes obvious, actual construction and testing of systems that might have the potential to deflect or mitigate a threat may be deferred because technology systems will improve."[44]

Following the Erice meeting, the 1,200-page *Hazards Due to Comets and Asteroids* was published as the 24th volume of the Space Sciences Series, edited by Tom Gehrels. The book was intended to provide a comprehensive answer, based on the latest research, to the question "What can be done if a dangerous object is identified?" and Gehrels sought to strike a balance to keep the various strong opinions in check. "At the early planning stage of the book, I was anxious to have a quorum," he wrote to JPL's Paul Weissman, one of the chapter authors.[45] Despite his initial optimism, the process was contentious. When two papers—one by Sagan and Steve Ostro on the dangers of deflection and the other by Edward Teller and William Tedeschi proposing a detailed program of nuclear development and testing—were declined by their referees, correspondence between the authors and Gehrels heated up and accusations of bias toward one party or the other flew. Gehrels got the volume

44. David Morrison and Edward Teller, "The Impact Hazard: The Issues for the Future," in *Hazards Due to Comets and Asteroids*, ed. Tom Gehrels (Tucson, AZ: University of Arizona Press, 1994), pp. 1135–1143; S. Nozette, personal communication dated 22 October 2016; Pete Worden, interview by Yeomans and Conway, 7 October 2016, copy in NEO History Project collection.

45. Gehrels to Weissman, letter dated 5 January 1994, folder 20, box 8, Gehrels papers (MS 541), University of Arizona Special Collections.

published in early 1994, but the dustup did nothing to bring the two sides closer to agreeing on a long-term mitigation strategy.

On the Political Impacts of Comets

Two comet visitations in the early 1990s helped keep the risk of impacts in the public eye, facilitating a political response. On 15 October 1992, shortly before the Erice meeting, a significant amount of media attention erupted around an announcement of a possible Earth collision of Comet 109P/Swift-Tuttle on 14 August 2126.[46] Swift-Tuttle had been discovered in 1862 and then lost; it was recovered by U.S. Naval Observatory astronomers on 30 September 1992 and reported to the Minor Planet Center.[47] Brian Marsden, Director of the Minor Planet Center, then had difficulties in reconciling the 1862 and 1992 observations, which suggested to him that the comet might be subjected to significant accelerations from the comet's outgassing (i.e., rocket-like cometary thrusting or so-called nongravitational effects). Furthermore, these accelerations might allow Marsden's estimate for the comet's next predicted perihelion passage time (11 July 2126) to be early by 15 days—enough time to allow a subsequent Earth collision on 14 August 2126. By calculating the ratio of the time the comet would sweep past Earth to the uncertainty in the comet's perihelion passage time, Marsden computed an impact probability of about one in ten thousand, and this was widely reported by David Chandler in the *Boston Globe* and by William Broad in the *New York Times*.[48]

Upon receiving Marsden's announcement of a possible 2126 Earth impact, Yeomans at JPL was able to fit the 1992 and 1862 data, as well as the time of a 1737 perihelion passage recovered from Chinese texts, to within 24 hours, without the so-called nongravitational parameters that are often used to represent the motions of active comets. These computations, including a more

---

46. *International Astronomical Union Circular (IAUC)* 5636, dated 15 October 1992; B. G. Marsden, "Comet Swift-Tuttle: Does It Threaten Earth?" *Sky and Telescope* (January 1993): 16–19; Sharon Begley, "The Science of Doom," *Newsweek* (23 November 1992): 56–60. Comet Swift-Tuttle is the source of the Perseid meteor shower.

47. IAUC 5627, 2 October 1992.

48. David Chandler, "Don't Look Now, But a Comet's Coming in 134 Years—Maybe," *Boston Globe* (17 October 1992): 3; William Broad, "Scientists Ponder Saving Planet from a Distant Comet," *New York Times* (3 November 1992): 39, 48.

sophisticated consideration of the uncertainties, indicated that there was no chance of an Earth impact in 2126. Yeomans concluded that 12 observations from the Cape of Good Hope taken in September–October 1862 were discordant, and an orbit calculated without them precluded an Earth collision in 2126.[49] In a letter to William Broad of the *New York Times* on 3 November, Yeomans noted these computations and confidently stated that the comet's next perihelion passage on 12 July 2126 was known to within 24 hours. Furthermore, "the comet will pass no closer to the Earth than 60 lunar distances on August 5, 2126. Then it will be a very bright naked eye object but certainly not a threat to Earth."[50] While the *New York Times* did not publish this response, Yeomans also sent this information on to Sharon Begley, a science writer for *Newsweek*, in response to her enquiries. A *Newsweek* cover story entitled "The Science of Doom" noted Marsden's 1-in-10,000 impact prediction for 2126 and Yeomans's dismissal of the same threat.[51]

The tidal wave of media attention that accompanied Marsden's impact probability computation could have been avoided by postponing the announcement for a few days while additional analyses were carried out. Had this been done, it would have been clear that the comet represented no threat in 2126. In 12–14 November 1992 e-mail correspondence with Gene Shoemaker, Don Yeomans, and Clark Chapman, Marsden suggested that the 2126 Earth impact that he then thought possible had the "benefit" of drawing much-needed attention to NASA's NEO detection efforts.[52]

In the end, computations by both Marsden and Yeomans arrived at the same predicted perihelion passage time in 2126.[53] There is no possibility of

49. IAUC 5671, 5 December 1992; B. G. Marsden, G. V. Williams, G. W. Kronk, W. G. Waddington, "Update on Comet Swift-Tuttle," *Icarus* 105 (1993): 420-426; K. Yau, D. K. Yeomans, and P. Weissman, "The Past and Future Motion of Comet P/Swift-Tuttle," *Monthly Notices, Royal Astronomical Society* 266 (1994): 305–316.

50. Donald K. Yeomans to William J. Broad, 3 November 1992, copy in "Swift Tuttle dust up doc scan 2-20-2016.pdf," folder Don Yeomans materials, NEO History Project collection, courtesy of Donald K. Yeomans.

51. Begley, "The Science of Doom," *Newsweek* (23 November 1992): 56–60.

52. Marsden to Shoemaker, e-mail dated 12 November 1992; Marsden to Yeomans, e-mail dated 13 November 1992; Marsden to Clark Chapman, e-mail dated 14 November 1992. All in Swift-Tuttle e-mail collection, NEO History Project collection, courtesy of Donald K. Yeomans.

53. Marsden et al., "Update on Comet Swift-Tuttle"; Yau et al., "The Past and Future Motion of Comet P/Swift-Tuttle."

an Earth impact in 2126, and the observed returns of Comet Swift-Tuttle in 1992–93 and 1862, as well as Chinese observations of 1737, 188, and 69 BCE, are consistent with no obvious outgassing accelerations over the entire observed interval. Yeomans pointed out that the nucleus of 109P/Swift-Tuttle was likely to be unusually large and massive, so the subtle thrusting due to cometary outgassing would have had very little effect upon this massive object's motion.[54] From mid-infrared observations of the nucleus and dust, Ames Research Center scientist Marina Fomenkova and colleagues provided an estimate of 30 kilometers for the comet's diameter, making it likely to be about 30 times more massive than the nucleus of Comet Halley.[55]

The more spectacular comet visitation that helped promote a political response in the United States was the destruction of Comet Shoemaker-Levy 9 in a collision with Jupiter in July 1994. This went a long way toward dispelling the giggle factor that Tagliaferri and others had experienced during the earlier discussions concerning the threat of NEOs. However, it did little to bolster confidence in scientists' ability to predict impact consequences. Some specialists anticipated the event to be a fizzle, with nothing much visible from Earth, while others expected to see dramatic fireballs and debris clouds rise into view.[56] The excitement generated by the impending impact developed from a serendipitous discovery some 15 months earlier.

On the evening of 25 March 1993, just before the Erice meeting and during their monthly search program using the Palomar Mountain 46-centimeter Schmidt telescope, Gene Shoemaker, Carolyn Shoemaker, and David Levy were dealing with two problems: films that had inadvertently been partially exposed and cloud cover that interfered with their observations. Rather than give up and call it a night, they exposed a few films in the neighborhood of Jupiter, and Carolyn began to scan them with the stereomicroscope, looking

---

54. Yau et al., "The Past and Future Motion of Comet P/Swift-Tuttle."

55. M. N. Fomenkova, B. Jones, R. Pina, R. Puetter, J. Sarmecanic, R. Gehrz, and T. Jones, "Mid-Infrared Observations of the Nucleus and Dust of Comet P/Swift-Tuttle," *Astronomical Journal* 110 (1995): 1866–1874.

56. Paul Weissman, "The Big Fizzle Is Coming," *Nature* 370 (1994): 94–95; John R. Spencer and John H. Rogers, "The Great Crash," in *The Great Comet Crash*, ed. John R. Spencer and Jacqueline Mitton (Cambridge, U.K.: Cambridge University Press, 1995), pp. 55–96; Mark B. Boslough and David A. Crawford, "Impact Modellers Not Surprised," *Nature* 373, no. 6509 (5 January 1995): 28; Kevin Zahnle and Mark-Mordecai Mac Low, "The Collision of Jupiter and Comet Shoemaker-Levy Nine," *Icarus* 108 (1994): 1–17.

Figure 5-2. Comet Shoemaker-Levy 9 discovery image.
(Image courtesy of Palomar Observatory)

for points that appeared to "float" above the background stars—a sign of a possible near-Earth object. Carolyn soon noted what she described as a bar-shaped, squashed comet image in a region about 4 degrees from Jupiter.[57] Cloud cover at the Palomar Observatory prevented further observations there, so they sent positions off to the Minor Planet Center and made a confirmation request to Jim Scotti at the Spacewatch telescope in Arizona. Scotti confirmed Carolyn's squashed comet ("Do you three ever have a comet!"); subsequent follow-up observations at Kitt Peak, Arizona, and Maunakea, Hawai'i, revealed a linear collection of several cometary fragments, each surrounded by dust. Comet Shoemaker-Levy 9 looked like a dusty string of pearls.[58]

Initial orbital computations by Brian Marsden established that the comet had made a Jupiter close approach in July 1992 and that it appeared to be

---

57. The comet was evident on earlier images taken by others, including an image that Eleanor Helin had taken on the same telescope on 19 March, but Carolyn Shoemaker was the first to recognize and report the image as a comet.

58. David Levy, *Shoemaker by Levy: The Man Who Made an Impact* (Princeton, NJ: Princeton University Press, 2000); Carolyn S. Shoemaker and Eugene M. Shoemaker, "A Comet Like No Other," in *The Great Comet Crash*, ed. John R. Spencer and Jacqueline Mitton (Cambridge, U.K.: Cambridge University Press, 1995), pp. 7–12; C. Shoemaker, interview by Rosenburg, 7 February 2017, transcript in NEO History Project collection. The number of fragments did not remain constant, with a few disintegrating into dust clouds and a few splitting into sub-fragments.

in orbit about Jupiter.[59] In early May 1993, computations by the Japanese dynamicist Syuichi Nakano began to indicate a possible Jupiter impact the following July. JPL dynamicists Paul Chodas and Don Yeomans then computed impact probabilities and orbital computations as more and more observations were provided. They computed the probability of a July 1994 Jupiter impact to be 50 percent on 21 May, and each week this confidence rose; they reported a 64 percent probability on 28 May and 95 percent a week later. The impacts would occur just behind the limb of Jupiter, but the planet's rotation would soon bring the impact points into view for Earth-based observers and for the Hubble Space Telescope in orbit around Earth.[60] The Galileo spacecraft, on its way to observe Jupiter, was only 1.6 au away and would have a direct view of the impact points. Chodas and Yeomans provided a series of updated times for each of the fragments, which had been lettered A through W in the order of their expected impact times over the week of 16–22 July. (Fragment A would be first to collide with Jupiter, followed by fragment B, and so on.) It became clear that a series of unexpected Jupiter impact events was coming, and the international community of astronomers had more than a year to make observing plans.[61]

59. Brian Marsden, "The Path to Destruction," in *The Great Comet Crash*, ed. John R. Spencer and Jacqueline Mitton (Cambridge, U.K.: Cambridge University Press, 1995), pp. 13–18; *IAU Circulars* 5744–5745, dated 3 April 1993. In 1992, Gonzalo Tancredi of the Universidad de la Republica, Uruguay, had argued that Jupiter orbit was a good place to look for comets as it had a high likelihood of temporarily capturing them. See Gonzalo Tancredi and Mats Lindgren, "The Vicinity of Jupiter: A Region to Look for Comets," in *Asteroids, Comets, Meteors 1991* (Houston: Lunar and Planetary Society, 1992), p. 601.

60. Heidi B. Hammel, "HST Imaging of Jupiter Shortly After Each Impact: Plumes and Fresh Sites," in *The Collision of Comet Shoemaker-Levy 9 and Jupiter*, ed. Keith S. Noll, Harold A. Weaver, and Paul D. Feldman, Space Telescope Science Institute Symposium Series, no. 9 (Cambridge, U.K.: Cambridge University Press, 1996), pp. 111–120.

61. A pre-impact summary of planned activities was provided by JPL scientist Paul Weissman and is available at *https://onlinelibrary.wiley.com/doi/epdf/10.1111/j.1945-5100.1994.tb00667.x* (accessed 16 March 2021). A complete chronology of predictions for the comet's orbital characteristics is given in Paul W. Chodas and Donald K. Yeomans, "The Orbital Motion and Impact Circumstances of Comet Shoemaker-Levy 9," in *The Collision of Comet Shoemaker-Levy 9 and Jupiter*, ed. Keith S. Noll, Harold A. Weaver, and Paul D. Feldman, Space Telescope Science Institute Symposium Series, no. 9 (Cambridge, U.K.: Cambridge University Press, 1996), pp. 1–30.

As the comet fragments approached Jupiter, observations were taken in the ultraviolet and visible wavelength regions, but because of the hot, dark dust and gases produced by the impacts, most of the action took place in the infrared. One of the first indications that the impacts would be far from a fizzle arrived when observers at Calar Alto Observatory in southern Spain reported that a small dot had appeared on the limb of Jupiter where the fragment A impact point was predicted to be.[62] The dot brightened quickly, becoming as bright as the innermost satellite Io before fading again. When the impact point rotated into view of ground-based astronomers again, a small but persistent dark spot appeared at visible wavelengths. It was immediately clear that the coming impacts would put on a show—and fragment A was one of the smaller fragments! After the largest fragment (G) hit, Vikki Meadows, who had observed in Australia, recalled that they had to partly mask the mirror of the Anglo-Australian telescope, effectively shrinking its aperture from 3.9 meters to only 1.9 meters, because the brightness had saturated their infrared detector.[63]

Each impact event had three stages:[64]

1. An early bright, high-temperature flash appeared, indicating either the initial entry of the fragment into the atmosphere of Jupiter or possibly the expanding fireball after the fragment had deposited its energy into the atmosphere, or both. The impact kinetic energy of a kilometer-sized fragment was estimated to be equivalent to 24,000 megatons of TNT explosives—roughly 2,400 times more energetic than the event 50,000 years ago that ripped a 1.2-kilometer crater (Meteor Crater) in the Arizona desert.

2. A plume of dusty material rose above Jupiter's limb to more than 3,000 kilometers.

3. Several minutes later, dark gas and dust re-impacted Jupiter over a region 10,000 kilometers across. The dust deposited on Jupiter's cloud tops was either condensed silicates and metal oxides from the comet fragments or

---

62. The predicted impact times for each of the fragments, provided by Chodas and Yeomans, were correct to within about 7 minutes. See Chodas and Yeomans, "The Orbital Motion and Impact Circumstances."

63. Michael F. A'Hearn, "The International Observing Campaign," in *The Great Comet Crash*, ed. John R. Spencer and Jacqueline Mitton (Cambridge, U.K.: Cambridge University Press, 1995), p. 43.

64. H. Jay Melosh, "Wiser After the Event?" in *The Great Comet Crash*, ed. John R. Spencer and Jacqueline Mitton (Cambridge, U.K.: Cambridge University Press, 1995), p. 100.

perhaps dark organic mixtures synthesized in the comet fireballs. These dark spots, some of which were larger in area than Earth's disk, were particularly notable in Hubble Space Telescope infrared imagery, but they could also be seen through modest telescopes by amateur astronomers. They compared in size and prominence to the Red Spot and the most prominent features ever observed in Jupiter's atmosphere.

Before the July 1994 impact, the parent comet had undergone tidal splitting during a very close Jupiter approach; it had passed only 0.3 Jupiter radii above the surface in July 1992. Even so, the tidal force that had fragmented the comet was so weak that one impact specialist likened the comet's strength to less than that of a fluffy soufflé. The comet was clearly already friable and extremely weak before its breakup, and this perception was reinforced when several fragments themselves broke apart for no apparent reason after the July 1992 close approach.[65] Several attempts to estimate the pre-breakup diameter of the parent comet were made, but the uncertain estimates spanned a large range from about 1 kilometer to 10 kilometers. The time of capture into Jupiter orbit is likewise uncertain due to the chaotic nature of the comet's motion. JPL scientist Lance Benner and Washington University's William McKinnon integrated the motions of several fragments back in time and suggested that the comet may have been captured by Jupiter around the turn of the 20th century.[66] Chodas and Yeomans followed the chaotic motion of the comet's central fragment K backward in time using a statistical Monte Carlo technique that, given the orbital uncertainties, traced the motions of 1,000 points that represented the comet's possible motion. They concluded that the comet had likely transitioned from a heliocentric short-period comet orbit interior to that of Jupiter to an orbit about Jupiter itself in 1929, with an uncertainty of about 9 years.

Clearly the parent comet had been extremely weak and perhaps held together by only the self-gravity of the constituent fragments. Attempts to use spectroscopy to determine the comet's gases were foiled because the impacts themselves were so energetic that most molecules of the impactor

---

65. Ibid., p. 99.

66. Lance Benner and William B. McKinnon, "On the Orbital Evolution and Origin of Comet Shoemaker-Levy 9," *Icarus* 118 (1995): 155–168.

were dissociated, erasing any chemical memory.[67] While the impacts of the Shoemaker-Levy 9 comet with Jupiter shed more light on the dynamics and chemistry of Jupiter's atmosphere than upon the chemical constituents of the comet, the impacts themselves dramatically drove home the point that the planets, including Earth, run their courses about the Sun in a shooting gallery of comets and asteroids.

Clark Chapman pointed out later that "the impacts were witnessed as they happened by the public and scientific communities via the newly popular World Wide Web that showed images from ground-based telescopes as well as the Hubble Space Telescope." They "changed the impact hazard from a theoretical possibility into a very distinct possibility."[68] The discovery of the Chicxulub impact crater in 1990 and the impact of Comet Shoemaker-Levy 9 with Jupiter four years later went a long way toward removing the so-called giggle factor that had long been associated with the impact hazard from near-Earth objects.

The 1995 Shoemaker NEO Survey Working Group Report

Even before the fragments of Comet Shoemaker-Levy 9 had finished hitting Jupiter in July 1994, the U.S. House of Representatives Committee on Science, Space and Technology provided additional direction to NASA: "To the extent practicable, the National Aeronautics and Space Administration, in coordination with the Department of Defense and the space agencies of other countries, shall identify and catalog within 10 years the orbital characteristic of all comets and asteroids that are greater than 1 km in diameter and are in

---

67. Jacques Crovisier, "Observational Constraints on the Composition and Nature of Comet D/Shoemaker-Levy 9," in *The Collision of Comet Shoemaker-Levy 9 and Jupiter*, ed. Keith S. Noll, Harold A. Weaver, and Paul D. Feldman, Space Telescope Science Institute Symposium Series, no. 9 (Cambridge, U.K.: Cambridge University Press, 1996), p. 31.

68. Clark Chapman, "History of the Asteroid/Comet Impact Hazard," online essay, 1999, *http://www.boulder.swri.edu/clark/ncarhist.html* (accessed 16 March 2021). More than two and a half million users accessed NASA's internet resources for the weeklong impact events, likely making this the world's first major web-based event. The JPL website, established by Ron Baalke, handled more than a million file requests from 59 different countries.

an orbit around the Sun that crosses the orbit of the Earth."[69] In response, NASA then formed the Near-Earth Objects Survey Working Group, with Gene Shoemaker once again as Chair. In their report dated June 1995, the Near-Earth Objects Survey Working Group recommended a program that would be expected to discover 60–70 percent of short-period NEOs larger than 1 kilometer in 10 years (by the end of 2006, for funding beginning in FY 1996). They recommended the construction of two dedicated 2-meter aperture telescopes and one or two 1-meter telescopes with advanced focal plane detectors. Also recommended was access to larger-aperture telescopes on demand for the physical characterization of near-Earth objects. These assets would then be expected to extend the survey's completeness to 90 percent over the following five years (by 2011), but anticipated cooperation from the U.S. Air Force and international programs could shift the attainment of the 90 percent goal forward to 2006.[70]

The estimated five-year development cost to NASA would be $24 million (in FY 1995 dollars) with annual operations costs of $3.5 million. The Near-Earth Objects Survey Working Group Report was delivered to Congress in August 1995, but due to NASA's budgetary pressures, the cover letter provided by NASA recommended against an initiation of the recommended program. Not surprisingly, the funding wasn't appropriated.[71] This would not be the last time that NASA management would be reluctant to accept the mantle of planetary defense against asteroid impacts. Even the term "planetary defense" would come, not from NASA, but from a member of the U.S. Air Force.

Who Should Be in Charge of Planetary Defense?

The youngest of seven children in a Kansas farm family, Lindley N. Johnson would play a key role in NASA's planetary defense program, but only after having little success in trying to interest the Air Force in accepting this role. After four years of Air Force Reserve Officers' Training Corps (ROTC) and

---

69. Report of the Near-Earth Objects Survey Working Group (NASA Ames Research Center, June 1995), p. 3, *https://permanent.access.gpo.gov/lps19279/neosurvey.pdf* (accessed 8 March 2022). This PDF carries an incorrect date stamp.

70. Due to the constraints of time and resources, the Working Group deferred, for future study, the consideration of surveys from space and the hazards posed by comets—particularly long-period comets. NEO Survey Working Group Report, p. 38.

71. Chapman, "History of the Asteroid/Comet Impact Hazard," p. 17.

an astronomy major at the University of Kansas, Johnson received his commission as a second lieutenant in the U.S. Air Force in 1980. In the early 1990s, while assigned to the Air Force Phillips Laboratory in Albuquerque, New Mexico, he interacted with Tom Gehrels, who was anxious to gain access to the CCD detector development that the Air Force was supporting at MIT's Lincoln Laboratory in Lexington, Massachusetts. These CCD detectors were being developed to improve the Air Force's capability to observe and track Earth-orbiting spacecraft using ground-based optical telescopes—an activity called space surveillance. Gehrels, who actively solicited funds and new technology from a number of sources, wished to utilize modern CCD detectors for his Spacewatch near-Earth asteroid survey program. At the Lincoln Lab, the person in charge of the state-of-the-art, extremely fast-readout CCD detectors was Grant Stokes, who would later become the Principal Investigator for the extremely successful, NASA-supported Lincoln Laboratory Near-Earth Asteroid Research (LINEAR) program that operated (1996–2013, 2015–17) near Socorro, New Mexico. LINEAR was the first survey to utilize sensitive and rapid-readout CCD detectors for near-Earth asteroid discoveries.[72] Prior to the use of CCD detectors for space surveillance, the Air Force had used slower and less-sensitive analog video detectors in their so-called Ground-based Electro-Optical Deep-Space Surveillance (GEODSS) systems that were operational near Socorro; on Haleakalā, Maui, Hawai'i; and on Diego Garcia in the Indian Ocean.[73]

With the end of the Cold War, support for Air Force space surveillance assets had dwindled, and Lindley Johnson thought that if the Air Force Space Command were to adopt asteroid planetary defense as a mission, then the support for space surveillance might rebound. He saw planetary defense as an extension of ongoing space surveillance Air Force activities rather than a completely new mission. The Air Force's GEODSS routinely captured near-Earth asteroids in its imagery, but GEODSS operators simply ignored them

---

72. For the discovery of near-Earth asteroids, the survey that can search the most sky in a given time will dominate other surveys with comparable telescope apertures. Hence LINEAR, with its extremely fast data readout CCD design, was the dominant survey for a number of years after it began routine operations in March 1998. See R. Jedicke, M. Granvik, M. Michelli, E. Ryan, T. Spahr, and D. K. Yeomans, "Surveys, Astrometric Follow-up, and Population Statistics," in *Asteroids IV*, ed. P. Michel, F. E. Demeo, and W. F. Bottke (Tucson, AZ: University of Arizona Press, 2015), pp. 795–813.

73. The NEAT and LINEAR programs are discussed in more detail in chapter 6.

as asteroid discovery was not their job. A policy change and some software to more efficiently extract NEAs from GEODSS imagery could make GEODSS part of a planetary defense system.

As part of an Air Force future planning study called "SpaceCast 2020," undertaken by students in the 1994 class of the Air Command and Staff College, Johnson was lead author on a 1994 paper that looked at what the Air Force's space-related capabilities should be in 2020, then a quarter century into the future. This paper, entitled "Preparing for Planetary Defense," called for a more closely consolidated, coordinated, and expanded international effort to search for, track, characterize, and mitigate near-Earth asteroids that presented potential threats. Johnson, who coined the term "planetary defense," also drew attention to the similarities between the assets required for space surveillance and those needed for asteroid detection and characterization. His report ended with a plea to form an international effort, perhaps under the auspices of the United Nations, to deal with a threat that could be catastrophic for humanity.[74]

While the research head of the SpaceCast 2020 study, Colonel Richard Szafranski, and a few others involved were intrigued and supportive of Johnson's proposed Air Force Space Command mission, the general reception by the Air Force leadership (Johnson called them the "corporate Air Force") was less enthusiastic.[75] Johnson noted typical comments including "…you're so far out there[,] Johnson. This isn't something that the Air Force needed to be interested in." Johnson then noted,

> …but it was only a week or so after we started talking about this that the discovery of Comet Shoemaker-Levy 9 was announced and it would be impacting Jupiter the following summer, and suddenly, I was this great prognosticator of the future. The papers we wrote for SpaceCast 2020 were kind of considered

---

74. Lindley Johnson, Jeffrey L. Holt, Gregory J. Williams (majors, U.S. Air Force [USAF]), "Preparing for Planetary Defense, Detection and Interception of Asteroids and Comets on Collision Course with Earth" (paper presented at SpaceCast 2020, 1995), *https://commons.erau.edu/space-congress-proceedings/proceedings-1995-32nd/april-25-1995/18/* (accessed 16 May 2019).

75. Lindley Johnson's wife, Brandy Johnson (major, USAF), who had also taken astronomy courses and Air Force ROTC at the University of Kansas, was working for Colonel Szafranski and was involved with the overall integration and management of the SpaceCast 2020 effort. Lindley Johnson interview, 29 January 2016.

thesis-level endeavors, and I like to say that I'm probably one of the few students that could claim that his thesis had been saved by an act of God.[76]

Even so, the Air Force was only too glad to leave the asteroid planetary defense mission to NASA. In a 2016 interview, Johnson summed up the Air Force position toward asteroid planetary defense as "…no, no, we're not interested in that. We've got enough missions as it is."[77]

By the mid-1990s, the Air Force had deferred the asteroid planetary defense mission to NASA, and while NASA had been providing a modest level of support for a few asteroid discovery surveys out of its planetary science program, it too did not seem to welcome an asteroid planetary defense mission. The arrival of near-Earth asteroid 1997 XF11 and the (incorrect) prediction of its possible Earth impact in 2028 changed the game, helping to push NASA management into finally accepting the mantle of asteroid planetary defense.

Would Asteroid 1997 XF11 Collide with Earth in 2028?

Six years after the brief brouhaha surrounding the Swift-Tuttle misprediction, a media frenzy accompanied another incorrect Earth impact prediction, for a collision of asteroid (35396) 1997 XF11 on 26 October 2028. Scotti, using the Spacewatch telescope on Kitt Peak, Arizona, discovered the asteroid on 6 December 1997. The asteroid was well observed for another two months but then went largely unobserved in February. When Peter Shelus at the McDonald Observatory in Texas picked it up again on the nights of 3 and 4 March, his four observations extended the data interval to 88 days, allowing a significantly improved orbit estimate. On 11 March, Brian Marsden, director

---

76. Lindley Johnson, interview by Yeomans and Conway, 9 May 2016. In February 2003, Johnson retired from the Air Force; two months later, he was hired by NASA, where he immediately became the Program Officer for the Near-Earth Object and Planetary Astronomy programs and, soon thereafter, the Program Executive for the Deep Impact mission. In January 2016, he was appointed Planetary Defense Officer (perhaps the coolest job title ever) in charge of NASA's newly formed Planetary Defense Coordination Office.

77. Ibid. Another Air Force planetary defense proposal was led by John M. Urias in 1996. In this ambitious proposal, the authors suggested a three-tiered system whereby Earth-crossing object detection systems would employ detectors (optical, infrared, radar) on Earth, between Earth and the asteroid belt, and then use constellations of remote sensing satellites within or around the main asteroid belt itself.

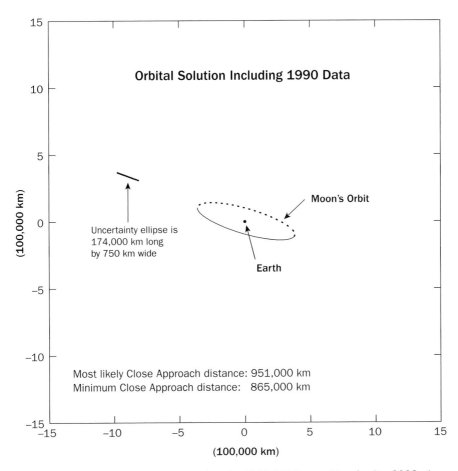

Figure 5-3. A diagram of the uncertainty in 1997 XF11's position for its 2028 close approach to Earth, as calculated in 1998 by Paul Chodas. Note that the uncertainty is primarily in the object's direction of motion. The 1990 data mentioned in the upper right is from pre-discovery ("precovery") images found by Lawrence and Helin of JPL. (Image courtesy of Paul Chodas, JPL/Caltech)

of the Minor Planet Center, announced in an *International Astronomical Union (IAU) Circular* that the prediction for the Earth-miss distance in 2028 was remarkably small (0.00031 au), which was less than 15 percent of the distance to the Moon. The *Circular* noted that "error estimates suggest that passage within 0.002 au was virtually certain." In an accompanying Press Information Sheet, Marsden stated, "The chance of an actual collision is small, but one is not entirely out of the question."[78] Although Marsden never quoted a quantitative impact probability, a few hours after the release of the 11 March *IAU Circular*, Paul Chodas and Don Yeomans at JPL computed the impact probability as zero. Just after 5 p.m. on 11 March, this result was e-mailed to Brian Marsden and a number of other astronomers.

In a subsequent 11 March e-mail to Marsden and others, Yeomans pointed out that, at the time of closest approach, the extremely elongated error ellipsoid, a region in space within which the asteroid is likely to be at a given time, was oriented in the direction of the asteroid's motion, and while this very narrow, pencil-shaped error ellipsoid would come very close to Earth, it would not include Earth.[79] Yeomans wrote, "Probability of impact: 0 (that's zero folks)."[80] Marsden refused to publish this result, noting that an *IAU Circular* should not be used to correct text in his Press Information Sheet. However, a JPL press release that was issued late on 12 March 1998 noted that "Asteroid 1997 XF11 will pass well beyond the Moon's distance from Earth in October 2028 with a zero probability of impacting the planet."[81]

Subsequent independent analyses by the Finnish dynamicist Karri Muinonen supported this conclusion.[82] The conclusion was agreed to by Marsden himself in an *IAU Circular* published on 18 April 1998. Kenneth Lawrence's subsequent identification of pre-discovery (or "precovery") 1990 observations of the asteroid taken by Eleanor Helin and himself with the Palomar Schmidt 46-centimeter telescope on 12 March made it even more clear that an Earth collision in 2028 was ruled out.[83] Nevertheless, the international media circus following the impact prediction for 1997 XF11 was

78. IAUC 6837, 11 March 1998; Minor Planet Center Press Information Sheet, 11 March 1998. The Earth close approach on 26 October 2028 will actually be 0.00621 au—more than three times the 0.002 au noted in the circular.

79. Donald K. Yeomans to Alan W. Harris et al., 11 March 1994 re: 1997 XF11; Alan W. Harris to Richard Binzel et al., 11 March 1998, re: Brian; copies in "1997 XF11 doc scan 02-20-2016.pdf," NEO History Project collection.

80. Donald K. Yeomans to Paul Chodas et al., 11 March 1994, copy in "1997 XF11 doc scan 02-20-2016.pdf," NEO History Project collection.

81. "Asteroid Will Miss Earth by 'Comfortable Distance' in 2028," JPL Release, 12 March 1998, *https://www.jpl.nasa.gov/news/news.php?feature=5185* (accessed 7 May 2018).

82. K. Muinonen, "Asteroid and Comet Encounters with the Earth," in *Proceedings of the Dynamics of Small Bodies in the Solar System: A Major Key to Solar System Studies*, ed. A. E. Roy and B. Steves (Dordrecht: Springer, 1999), pp. 127–158.

83. IAUC 6839, 12 March 1998; IAUC 6879, 18 April 1998.

even more intense than it had been for Comet Swift-Tuttle six years earlier.[84] All four of the United Kingdom national papers (the *Times*, the *Guardian*, the *Independent*, and the *Daily Telegraph*) carried the story on their front pages, as did the *New York Times* in the United States. The *Washington Post*, the *Wall Street Journal*, and many local papers carried the story on their inside pages.[85] The *New York Post* front-page headlines first read "Kiss Your Asteroid Goodbye!" and then "NASA Needs a 'Crash' Course in Math."[86] As was the case for his announced impact possibility for Comet Swift-Tuttle in 2126, Marsden's stated reason for noting the possible impact on the Press Information Sheet was to motivate new observations, including searches for any unreported past observations, in order to refine its orbit and the close-approach prediction.[87] Most astronomers took a different view of the announcement and were concerned that announcing unfounded possibilities of an Earth impact would erode the credibility of future warnings. Clark Chapman, who had been working in asteroid research for nearly fifty years, noted: "Astronomers dare not appear to be Chicken Little and lose credibility in an arena in which they conceivably might someday have to forecast an event that would deserve to be taken seriously at the highest public and governmental levels."[88]

Three months after the 1997 XF11 story first hit the news, Marsden announced on an Internet discussion group evidence that, prior to the discovery of the 1990 observations, "there was in fact a small, but real, possibility of

---

84. The intense media attention given to near-Earth objects in the summer of 1998 was heightened by the release of two Hollywood blockbuster movies involving Earth impacts. The movie *Deep Impact* dealt with the imminent threat of a comet, while the film *Armageddon* dealt with an asteroid impact threat. While the former movie was more scientifically accurate, the latter was commercially more successful and was the highest-grossing film in 1998.

85. Felicity Mellor, "Negotiating Uncertainty: Asteroids, Risk and the Media," *Public Understanding of Science* 19 (2010): 16–33.

86. Clark Chapman, "The Asteroid/Comet Impact Hazard," online essay, 19 August 1999, *http://www.boulder.swri.edu/clark/ncar.ps* (accessed 16 April 2017). Chapman provides a rather detailed, no-holds-barred case study of the 1997 XF11 affair and the issues that it raised.

87. Gretchen Vogel, "Asteroid Scare Provokes Soul-Searching," *Science* 279 (20 March 1998): 1843–1844; B. G. Marsden, "Comets and Asteroids: Searches and Scares," *Advances in Space Research* 33 (2004): 1514–1523.

88. Chapman, "History of the Asteroid/Comet Impact Hazard."

collision" in the decade or so after 2028. He noted that the asteroid's descending nodal crossing was outside Earth's orbit in 2028 but that planetary perturbations would cause the nodal distance to decrease in subsequent years and, in fact, cross Earth's orbit around 2037.[89] Furthermore, the Earth close approach in 2028 could change the orbital period of the asteroid from 1.73 years to anything between 1.58 and 1.99 years, depending on the conditions of the encounter. For every rational number between these limits, there is a corresponding trajectory that would bring the asteroid back for another close approach to Earth in a later year. For example, if the asteroid were to pass about 210,000 kilometers from Earth in 2028, its period would change to 1.80 = 9/5 years, which would bring it back to Earth in 2028 + 9 = 2037, during which time the asteroid would have completed five orbits. In a mathematical exercise, Marsden provided an example orbit, consistent with the earlier 88-day set of observations, which he claimed would actually result in an impact in 2037.

The day after Marsden's claim, Chodas performed a preliminary analysis of the 2037 close approach using an orbital error analysis involving a Monte Carlo technique.[90] His conclusions disagreed with Marsden's. He found that an impact in 2037 was essentially impossible but that an extremely close approach of only 0.5 Earth radius above the surface was possible. In a 10 June e-mail to colleagues summarizing the results, Chodas noted that "about 40 cases out of 20,000 passed through a keyhole in the 2028 error ellipse to arrive within 1.5 million km of Earth in 2037." The descriptive term "keyhole," coined by Chodas, refers to a particularly small region of space near Earth. If the approaching object happens to pass through this small region, Earth's gravity can perturb it onto a trajectory that returns for a later close approach. The 2028 keyhole passage leading to the 2037 Earth encounter did not seem to allow an impact in that year, but what about other keyhole passages that would lead to Earth close approaches in other years? Marsden later identified another impacting scenario based on the short 88-day arc, this one having a post-2028 period of 12/7 = 1.71 years and leading to Earth

89. That is, 1997 XF11 would pass, north to south, through Earth's orbital plane and at Earth's actual heliocentric distance.

90. In the Monte Carlo technique, thousands of test points that represent an object's possible positions and velocities within its error ellipsoid at a particular epoch time are numerically integrated forward in time to obtain the error ellipsoid at a later time. It is equally possible to propagate backward to positions in the past.

impact in 2040. Chodas confirmed Marsden's result that, based upon an early observation set of only 88 observations, the 2040 impact could not be ruled out.[91] Subsequent observations that were included in orbit updates ruled out any possibility of an Earth impact for at least the next 100 years, underlining the importance of numerous observations. Nevertheless, Marsden's reasoning had opened up a rich field of investigation whereby an object could closely pass Earth, enter into a specific narrow keyhole in space, and return for a subsequent Earth impact.

For both Comet 109P/Swift-Tuttle and near-Earth asteroid (35396) 1997 XF11, sensational reports of a possible Earth impact—in reality of no credibility—were made without any sort of peer review by other orbital specialists. Furthermore, such a vetting process was not even possible, since the observations necessary to undertake the analysis were not available until after the announcements had been made. In his 11 March correspondence, Yeomans commented, "I find it more than a little disconcerting that Brian [Marsden] issued a press release before issuing the astrometric data to the small community of orbit computers."[92] This preserved Marsden's scientific priority but also prevented others from checking his work. Although the MPC did eventually provide the observations after a specific request was made following the announcement, it was clear that NEO observations had to be routinely distributed to the scientific community in a more timely fashion. In his own online critique of the NEO community's handling of this event, Clark Chapman criticized Marsden for rushing his 1997 XF11 prediction into the press for a subject that lacked urgency—a prediction of an impact that would not occur for 30 years could have reasonably waited a few days for peer review. The need to obtain more observations did not justify the alarm or the loss of credibility that occurred when the prediction had to be rather publicly corrected.[93]

91. P. W. Chodas and D. K. Yeomans, "Predicting Close Approaches and Estimating Impact Probabilities for Near-Earth Objects" (paper AAS99-462, presented to the American Astronautical Society, Astrodynamics Conference, Girdwood, Alaska, 16 August 1999).

92. Donald K Yeomans to Alan W. Harris et al., 11 March 1998, re: 1997 XF11, copy in "1997 XF11 doc scan 02-20-2016.pdf," NEO History Project collection.

93. Clark R. Chapman, "The Asteroid/Comet Impact Hazard," 19 August 1999, *http://www.boulder.swri.edu/clark/ncar799.html* (accessed 7 May 2018). This dispute within the community did not end right away. Irwin Shapiro, director of the Harvard Smithsonian Center for Astrophysics, criticized Yeomans's comments about Marsden as well. See Irwin Shapiro to D. K. Yeomans, 21 March 1998 et seq., re: Letter, in

Finally responding to the congressional pressure to include detection of hazardous NEOs in its mission, NASA convened a meeting on 17 March 1998 at the Lunar and Planetary Institute in Houston, Texas. The issues raised by concerned scientists and media representatives over the very recent 1997 XF11 affair were very much on the minds of the meeting participants. Carl Pilcher, Science Director for NASA's Solar System Exploration program, chaired the meeting. This workshop established a set of roles and responsibilities regarding NEO data releases and public announcements in the event of cases where a future Earth impact could not be ruled out. These guidelines included the provision that the Minor Planet Center would release NEO astrometric data within 24 hours of its arrival there via the *MPC Daily Electronic Circulars (MPECs)*, which it has done ever since. Before any orbit computation specialist or team could publicly announce a possible future impact, other specialists in the field would verify their computations. This verification period was expected to last up to 48 hours, and NASA's Office of Space Science asked to be informed at least 24 hours in advance of any public report of a Potentially Hazardous Object. Thus there could be a delay of up to 72 hours.[94]

In his May 1998 testimony to the House Subcommittee on Space and Aeronautics, Carl Pilcher committed NASA to achieving the goals of the 1992 Spaceguard Survey—discover 90 percent of Earth-approaching asteroids larger than 1 kilometer within 10 years and physically characterize a representative sample of these objects.[95] Pilcher explained later that the years of delay between Congress's direction in the Agency's 1991 Authorization language and its acceptance of the Spaceguard Survey goal in 1998 were due to the fact that the Office of Space Science did not consider asteroid hunting to qualify as science. "We're not interested in butterfly collecting, was the attitude at the time. Knowing that [they're] there and counting them up just

---

"1997 XF11 doc scan I Shapiro dust up part 1 02-20-2016.pdf" and "1997 XF11 doc scan I Shapiro dust up part 2 02-20-2016.pdf," copies in NEO History Project collection.

94. In late 2000, this 72-hour restriction was eliminated. See chapter 6. Interim communications guidelines can be found in "Interim Roles and Responsibilities for Reporting on Potentially Hazardous Objects," in "1997 XF11 doc scan 02-20-2016. pdf," NEO History Project collection.

95. Summary of Activities of the Committee on Science, H. Rept. 105-847, 105th Cong., 2nd sess., 2 January 1999, *https://www.congress.gov/congressional-report/105th-congress/house-report/847/1* (accessed 29 May 2019).

wasn't a priority for the science office in NASA." But he talked the Associate Administrator for Science, Wesley Huntress, into accepting the task. Pilcher recalls contending, "We shouldn't be resisting Congress. We've been resisting Congress for eight years. What's the benefit?"[96] Huntress agreed, and they combed through the Agency science budget to find about $3.5 million from various other programs. Huntress then convinced NASA Administrator Daniel Goldin to approve the new program, which was named the Near-Earth Objects Observations Program (NEOOP). The NASA FY 1998 and 1999 authorization bills contained $3.4 million for the new program, though the House Committee on Science the following year already thought this was too little: "[T]he Committee notes with great concern NASA's failure to submit a budget for the Near Earth Object Survey that is sufficient to achieve the Shoemaker metric to which NASA has repeatedly committed."[97] They thought the budget should be $10.5 million.

On 6 July 1998, NASA established a Near-Earth Object Program Office at JPL to coordinate and monitor the discovery of NEOs and their future motions, to compute close Earth approaches and, if appropriate, their Earth impact probabilities. Yeomans was appointed manager and remained in that post until his retirement in early 2015, at which point Paul Chodas became the manager. The name of the office was subsequently changed to the Center for Near-Earth Object Studies.

---

96. Carl Pilcher, interview by Yeomans and Conway, 28 February 2016, NEO History Project collection.

97. National Aeronautics and Space Administration Authorization Act of 1999," H. Rept. 106-145, 106th Cong., 1st sess., 18 May 1999, *https://www.congress.gov/congressional-report/106th-congress/house-report/145/1* (accessed 16 May 2019).

CHAPTER 6

AUTOMATING NEAR-EARTH
OBJECT ASTRONOMY

Recognition that near-Earth asteroids and comets represented a new kind of environmental risk led to the establishment of NASA's Near-Earth Object Program Office at JPL in 1998, providing a locus of quantitative risk assessment. An important aspect of that task was the need to refine knowledge about the numbers and sizes of actual asteroids. The Spaceguard goal of finding 90 percent of the 1-kilometer- or larger-diameter asteroids had been based upon modeling that relied upon a very small sample size—only 128 Earth-crossing asteroids were known when the Spaceguard Survey report was published, and only 61 of those had well-established orbits.[1] This small sample size meant a large uncertainty, which could be reduced only by finding a larger fraction of the real population.

The obvious solution to the challenge of finding a large fraction of the near-Earth object population was automated discovery surveys, similar to the work that Tom Gehrels's Spacewatch was already starting to do. The Spaceguard Survey committee had recommended a "survey network" of six 2.5-meter telescopes designed to cover about 6,000 square degrees of sky per month. While that network was never built, during the 1990s, other automated surveys were developed in its place. Two would use U.S. Air Force telescopes built for tracking satellites; two others would be built around older telescopes that had fallen into disuse. And Gehrels's Spacewatch would finally build its

1. David Morrison, ed., "The Spaceguard Survey: Report of the NASA International Near-Earth-Object Detection Workshop" (Moffett Field, CA: NASA Ames Research Center, 1992), *http://ntrs.nasa.gov/search.jsp?R=19920025001* (accessed 8 March 2021), p. 15.

dedicated 1.8-meter telescope, after a good bit of wrangling to get an old mirror returned.

The advent of many automated asteroid surveys during the mid- to late 1990s meant an overwhelming flow of new observations into the Minor Planet Center and other data repositories, necessitating transformations in their processing procedures. But it also meant that each new discovery arrived with some risk that the discovered asteroid would turn out to be a potential hazard—a "PHA," or Potentially Hazardous Asteroid—that might collide with Earth in the future. To keep these objects from slipping through the cracks, astronomers also developed automated warning systems to help identify which of the many thousands of newly discovered objects actually presented a threat. These were designed to help target additional observations to quickly reduce errors in orbit knowledge while also serving as public communications tools.

Survey and Population Modeling

Developing an understanding of what it would take to meet the Spaceguard goal involved modeling the near-Earth object population and evaluating the ability of various search strategies to find objects of different sizes. In 1979, for example, Glo Helin and Gene Shoemaker had made initial estimates of near-Earth and near-Mars asteroids based on their Palomar Planet-Crossing Asteroid Survey. Between 1973 and 1978, they had identified five Earth-crossing and seven Mars-crossing asteroids. Of the Earth-crossers, three were "Aten"-class asteroids whose orbits lay primarily sunward of Earth's, while two were "Apollo"-class asteroids. (See appendix 2.) Other surveys had found four more Aten and Apollo-class asteroids, bringing the known total to nine. Accounting for the limiting magnitude of their 46-centimeter Schmidt telescope and using several different statistical methods to extrapolate from this tiny number of asteroids, they arrived at an estimate of 800 ± 300 near-Earth asteroids of these classes. This projection fell within the same order of magnitude as an earlier estimate of Shoemaker's, which had been based on the numbers and ages of large impact craters.[2]

2.   E. F. Helin and E. M. Shoemaker, "The Palomar Planet-Crossing Asteroid Survey, 1973–1978," *Icarus* 40, no. 3 (1 December 1979): 321–328, doi:10.1016/0019-1035(79)90021-6.

The limiting magnitude of the Palomar Schmidt, which they assessed to be about 18, implied that the asteroids they could see were likely to be quite large. The size of the asteroids could not be directly determined because the telescope only received reflected light; any given asteroid's albedo affects the amount of light it reflects, so an asteroid of a given brightness could be large and dark or small and bright. As a practical matter, therefore, size estimates depended upon albedo estimates.[3] In their 1979 paper, Helin and Shoemaker avoided discussion of asteroid size altogether, instead focusing on the likely populations of various classes of asteroids as defined by their orbits.[4]

For his unpublished 1981 report, Shoemaker compiled data for the 57 known near-Earth asteroids, of which 8 also had albedo estimates.[5] He also identified eight asteroids that would likely now be referred to as "potentially hazardous," with their orbits passing within 0.02 au of Earth's orbit. But he didn't revisit the issue of asteroid populations in the report. He and Carolyn Shoemaker, with Ruth Wolfe of the USGS, picked up the subject again for the 1988 "Global Catastrophes in Earth History" meeting (aka Snowbird II).

---

3. Telescopic visual brightness estimates of asteroids, called apparent magnitudes, are routinely determined for all well-observed asteroids. Using these estimates, in addition to the known asteroid distances from Earth and the Sun as well as the asteroid-centered angle between Earth and the Sun (phase angle) at the time of the observations, the asteroid's absolute magnitude can be determined. An asteroid's absolute magnitude (H) is defined as its visual magnitude seen at zero phase and one au from both the Sun and Earth, and they are available for all well-observed asteroids. If an asteroid's geometric albedo (p_v) or reflectivity is known or assumed, then the object's diameter (D) in kilometers can be expressed in the form $D = 1329\, p_v^{-0.5}\, 10^{-H/5}$. This relationship is given in M. Delbo, M. Mueller, J. P. Emery, B. Rositis, and M. T. Capria, "Asteroid Thermophysical Modeling," in *Asteroids IV*, ed. P. Michel, F. E. DeMeo, W. F. Bottke (Tucson, AZ: University of Arizona Press, 2015), pp. 107–128. In 1985, the International Astronomical Union adopted an asteroid photometric system developed by Edward Bowell and colleagues that provided an expression giving an asteroid's predicted apparent magnitude as a function of its absolute magnitude, its distances from the Sun and Earth, and its phase angle: E. Bowell, B. Hapke, D. Domingue, K. Lumme, J. Peltoniemi, and A. W. Harris, "Application of Photometric Models to Asteroids," in *Asteroids II*, ed. R. P. Binzel, T. Gehrels, and M. S. Matthews (Tucson, AZ: University of Arizona Press, 1989), pp. 524–556.

4. E. F. Helin and E. M. Shoemaker, "The Palomar Planet-Crossing Asteroid Survey, 1973–1978," *Icarus* 40, no. 3 (1 December 1979): 321–328, doi:10.1016/0019-1035(79)90021-6.

5. Snowmass report (unpublished draft, dated July 1981), pp. 8A and 8B, courtesy of Donald K. Yeomans, copy in NEO History Project collection.

Based on a known population of 55 Earth-crossing asteroids, they estimated a population of about 1080 ± 500 down to a limiting magnitude of 17.7, having revised the 46-centimeter Schmidt's capabilities slightly since the 1979 work.[6] They also concluded that the existing survey efforts had probably already discovered all of the Earth-crossing asteroids of absolute magnitude 13 and brighter (and therefore larger than about 7 kilometers in diameter for "bright" asteroids, and 14 kilometers for "dark" ones).

When the Spaceguard Survey committee convened in 1991 in response to congressional direction, an important topic of interest was the development of an optimum search strategy for finding the most hazardous Earth-crossing asteroids and comets. Edward "Ted" Bowell of the Lowell Observatory in Flagstaff, Arizona, and a colleague, Karri Muinonen of the Helsinki Observatory, developed a simulation for that purpose. They constructed a model asteroid population of 320,000 objects using a set of power laws derived from the small number of known Earth-crossing asteroids and from cratering rates, and they explored various approaches to discovering the largest number in the shortest time. Key issues were the choice of limiting magnitude (which affected telescope design), how much of the sky needed to be scanned during each lunation, how often a part of the sky should be rescanned, and how many observations would have to be processed each day to achieve a real-time survey. Since new discoveries had to be reobserved shortly after being detected in order to provide sufficient information for orbit determination, a useful survey had to have near-real-time data-processing capability to ensure that asteroids would not be "lost" due to lack of follow-up observations.

This desire for a real-time survey ruled out the kinds of photographic surveys that the Shoemakers, Helin, and some others had been performing for decades. Bowell and Muinonen wrote, "[T]here is no feasible way, either by visual inspection or digitization of the films, to identify and measure the images in step with the search. A photographic survey would fail for lack

---

6.   Eugene M. Shoemaker, Carolyn S. Shoemaker, and Ruth F. Wolfe, "Asteroid and Comet Flux in the Neighborhood of the Earth," in *Global Catastrophes in Earth History: An Interdisciplinary Conference on Impacts, Volcanism, and Mass Mortality*, ed. Virgil L. Sharpton and Peter D. Ward (Boulder, CO: Geological Society of America, 1990), pp. 174–176.

of adequate data reduction and follow-up."[7] Early in their ideal survey, they anticipated that thousands of objects would be seen every lunation, with about one thousand of those turning out to be new Earth-crossing asteroids.

The two first explored an idealized "whole sky" survey, in which the entire sky was surveyed monthly. In reality, this was impossible to achieve in a single lunation because the Moon brightens the sky too much to find faint asteroids and comets over part of the month, but simulating it nevertheless proved a useful exercise to help them understand all of the constraints on a real survey. At the limiting magnitude of the Palomar 46-centimeter Schmidt that the Shoemakers and Helin had been using, even a 25-year-long whole-sky survey would not find even half of the Earth-crossing asteroids larger than 1 kilometer in diameter. At a limiting magnitude of 20, slightly worse than that of the Spacewatch 0.9-meter telescope, a 25-year survey would find around 70 percent of the 1-kilometer or larger objects. Higher limiting magnitudes were clearly beneficial; at 22nd magnitude, a survey system could reach 90 percent completion in five years.[8] This high yield occurs because the more light a telescope can collect, the farther away it can see any given asteroid. In other words, fainter performance meant peering farther into space, increasing the volume of space being scanned.

The Spacewatch telescope itself would not be able to achieve these results, however, as the drift scan technique severely limited the amount of sky it could access during the dark time of each month. That technique, of course, had been developed to overcome the slow readout of the telescope's detector. The simulations Bowell and Muinonen performed showed that a survey system more efficient than the Spacewatch 0.9-meter telescope would be needed to scan the entire dark sky monthly to discover all of the 1-kilometer-class Earth-crossing asteroids within 25 years. Such a survey still would not discover all the Earth-crossing comets because long-period comets (by definition, with orbit periods greater than 200 years) would not necessarily be visible from Earth during such a "short" survey period. The long-period comets would require surveillance in perpetuity. But comets aside, it was now possible to assess the risk from asteroids in a reasonable period of time.

---

7. Edward Bowell and Karri Muinonen, "Earth-Crossing Asteroids and Comets: Groundbased Search Strategies," in *Hazards Due to Comets and Asteroids*, ed. Tom Gehrels (Tucson, AZ: University of Arizona Press, 1994), p. 181.

8. Morrison, "Spaceguard Survey," p. 30.

These simulation studies led the Spaceguard committee to recommend a survey system based upon the Spacewatch 2,048- by 2,048-pixel sensor with faster readout and data-processing capabilities and seven telescopes 2 meters in diameter, with four detectors each. Fewer, larger telescopes could be substituted, but they needed to be distributed between both hemispheres.[9] A coordination center would be necessary to assure monthly sky coverage, and of course it would need some sort of data system. They expected this Spaceguard Survey system to cost about $50 million to build and $10 to $15 million per year to operate.[10]

Nothing like the Spaceguard Survey system would actually be built. Instead, when the NASA Near-Earth Objects Observations Program was established in 1998, it funded renovations to existing telescopes to enable reaching the Spaceguard goal.

Automating Surveys

Tom Gehrels had long sought to create a dedicated facility for assessing the true population of near-Earth asteroids. The 0.9-meter Newtonian telescope his Spacewatch team had adapted during the late 1980s had never been his end goal. He still intended to build a larger, dedicated 1.8-meter telescope on Kitt Peak in Arizona. He even had a mirror for it—at least, he had the promise of one. During the 1960s, the U.S. Air Force had built several lightweight mirrors for a project known as the Manned Orbiting Laboratory, which was intended to be, in essence, a crewed surveillance satellite.[11] But the program was canceled in 1969. Aden Meinel, director of the University of Arizona's Optical Sciences Center, proposed using six of these mirrors in a "multiple mirror telescope" in 1970.[12] When given the go-ahead, he allocated one of

---

9. Ibid., p. 43.

10. Ibid., p. 51.

11. Carl Berger, "History of the Manned Orbiting Laboratory Program," Department of the Air Force, February 1970, *https://www.nro.gov/FOIA/MOL/* (accessed 14 September 2021).

12. His proposal was called Project COLT initially, and for some reason the National Reconnaissance Office's declassified copy redacts Meinel's affiliation as well as the name of the telescope he was proposing. See Lawrence E. Pence to Dr. McLucas, Subject Project COLT, 19 January 1971, National Reconnaissance Office, Washington, DC, released 1 July 2015, document 816 at *https://www.nro.gov/FOIA/MOL/* (accessed 14 September 2021).

the mirrors to Gehrels's asteroid telescope and six others to the Multi-Mirror Telescope (MMT) project.[13]

Funding for Gehrels's asteroid telescope did not come through during the 1970s or the 1980s. But the Multi-Mirror Telescope was more successful. It was funded and built on Mount Hopkins outside Tucson. During its construction, though, one of its six mirror blanks was damaged, and Gehrels loaned the project his mirror until the other could be repaired to help keep the project on schedule. The MMT was finished in 1979, and his mirror was removed and put in storage—where it would remain until January 1993. Getting the mirror returned was a considerable struggle, beginning in 1990. As Gehrels's loan had never been properly documented, without proper paperwork, he kept running into bureaucratic barriers. He resorted to getting testimonials from Aden Meinel and others to convince the Steward Observatories of the University of Arizona (which operated the MMT) and the Smithsonian Institution (which owned the telescope) to transfer the mirror to Spacewatch—which was itself, of course, also a unit of the University of Arizona. The deal that was finally struck between the MMT and Spacewatch late in 1991 was that Gehrels would get one of the mirrors (ultimately, he acquired MMT-5) once full funding for his telescope was in hand.[14] In further irony, by the time the transfer took place, the MMT was being converted to a single large mirror configuration (a 6.5-meter mirror for it was cast in 1992), and all of its original mirrors would soon be removed and boxed, too.

The successful operation of the 0.9-meter Spacewatch telescope and the growing interest in the near-Earth object threat enabled Gehrels to raise the funds he needed to finally build the larger telescope. As had been true of the 0.9-meter telescope modification, funds came from a blend of

---

13. See W. Patrick McCray, *Giant Telescopes: Astronomical Ambition and the Promise of Technology* (Cambridge, MA: Harvard University Press, 2004), pp. 63–69.

14. See, e.g., Tom Gehrels to F. H. Chaffee, 3 October 1990, in folder 15 Spacewatch 72" telescope 1990–1994, box 22, Gehrels papers (MS514), University of Arizona Special Collections; Irwin Shapiro to Tom Gehrels, 17 December 1990, "Re: Mirror, Mirror in the Box," folder 15 Spacewatch 72" telescope 1990–1994, box 22, Gehrels papers, (MS514), University of Arizona Special Collections; Aden B. Meinel to Tom Gehrels, 19 April 1991, folder 15 Spacewatch 72" telescope 1990–1994, box 22, Gehrels papers, (MS514), University of Arizona Special Collections; Michael A. Cusanovich and Ed McCullough to Frederic Chaffee et al., 1 November 1991, "Re: The '7th mirror,'" folder 15 Spacewatch 72" telescope 1990–1994, box 22, Gehrels papers (MS514), University of Arizona Special Collections.

private and public sources. The telescope itself was constructed with donations from the University of Arizona Foundation, the David and Lucille
Packard Foundation, a John Nitardy of Seattle, and grants from NASA and
the Ballistic Missile Defense Organization.[15] Jurgen Rahe, head of NASA's
Planetary Astronomy Program, was the source of NASA's funds, while within
the Air Force, Simon "Pete" Worden was Gehrels's chief patron. Rahe also
funded the camera and data system, and both NASA and the U.S. Air Force
Office of Scientific Research supported salaries. In a 1994 memo to Hans
Mark, Gehrels remarked that Spacewatch's operating funds were 60 percent
Air Force, 40 percent NASA.[16]

Gehrels's 1.8-meter telescope (often called Spacewatch II) saw "first light" in
2001. By that time, Spacewatch was no longer the sole semiautomated asteroid
survey. Grant Stokes at the MIT Lincoln Laboratory, a Defense Department–
supported Federally Funded Research and Development Center, had proposed using a CCD camera with very-high-speed readout capabilities on one
of the U.S. Air Force's experimental space surveillance telescopes in New
Mexico for this purpose. Eleanor Helin made arrangements to put another
CCD-based camera on an operational Air Force space surveillance telescope
on Haleakalā, Maui, Hawai'i. Ted Bowell at the Lowell Observatory began a
survey on a 56-centimeter Schmidt telescope in 1998. And another group at
the University of Arizona's Lunar and Planetary Laboratory was renovating an
observatory on nearby Mt. Lemmon to discover asteroids from there.

The first of these second-generation automated surveys to get off the
ground was Helin's, and it was known as the Near-Earth Asteroid Tracking
program, or NEAT. NEAT was based upon the use of 1-meter Ground-based
Electro-Optical Deep Space Surveillance (GEODSS) telescopes in Hawai'i.
The GEODSS system had been developed for the U.S. Air Force in the 1980s,
and while it was not CCD-based, it was an automated wide-field search and
tracking system. In the mid-1980s, Helin, like Gehrels, had started thinking about moving away from film. JPL's Raymond Bambery, one of her early

---

15. See, e.g., Wilbur S. Coburn to Tom Gehrels, 29 June 1995, "Re: Grant #95-1390,"
     folder 16 Spacewatch Program 1995, box 22, Gehrels papers (MS514), University of
     Arizona Special Collections; Tom Gehrels to Colonel S. P. Worden, 15 November
     1993, folder 30 Worden SP, box 8, Gehrels papers (MS514), University of Arizona
     Special Collections.

16. Tom Gehrels to Hans Mark, 2 March 1994, folder 22, box 5, Gehrels papers (MS514),
     University of Arizona Special Collections.

Figure 6-1. Robert S. McMillan in the Spacewatch control center in November 2011. (Photographer Ron Mastaler, courtesy of Robert S. McMillan)

collaborators on the NEAT project, recalled that in 1988 they had proposed to JPL's Directors Discretionary Fund a simple proof of concept that scanned plate pairs to find asteroids.[17] Helin's first CCD camera, though, got its start due to the cancellation of the Comet Rendezvous Asteroid Flyby mission (generally known as CRAF). CCDs for one of CRAF's instruments had already been bought and suddenly had no purpose; Helin and others at JPL wrote a series of proposals over the ensuing few years to build a camera intended for the 18-inch telescope she had been using on Palomar. But this plan presented a new problem: the 18-inch telescope on Palomar was ancient, and upgrading it for use with a digital camera would be more expensive than building a new telescope. And NASA had no interest in funding a new telescope.

Access to the GEODSS telescope came about after a conversation within the Air Force Research Laboratory (AFRL) in 1992. Paul Kervin of the Air Force Research Laboratory in Hawai'i recalls that he was asked by Janet

---

17.   Raymond Bambery, interview by Yeomans and Conway, 16 February 2017, transcript in NASA History Division HRC.

Fender at the Air Force's facility in Albuquerque to talk to Helin about possibly cooperating on an observing program using the 1.2-meter Air Force Maui Optical Station (AMOS) telescope, which had recently been transferred to the AFRL's control. Helin and Yeomans went to Maui, where the AMOS telescope was situated on Haleakalā, to discuss the opportunity. In March 1993, Helin gained permission to use the AMOS telescope to do follow-up observations of asteroids.[18]

Later, Helin pursued access to one of the GEODSS telescopes for the CCD camera, too, which was ultimately built for JPL at San Diego State University.[19] There were three GEODSS stations in operation, each with two 1-meter telescopes and a smaller, auxiliary wide-field telescope: in Socorro, New Mexico; on Haleakalā; and on the atoll Diego Garcia. Kenneth Lawrence, who followed Helin from her film-based survey to NEAT, recalled much later that she preferred the Maui site to Socorro, in part due to the monsoon season in New Mexico and in part because Socorro was too close to Tucson and Spacewatch.[20] A different piece of sky could lead to finding more new objects.

Quite a lot of effort had to go into making the GEODSS telescopes available because—unlike the AMOS telescope—GEODSS was an operational system belonging to Air Force Space Command, not to the Air Force Research Laboratories. It was a challenge convincing Air Force officers to spend some of the systems' time finding near-Earth asteroids instead of satellites (and, perhaps more importantly, assuring them that classified satellites would not be inadvertently revealed). But with the Cold War over, GEODSS was relatively underutilized, and eventually Helin's group was granted 18 nights per month at the Maui site for their CCD camera.[21] Lindley Johnson, then assigned to Air Force Space Command, was part of the effort to persuade his superiors to

18. Paul Kervin to Donald K. Yeomans, e-mail dated 30 January 2017, copy filed as "PKervin_30 Jan 2017.pdf," NEO History Project collection. Also see James C. Mesco, "Watch the Skies," *Quest* 6, no. 4 (1998): 35–40.

19. Steven H. Pravdo, David L. Rabinowitz, Eleanor F. Helin, Kenneth J. Lawrence, Raymond J. Bambery, Christopher C. Clark, Steven L. Groom, et al., "The Near-Earth Asteroid Tracking (NEAT) Program: An Automated System for Telescope Control, Wide-Field Imaging, and Object Detection," *Astronomical Journal* 117, no. 3 (1999): 1616.

20. Kenneth Lawrence, interview by Conway, 27 January 2017, transcript in NASA History Division HRC.

21. Kervin to Yeomans, 30 January 2017.

allow this research, and Helin named asteroid 1989 CJ1 for Johnson for his support in 2005.[22]

The kind of "hands-on" observing that was done at Palomar and at Spacewatch on Kitt Peak was infeasible here, owing to the great distance between JPL and Haleakalā. Instead, the system developed by Steven Pravdo, Ray Bambery, David Rabinowitz (who joined the NEAT team to help build the software), and others was designed to be run remotely from Pasadena. GEODSS personnel in Maui only had to mount the camera to the telescope's prime focus, turn on the local computer, and open and close the dome. A Sun Sparc 20 dual–central processing unit computer handled both telescope control and on-site data reduction. It also identified likely asteroid candidates and transmitted their information back to JPL. Helin's team wrote and uploaded a script to the telescope control computer each observing day for that night; back in Pasadena, they also reviewed the asteroid candidates each morning to weed out false identifications. The job of reviewing candidates and sending identifications to the Minor Planet Center fell mostly to Kenneth Lawrence.[23]

NEAT began observing on Haleakalā in December 1995. In its first 15 months, the program identified 5,637 new asteroids, including 14 near-Earth asteroids and 2 long-period comets.[24] In an article she wrote for the New York Academy of Sciences, Helin commented on the different observing strategies of NEAT and Spacewatch. NEAT covered about 10 times as much sky per night, but only to a lower limiting magnitude of 20, compared to 22 for Spacewatch.[25] It achieved wider coverage with less depth. In mid-1996, the camera was upgraded to a commercially available 4k by 4k CCD,

---

22. 5905 Johnson. See *https://ssd.jpl.nasa.gov/sbdb.cgi?orb=1;sstr=5905* (accessed 17 September 2019).

23. Lawrence interview, 27 January 2017; Raymond Bambery, interview by Yeomans and Conway, 16 February 2017; Steven Pravdo, interview by Yeomans, 7 March 2017; David Rabinowitz, interview by Conway, 27 January 2017 (transcripts for all of these in NEO History Project collection); Steven H. Pravdo, David L. Rabinowitz, Eleanor F. Helin, Kenneth J. Lawrence, Raymond J. Bambery, Christopher C. Clark, Steven L. Groom, et al., "The Near-Earth Asteroid Tracking (NEAT) Program: An Automated System for Telescope Control, Wide-Field Imaging, and Object Detection," *Astronomical Journal* 117, no. 3 (1999): 1616.

24. Eleanor F. Helin, Steven H. Pravdo, David L. Rabinowitz, and Kenneth J. Lawrence, "Near-Earth Asteroid Tracking (NEAT) Program," *Annals of the New York Academy of Sciences* 822, no. 1 (1997): 6–25.

25. Ibid., pp. 6–25.

and the computers were upgraded as well. Further improvements in 1997 and 1998 reduced their "cycle time," the time delay between one 20-second exposure and the next, from 160 seconds to 45 seconds, increasing their nightly sky coverage. NEAT's allocation of observing time on Haleakalā was reduced while they were upgrading the system; by 1998, they were down to six nights per month prior to the new Moon. The reason was simply that Air Force Space Command, which owned the telescopes, had growing needs for them, and asteroids were not part of its mission. In 1998, the Air Force Space Command decided to upgrade the GEODSS data infrastructure in ways that rendered the NEAT system incompatible. Ultimately, NEAT moved back to the 1.2-meter AMOS telescope, which came under the Air Force Research Laboratory's jurisdiction. The AMOS telescope needed upgrades too (that the Air Force paid for) to accept the NEAT system,[26] and NEAT stopped observing between 1999 and 2000 to accommodate this work.

NEAT also gained an opportunity to use the 1.2-meter Oschin Schmidt at Palomar Observatory in California. That telescope had a much wider field of view, and the team designed a new camera for it that would cover part of its focal plane with three 4k by 4k CCDs. The new design (called the "three-banger" for the three CCDs) also had better cooling and faster readout speeds. They built two of these, one for Palomar and one for the AMOS telescope. The team also renovated the Oschin Schmidt telescope for the same kind of robotic control that was used in Hawai'i. For all of these reasons, NEAT was not generating new discoveries in 1999.[27] They began observing again on the Maui Space Surveillance telescope during the spring of 2000, though that year witnessed very poor observing weather in Hawai'i that undermined their productivity. They also experienced technical problems related to the humidity of the site. As a result, they were increasingly seen as unproductive compared to the other surveys.[28] The NEAT project's upgrades to the Oschin telescope on Palomar were completed late in 2000, and observations with its three-banger camera began in spring 2001.

---

26. Lawrence interview, 27 January 2017.

27. Lawrence interview, 27 January 2017; Steven Pravdo to Dennis L. Sparrow, e-mail dated 2 June 1999, "Re: NEAT Operations Post-GMP," provided by D. Yeomans, copy in NEO History Project collection, Don Yeomans materials.

28. Yeomans notebook, "NEO Program Office," entries for 6 March 2000, 10 July 2000; 11 September 2000, and 13 November 2000, courtesy of Don Yeomans.

The second new survey to begin in the 1990s was also a cooperative program with the Air Force, known as LINEAR—the Lincoln Near-Earth Asteroid Research program. Grant Stokes of the MIT Lincoln Laboratory in Lexington, Massachusetts, had proposed using Lincoln-developed fast-readout CCD cameras on the GEODSS telescopes at the Laboratory's Experimental Test Site (ETS) at White Sands Missile Range in New Mexico for an asteroid survey. Unlike the operational GEODSS telescopes that NEAT shared with other Air Force efforts in Maui, the ETS telescopes were for research; the Lincoln Labs used them to try out new sensor technologies. The large-format, fast-transfer CCD cameras that Stokes intended to use had been specifically developed as a potential upgrade for the operational GEODSS telescopes, so compatibility with the telescopes was never an issue. A prototype of the new camera and processing system was tested in August 1995 and July 1996 at ETS, just a few months before NEAT began observing on Haleakalā. The July 1996 test series netted LINEAR its first NEO, 1996 MQ.[29]

The success of the test series garnered Air Force funding to develop the complete LINEAR system. In mid-1997, a nearly complete system with a smaller CCD underwent further trials at ETS; the large-format CCD was installed that summer, and in October of 1997, a 10-night observing campaign generated 52,542 observations and netted 9 confirmed new NEOs. Over the course of 1998, LINEAR became the dominant discoverer of near-Earth asteroids, and in 1999, it expanded further by equipping another of the ETS telescopes. It was so prolific that it overwhelmed the ability of follow-up observers to keep up with its new discoveries and had to reprogram some of its observing time to follow up its own discoveries, reducing its sky coverage to an extent.[30] But it remained the dominant survey through 2005. After 1998, though, Stokes had to begin drawing NASA funds from the NEO Observations Program to maintain operations. Developing new technologies was an Air Force objective, while discovering asteroids was not.

---

29. Grant Stokes, interview by Yeomans, 25 February 2016, transcript in NASA History Division HRC; Grant H. Stokes, Jenifer B. Evans, Herbert E. M. Viggh, Frank C. Shelly, and Eric C. Pearce, "Lincoln Near-Earth Asteroid Program (LINEAR)," *Icarus* 148 (2000): 21–28. The discovery credit went to astronomer Robert Weber. See MPEC 1996-MO4, *https://www.minorplanetcenter.net/mpec/J96/J96M04.html* (accessed 4 November 2019).

30. Follow-up observations were done mostly by amateurs at the time, though Spacewatch did them on occasion.

Two more surveys, both of which had been started in the film era and were later revamped in the early 1990s, were funded out of the small NASA NEO budget. One of these surveys was initiated at the Lowell Observatory by Ted Bowell. The Lowell Observatory NEO Survey (LONEOS) was developed around a 0.6-meter telescope and a wide-field camera. The Lowell Observatory had been involved in doing astrometric analysis of the films taken by the Shoemakers at Palomar and had built an automated film analyzer for that purpose. Bowell had also developed his own public database for near-Earth asteroids during the mid-1990s, so the group, including observer Brian Skiff and software engineer Bruce Koehn, had some expertise to build on. LONEOS was active over the 1993–2008 period, but technical problems getting the camera and telescope together hampered early efforts, and the group did not start discovering NEOs until 1998. Between 1998 and 2008, the LONEOS team discovered a total of 290 NEOs, including the "lost" asteroid 69230 Hermes and the second Atira asteroid to be found, 2004 JG_6. Ultimately, though, their modest aperture did not make them competitive with other operational surveys, and, after a period when they provided photometry of NEOs, they ceased operations in 2008.

Finally, a second survey wound up being established at the Lunar and Planetary Laboratory of the University of Arizona (just three floors above Spacewatch). Scotti, who had been hired to write Spacewatch's software, decided to learn how to find asteroids and comets the old-fashioned way, on film. There was an essentially unused 0.4-meter Schmidt on Mount Bigelow still configured for film, so he bought some film and recruited an undergraduate student, Timothy Spahr, to help him. Scotti wanted to try observing in the early-morning twilight near the Sun (where Spacewatch itself would never look) in hopes of finding a comet. Scotti participated for only a few months, but Spahr became an enthusiastic convert to asteroid and comet hunting and found a University of Arizona mentor in Steve Larson.[31]

Larson commented many years later that when he was preparing for Comet Hale-Bopp's appearance in 1997, the company that made the 2k by 2k CCDs used by the NEAT project was having a sale, and he acquired one

---

31. Jim Scotti, interview with Conway, 2 August 2016, transcript in NEO History Project collection.

to upgrade a camera he had first made for the Comet Halley Watch.[32] He used it on the Catalina 0.4-meter Schmidt to demonstrate its utility for wide-field surveys, completing the upgrade by the time Spahr received his Ph.D. from the University of Florida in 1998. Larson hired him back to write the detection software, and Spahr recalls that he and Carl Hergenrother had their software running by fall of 1999.[33] This survey became known as the Catalina Sky Survey, and it began reporting observations to the Minor Planet Center in 1998. There was also an unused 1.5-meter telescope on Mount Lemmon, and, after getting the Schmidt operating, Larson proposed converting that telescope for asteroid survey work as well.

Larson also wanted to tackle the problem of the lack of Southern Hemisphere observations. Between 1990 and 1996, an Anglo-Australian Near-Earth Object Survey was conducted by reviewing plates taken for another purpose on the 1.2-meter U.K. Schmidt telescope at the Siding Spring Observatory in New South Wales, Australia.[34] The Siding Spring Observatory also had a smaller, 0.5-meter Schmidt telescope known as the Uppsala Schmidt (after its original home in Sweden) that was largely unused. Larson e-mailed the Uppsala Observatory director at the time, discovered that he was actually visiting Tucson, and had a quick meeting to work out a memorandum of understanding. Once the extensive modifications necessary to host a CCD camera on the Uppsala Schmidt were completed, Robert McNaught became the lead observer, and, together with his coworker Gordon Garradd, they provided the only NEO observations from the Southern Hemisphere. These observations lasted until 2013, when budget considerations triggered the closure of the Siding Spring arm of the Catalina Sky Survey.

Larson was able to get NEO Observation Program funds to finance the upgrades to the 1.5-meter Mount Lemmon telescope and the Uppsala Schmidt in 2000; in effect, the Catalina Sky Survey stopped observing while the small team did all the upgrade work involved. He hired Edward Beshore—a technical writer by profession who had built his own automated telescope for his

---

32. Stephen Larson, interview by Conway, 17 October 2016, transcript in NASA History Division HRC.

33. Tim Spahr, interview by Yeomans and Conway, 26 January 2016, transcript in NASA History Division HRC.

34. "AANEAS: A Valedictory Report," *http://users.tpg.com.au/users/tps-seti/spacegd4.html* (accessed 24 May 2017).

backyard—to help him manage the effort, and they restarted observations in 2003.

By 2004, several NASA-funded near-Earth object discovery surveys were operating and discovering hundreds of new NEOs per year. These were accomplished on budgets that did not exceed $4 million per year, not counting the Air Force contributions of technology and time (to LINEAR and NEAT), or that of amateurs in carrying out follow-up observations for new detections. European scientists' efforts to stimulate interest were not nearly so successful. In 2000, the British National Space Centre commissioned a study of what the United Kingdom could provide to an international NEO discovery and characterization program.[35] It gained little immediate traction. European efforts to address the issue would not begin to crystallize until 2008, when the European Space Agency initiated a "Space Situational Awareness Programme" to develop a European space surveillance capability covering space weather, satellites and debris, and near-Earth objects.

In 2000, the NEAT team published a new estimate of the number of near-Earth objects larger than 1 kilometer based on their discovery statistics to date. Between March 1996 and August 1998, they had discovered 26 near-Earth asteroids larger than 1 kilometer, out of a total of 45 discoveries. They then generated a series of simulations of fictitious asteroid populations and evaluated what NEAT would have detected out of those artificial populations, given its known performance—amount of sky searched per month, limiting magnitude, and so forth. Comparing the simulation results to their actual results gave them an estimate of the likely completeness of their survey to date and thus the total number of large near-Earth objects. They contended that there were about 700 such objects (±200), not greatly different from the estimate developed by Shoemaker and Helin from their film survey in the 1970s. They also argued that the existing surveys would probably reach the goal of 90 percent discovery within 20 years, but probably not by the goal year of 2008. "Doubling the current world-wide detection rate would therefore lead to near completion in the next decade," they wrote.[36]

35. Harry Atkinson, Chair, *Report of the Task Force on Potentially Hazardous Near-Earth Objects* (September 2000), *http://spaceguardcentre.com/what-are-neos/task-force-pdf/* (accessed 24 March 2021).

36. David Rabinowitz, Eleanor Helin, Kenneth Lawrence, and Steven Pravdo, "A Reduced Estimate of the Number of Kilometre-Sized Near-Earth Asteroids," *Nature* 403, no. 6766 (2000): 165–166.

The Minor Planet Center

Despite these pessimistic evaluations of the automated surveys' performance, the scale of the data being produced by the automated surveys by the late 1990s—especially by LINEAR—caused the International Astronomical Union's Minor Planet Center (MPC) some growing pains. Its operating costs were primarily funded by NASA and by fees paid by subscribers to the *Minor Planet Circular*. Private donations provided its computing infrastructure. It was a very small operation, with only two or three people to run the place.

Brian Marsden had been its director since the center had moved from the Cincinnati Observatory to the Smithsonian Astrophysical Observatory in 1978. The MPC's charge was to be the community repository for the astrometric observations and orbits of comets and asteroids, and it also specified the rules for naming these bodies and for granting credit for new discoveries. Astronomers, professional or amateur, submitted observations to the center daily, and for objects that were seen on more than one night, the MPC generated orbits from the observation data. During the MPC's decades at the Cincinnati Observatory, Paul Herget had deployed punch-card–based calculating machines and later computers to calculate orbits.[37] His equipment, techniques, and banks of punch cards—Brian Marsden claimed almost 200,000 cards in 1980—then moved east with Conrad Bardwell, who had been Herget's assistant in Cincinnati and relocated to Cambridge with the MPC's move.[38] Bardwell retired in 1989 and was succeeded by Gareth Williams.

The 1990s were a decade of rapid transition for the MPC. At the beginning of the decade, observations were mostly submitted by telegram, telephone, or even on paper. By the end of the decade, nearly everything came in by e-mail, and the *Minor Planet Circular* had moved to the World Wide Web. In 1993, they developed a Minor Planet confirmation page to alert other astronomers about newly observed objects that needed follow-up observations to secure orbits and provided an electronic version of the *Minor Planet Circular*. Williams also worked with Jim Scotti of Spacewatch to develop a two-phase pipeline for Spacewatch's data, with "fast mover" objects being sent

---

37. Donald E. Osterbrock and P. Kenneth Seidelmann, "Paul Herget," *Biographical Memoirs* 57 (1987): 59.

38. Brian G. Marsden, "The Minor Planet Center," *Celestial Mechanics* 22 (1 July 1980): 63–71, doi:10.1007/BF01228757; Gareth Williams, interview by Rosenburg, 30 November 2016, transcript in NEO History Project collection.

and processed first, and the slower objects coming in later.[39] The scale of data being produced by LINEAR beginning in 1998, though, posed a big problem for the center—it could not keep up with the processing.

Timothy Spahr, whom Marsden had hired away from the Catalina Sky Survey in May 2000, remembered that when he had arrived at the MPC, about a year and a half of LINEAR data had only been partially processed. With the system that was then in place, it took 45 days to fully process a month's worth of LINEAR's output. So the MPC could never catch up.[40] Spahr remembers that he eventually encouraged Williams to change the algorithms that discriminated known objects from unknown ones to speed that process up dramatically.

Brian Marsden stepped down as director of the Minor Planet Center in 2006, at the same meeting of the International Astronomical Union in Prague that resulted in the demotion of Pluto to the status of "dwarf planet."[41] He was succeeded by Spahr. Marsden and Williams had obtained previous support from Steven Tamkin for MPC computer hardware, and Spahr approached Tamkin about funding a new computing infrastructure for MPC to help it keep up with the increasing flow of observations from the automated surveys. Tamkin agreed, and the conversion to a Linux-based system began in 2009.[42]

Modeling the Motions of Comets and Asteroids

The ability to turn observations into orbits and to be able to predict future positions and potential impact risks depends on numerical modeling. For all observed near-Earth objects, whether comets or asteroids, orbits are computed by personnel working within the Minor Planet Center and JPL's Solar System Dynamics Group, as well as European researchers. Using some or all of the available optical observations, they first compute a preliminary orbit. Then, from the preliminary orbit, they generate an ephemeris that provides predicted positions of the object at the times when the object was actually observed. These computed positions are then differenced with the actual,

---

39. Williams interview, 30 November 2016.

40. Spahr interview, 26 January 2016.

41. Gareth V. Williams and Cynthia Marsden, "Brian G. Marsden (1937–2010)," *Bulletin of the AAS* 43, no. 1 (1 December 2011), *https://baas.aas.org/pub/brian-g-marsden-1937-2010/release/1* (accessed 14 September 2021).

42. Spahr interview, 26 January 2016.

observed positions. These differences are then squared and summed together. The initial preliminary orbit is adjusted, or refined, until the sum of these squares is minimized. Modern computer-generated orbit determination processes also include the perturbing effects of the planets and some of the more massive asteroids and relativistic effects. Once a good orbit has been developed based upon optical observations alone, an accurate ephemeris for the object can often be used to predict range and Doppler measurements so that radar observations can be made. Radar observations, along with the optical observations, are sometimes available to provide more refined orbits along with ephemerides that are normally accurate well into the future.[43]

In the case of comets, today's procedures to define an orbit and account for nongravitational forces are rooted in ideas stretching back centuries. In the 19th century, the German astronomer Johann Encke studied the motion of a short-period comet—with an orbital period of only 3.3 years—over the interval from its discovery in 1786 through 1858. He noted that the observed times of perihelion passage were consistently a few hours earlier than his predictions. To explain the comet's early arrivals, Encke suggested that the comet's orbital period was decreasing as it moved through the interplanetary resisting medium that was thought to exist at that time. During the 1835 return of Comet Halley, Friedrich Bessel of the Königsberg Observatory had suggested that, alternatively, the activity of this comet's nucleus itself could introduce rocket-like deviations from its normal course.[44] In 1950, Fred Whipple's "dirty snowball" model for a cometary nucleus provided a likely solution to the so-called nongravitational motions of comets.[45] For a rotating cometary nucleus, the maximum vaporization of the cometary ices would take place at a position offset from the subsolar point (cometary noon). There would then be a nonradial component of thrust that would either introduce orbital energy

---

43. D. K. Yeomans, P. W. Chodas, M. S. Keesey, S. J. Ostro, J. F. Chandler, and I. I. Shapiro, "Asteroid and Comet Orbits Using Radar Data," *Astronomical Journal* 103, no. 1 (1992): 303–317; S. J. Ostro and J. D. Giorgini, "The Role of Radar in Predicting and Preventing Asteroid and Comet Collisions with Earth," in *Mitigation of Hazardous Comets and Asteroids*, ed. M. J. S. Belton, T. H. Morgan, N. Samarasinha, and D. K. Yeomans (Cambridge, U.K.: Cambridge University Press, 2004), pp. 38–65.

44. Donald K. Yeomans, *Comets: A Chronological History of Observation, Science, Myth, and Folklore* (New York: John Wiley and Sons, Inc., 1991).

45. Fred L. Whipple, "A Comet Model. I. The Acceleration of Comet Encke," *Astrophysical Journal* 111 (1950): 375–394.

for a nucleus in prograde rotation or subtract orbital energy for a retrograde rotator.[46] Brian Marsden and Zdenek Sekanina of JPL carried out a number of efforts to empirically model these so-called nongravitational forces acting on comets in the late 1960s and early 1970s. The model introduced by Marsden, Sekanina, and Yeomans in 1973, where the comet's rocket-like thrusting is based upon the vaporization of water ice as a function of heliocentric distance, was quite successful and is still in use more than 40 years later.[47]

Surprisingly, inactive asteroids are also affected by nongravitational forces. Because of their "thermal inertia"—their ability to retain heat—the motions of asteroids are affected by thermal reactive forces. This so-called Yarkovsky effect can cause the semimajor axis of an asteroid to drift inward or outward.

The Yarkovsky effect is named after Ivan Yarkovsky, a Polish civil engineer working in Russia who first introduced the concept in 1888.[48] This effect refers to the nongravitational force introduced by the thermal reradiation of sunlight from a rotating asteroid. Just as on Earth, midafternoon, rather than exactly noon, is the warmest part of the day, so it is for asteroids where the afternoon side of the asteroid is warmer than the subsolar point (noon). Absorbed solar radiation is re-emitted as infrared radiation (heat) with some delay. As a result, the reradiated solar energy introduces a component of thrust in the direction of the asteroid's motion (for an asteroid in prograde rotation) or a component of thrust opposite to the asteroid's motion (for an asteroid in retrograde rotation). Thus, the Yarkovsky effect can either introduce an outward drift in semimajor axis for asteroids in prograde rotation (orbital energy is added) or an inward spiral for asteroids in retrograde rotation (orbital energy is subtracted). In 2000, the Czech astronomer David Vokrouhlický and colleagues predicted that the well-observed, half-kilometer-sized asteroid 6489 Golevka would exhibit a noticeable Yarkovsky effect once the 2003

---

46. Prograde rotation is in the same sense as the object's motion about the Sun. Retrograde rotation is in the opposite sense. A detailed history of cometary motions is provided by Yeomans, *Comets: A Chronological History.*

47. A review of various cometary nongravitational models is provided by D. K. Yeomans, P. W. Chodas, G. Sitarski, G. Szutowicz, and M. Królikowska, "Cometary Orbit Determination and Nongravitational Forces," in *Comets II*, ed. M. C. Festou, H. U. Keller, and H. A. Weaver (Tucson, AZ: University of Arizona Press, 2004), pp. 137–151.

48. George Beekman, "I.O. Yarkovsky and the Discovery of 'His' Effect," *Journal for the History of Astronomy* 37 (2006): 72–86. Yarkovsky originally proposed the effect to explain why celestial bodies were not observed to lose orbital energy as they moved through the resistive interplanetary ether that was then thought to exist.

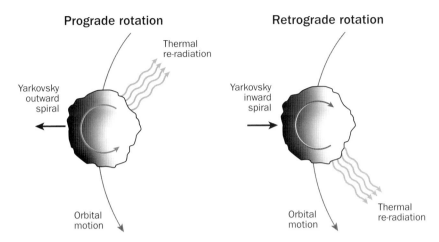

Figure 6-2. The Yarkovsky effect. Thermal radiation by a spinning asteroid can result in either acceleration or deceleration, slowly changing the orbit over time. It can also change the asteroid's spin, which is known as the YORP effect. See figure 6-3. (Republished from Yeomans, *Near-Earth Objects: Finding Them Before They Find Us*, p. 44)

optical and radar observations were available to refine its orbit. Subsequently, JPL's Steven Chesley and colleagues did indeed directly detect the Yarkovsky drift affecting this asteroid's motion and estimated the asteroid's bulk density as 2.7 grams per cubic centimeter.[49] There have been several subsequent Yarkovsky detections in the motions of other well-observed asteroids, including 1862 Apollo, 2062 Aten, 2340 Hathor, 101955 Bennu, and (85990) 1999 JV6. The motions of very small asteroids, only a few meters in diameter, can also be affected by solar radiation pressure since their surface-to-mass ratios are relatively large.

A related Yarkovsky-O'Keefe-Radzievskii-Paddack (YORP) effect can cause the rotation rate of an asteroid to increase or decrease.[50] This YORP effect is due to reradiation of sunlight from irregularly shaped asteroids, where either the morning or evening edge of the rotating asteroid is more effective than the other in catching and re-emitting solar radiation. Hence the re-emission of the sunlight in the infrared can introduce a small force

49. Steven R. Chesley, Steven J. Ostro, David Vokrouhlický, David Capek, Jon D. Giorgini, Michael C. Nolan, Jean-Luc Margot, Alice A. Hine, Lance M. Benner, and Alan B. Chamberlin, "Direct Detection of the Yarkovsky Effect by Radar Ranging to Asteroid 6489 Golevka," *Science* 302 (2003): 1739–1742.

50. "YORP" comes from the initials of four researchers who were responsible for describing the effect: Yarkovsky, John O'Keefe, V. V. Radzievskii, and Stephen Paddack.

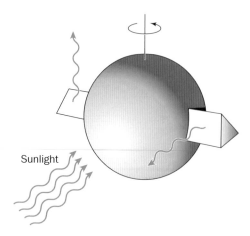

Figure 6-3. The YORP effect can affect an asteroid's spin when parts of an asteroid are more effective at absorbing and reradiating sunlight than others, which happens when an asteroid is irregularly shaped (as most smaller ones are). The YORP effect can alter both the rate of spin and the direction of the spin axis, so over time it can also change the Yarkovsky effect's influence on orbit. (Republished from Yeomans, *Near-Earth Objects: Finding Them Before They Find Us*, p. 45)

that either increases the asteroid's spin or slows it down. This effect can alter spin axis orientations as well as spin rates, so the YORP effect can dramatically affect the direction and rate of a Yarkovsky drift in orbital semimajor axis. Moreover, asteroids held together by only their own weak gravity, often referred to as "rubble pile asteroids," could change shape, shed mass, and fission to produce satellites if spun up by the YORP effect. By carefully visually monitoring the rotation rate of asteroid (54509) 2000 PH5 over a four-year span, Northern Ireland researcher Stephen Lowry and colleagues were able to determine an asteroid's increasing rotation rate, or spin-up. Over the same time period, Cornell University's Patrick Taylor and colleagues used visual and radar observations to confirm the spin-up of this asteroid, and several other YORP detections followed.[51]

---

51. Asteroid (54509) was subsequently, and appropriately, renamed 54509 YORP. S. C. Lowry, A. Fitzsimmons, P. Pravec, D. Vokrouhlický, H. Boehnhardt, P. A. Taylor, J.-L. Margot, A. Galad, M. Irwin, J. Irwin, and P. Kusnirak, "Direct Detection of the Asteroid YORP Effect," *Science* 316 (2007): 272–273; P. A. Taylor, J.-L. Margot, D. Vokrouhlický, D. J. Scheeres, P. Pravec, S. C. Lowry, A. Fitzsimmons, M. C. Nolan, S. J. Ostro, L. A. M. Benner, J. D. Giorgini, and C. Magri, "Spin Rate of Asteroid (54509) 2000 PH5 Increasing Due to the YORP Effect," *Science* 316 (2007): 274–277.

Celestial Pinball: The Evolution of Near-Earth Asteroids into Earth's Neighborhood

Over time, Earth and the Moon would clear their orbit around the Sun of these small bodies, except that the gravitational effects of Jupiter (mostly) keep scattering asteroids from the "main belt" between Mars and Jupiter in our direction. Many of these are fragments of the largest of those asteroids, Vesta. Vesta's unique spectral signature has been associated with the similar spectral characteristics of certain basaltic achondrite meteorites, the howardites, eucrites, and diogenites (HED).[52] However, it was considered dynamically very difficult for collision fragments of Vesta to reach one of two regions in the main belt where planetary perturbations could push them into Earth-crossing orbits. Two of the most likely escape hatches to the inner solar system for low-inclination Vesta fragments were the 3:1 resonance at a heliocentric distance of 2.5 au and the so-called nu6 (ν6) resonance region near the inner edge of the main belt at about 2.1 au.[53] Vesta was in between at a distance of 2.36 au, and asteroid collisions with Vesta could not be expected to send the resulting fragments as far as either escape hatch.

In 1985, MIT researcher Jack Wisdom studied the orbital behavior of asteroids at a heliocentric distance of 2.5 au, where they would enter a 3:1 resonance with Jupiter.[54] At that distance, each asteroid would make three orbits in 11.86 years and Jupiter would make one orbit in the same time. Thus, these asteroids would encounter Jupiter at the same orbital location every 11.86 years and suffer regular gravitational tugs in the same direction. Over a million or more years, these regular tugs could modify the asteroid's

---

52. Achondrites are a class of stony meteorites formed by igneous processes on or near the surfaces of large asteroids or planets. Unlike the most common meteorites, called chondrites, achondrites lack chondrules, which are millimeter-sized mineral spheres that form by the remelting of mineral grains in the solar nebula. HED meteorites, a subclass of the achondrites, originate from Vesta and make up about 5 percent of all meteorite falls. See Thomas B. McCord, John B. Adams, and Torrence V. Johnson, "Asteroid Vesta: Spectral Reflectivity and Compositional Implications," *Science* 168, no. 3938 (19 June 1970): 1445–1447.

53. A. Morbidelli, W. F. Bottke, Jr., Ch. Froeschle, and P. Michel, "Origin and Evolution of Near-Earth Objects," in *Asteroids III*, ed. W. F. Bottke, Jr., A. Cellino, P. Paolicchi, and R. P. Binzel (Tucson, AZ: University of Arizona Press, 2002), pp. 409–422.

54. J. Wisdom, "Meteorites May Follow a Chaotic Route to the Earth," *Nature* 315 (1985): 731–733.

orbital eccentricities in a chaotic fashion, allowing some to cross the orbit of Mars and then move into Earth's neighborhood.

Another route into near-Earth space is the so-called nu6 resonance first identified by JPL's James Williams in his 1969 Ph.D. thesis.[55] In this case, the gravitational perturbations of Jupiter and Saturn combine, over long time intervals, to increase an object's orbital eccentricity so that it ultimately crosses the orbit of Mars and reaches the inner solar system. This resonance is effective for low-orbital-inclination objects orbiting the Sun near the main belt's inner edge at about 2.1 au and for objects at somewhat larger distances with more highly inclined orbits.

In 1993, MIT astronomers Richard Binzel and his graduate student Shui Xu established that 20 small main-belt asteroids have distinctive optical spectral features similar to those of Vesta as well as to those of eucrite and diogenite meteorites.[56] Twelve of these "chips off Vesta" lay in the immediate vicinity of Vesta, and the rest bridged the space between the orbital distance of Vesta and the 3:1 resonance escape hatch into the inner solar system. Clearly, these "Vestoids" were capable of moving from Vesta's distance (2.36 au) to the 3:1 escape hatch at 2.5 au so that some could impact Earth. At the time, it was not clear just how this occurred, but within a few years, the recognition of the Yarkovsky effect provided the answer.[57] Once asteroid fragments are created via asteroid-asteroid collisions, the resulting fragments can slowly evolve either inward or outward via the Yarkovsky effect, with some reaching the 3:1 or nu6 escape hatches to the inner solar system.

A favorite analogy of the noted dynamicist William Bottke of the Southwest Research Institute in Boulder, Colorado, compares the Yarkovsky drift to something like playing pinball.[58] When the ball begins its slow descent down the table, it is rather like Yarkovsky drift slowly driving some asteroids toward resonances that push them into Earth's neighborhood. Once they reach Earth's neighborhood, the perturbations by Earth and Mars rapidly scramble

---

55. J. G. Williams, "Secular Perturbations in the Solar System" (Ph.D. diss., University of California, Los Angeles, 1969).

56. Richard P. Binzel and Shui Xu, "Chips off of Asteroid 4 Vesta: Evidence for the Parent Body of Basaltic Achondrite Meteorites," *Science* 260, no. 5105 (1993): 186–191.

57. P. Farinella, D. Vokrouhlický, and W. K. Hartmann, "Meteorite Delivery via Yarkovsky Orbital Drift," *Icarus* 132 (1998): 378–387.

58. Bottke's analogy was noted by Yeomans during the 1 July 2016 meeting of NASA's NEO Science Definition Team at the Applied Physics Laboratory in Maryland.

the asteroid's motion and the slow, long-term Yarkovsky drift is no longer important. Likewise, once the pinball reaches the active pedestals and the flippers at the bottom of the pinball machine, the slight decline of the pinball table is not very important.

Computing Earth Impact Probabilities

The orbits of comets and asteroids are not perfectly known; they have errors or uncertainties. For example, the astrometric data used to compute orbits are not exact; the masses of the perturbing planets are not known perfectly; and the dynamic model used to represent the object's motion will not include all perturbing bodies or effects (e.g., there may be perturbations by asteroids that are not included in the dynamic model). Each of the six orbital elements for a body at a particular instant in time (epoch) will have associated uncertainties, and these uncertainties will change as the body's motion is represented at times different from the epoch time.[59] These uncertainties are normally represented as an ellipsoid surrounding the most likely position of the object. Positions near the boundary of the uncertainty ellipsoid are far less likely to be correct but are still possible. The epoch time is usually chosen close to the times for which the object has been observed, so the orbital uncertainties will usually increase for times well before, or after, the epoch time. If there is a close Earth approach of the body, the body's uncertainty ellipsoid at that close approach time can be projected onto Earth's figure. If the body's position uncertainty ellipsoid is entirely within Earth's figure at the time of the closest approach, then the Earth impact probability is 100 percent—it will collide with Earth. If only a fraction of the object's position uncertainty ellipsoid touches Earth's figure, then the impact probability is equal to that fraction, and if the uncertainty ellipsoid does not intersect Earth's figure at

---

59. The orbital motion of an object, either forward or backward in time, is carried out on a computer by numerically integrating a second-order differential equation that represents the combined accelerations acting upon the object at any given time. This technique then provides position and velocity vectors and their uncertainties at selected time steps. At a given time, or epoch, an object's heliocentric position and velocity vectors in space (a total of six heliocentric position and velocity components) can be converted to a corresponding set of six orbital elements—and vice versa.

all (the most likely case), then no impact is possible.[60] Normally, the object's orbital size and orientation in space are far better known than its position along its orbit. So an object's uncertainty ellipsoid is most often shaped like an elongated cigar, rather than a sphere, centered on the nominal position of the body. The largest position uncertainty component (the long axis of the cigar) is generally in the direction of the object's motion, or in the so-called "along-track" direction. Misunderstanding this concept led to the incorrect prediction for a possible Earth impact in 2028 by asteroid 1997 XF11, as we saw in chapter 5.

The MPC's move online enabled the development of automated warning software based on these principles late in the 1990s. Steven Chesley developed what was called CLOMON, for Close Approach Monitoring System, and its public interface, the Near Earth Objects Dynamic Site (NEODyS), while on a North Atlantic Treaty Organization (NATO) fellowship to the University of Pisa. This worked by pulling data from the Minor Planet Center's daily orbit update pages and calculating orbit probabilities in real time. Chesley commented that the need for "automated monitoring came up at the Torino meeting in 1999 when the second object that we found that had virtual impactors was lost. It was 1998 OX4."[61] It had been discovered on 26 July 1998 by Spacewatch and then lost because nobody had realized that it was a potential impactor, and no follow-up observations had been arranged.

Chesley and his mentor at Pisa, Andrea Milani, had conceived a way to rule out a potential impact from a lost asteroid even if it was never recovered. Their "virtual impactor" method computed potential orbits that were compatible with the existing observations, projected them out for a century, and then figured out where in the sky the object would be at each of its future encounters with Earth. For the lost 1998 OX4, the object would be visible in particular patches of Earth's sky in 2001 and 2003 if it was on one of the

---

60. JPL's Paul Chodas is responsible for first applying to natural bodies the impact probability techniques that had been used for interplanetary spacecraft studies. D. K. Yeomans and P. W. Chodas, "Predicting Close Approaches of Asteroids and Comets to Earth," in *Hazards Due to Comets and Asteroids*, ed. T. Gehrels (Tucson, AZ: University of Arizona Press, 1994), pp. 241–258.

    These techniques would be successfully employed for the collision of Comet Shoemaker-Levy 9 with Jupiter in 1994. (See chapter 5.)

61. Steven Chesley, interview by Conway, 9 March 2017, transcript in NASA History Division HRC.

computed impact trajectories. If it was not on an impact trajectory, it would not be found in one of those spots. Thus one could rule out a future impact by looking at these computed spaces at the appropriate time. "Negative observations," as Chesley called them, were a simple, inexpensive way to retire risk. As it happened, 1998 OX4 was not observed during its potential encounter in 2001, and it was recovered by NEAT in August 2002.[62]

Chesley moved to JPL in late 1999 and developed a similar system to CLOMON/NEODyS called Sentry. While the two were similar in function, Sentry used different software and integration routines, so they were not identical. Instead, they served as checks on one another. They allowed almost immediate assessment of potential risk, overcoming some of the problems that the NEO community had encountered during the 1997 XF11 affair.

Less than a year after the establishment of NASA's Near-Earth Objects Observations Program in 1998, the discovery of near-Earth asteroid 1999 AN10 on 13 January 1999 by LINEAR set off an analysis of the object's future motion by Andrea Milani, Steven Chesley, and Giovanni Valsecchi that tested review guidelines adopted by NASA and the IAU after the 1997 XF11 incident.[63] The asteroid was observed until 20 February, when its position in the sky near that of the Sun prevented further observations. Milani and colleagues noted that the asteroid would make a close approach to Earth in August 2027 and that this close approach would introduce chaotic behavior such that one of the possible subsequent trajectories would bring the object to an Earth approach in 2034 and again in August 2039—and for this latter encounter, there remained a slight chance of an Earth impact. They were careful to note that the impact probability was about one in one billion and that this value was less than the probability of an impact by a similarly sized unknown object within the next few hours! In short, this was an interesting object from a mathematical point of view, but not of any particular concern. During the first two weeks of April 1999, others, including Paul Chodas at JPL, verified the basic analysis by Milani and his colleagues. This was the first case in which an impact prediction was made and then vetted by professional colleagues before any public announcement.

---

62. Chesley interview, 9 March 2017; A. Milani and S. R. Chesley, "Virtual Impactors: Search and Destroy," *Icarus* 145, no. 1 (May 2000): 12–24, doi:10.1006/icar.1999.6324.

63. Robert Lee Hotz, "How Should NASA Break It to Us When the World's Ending," *Los Angeles Times* (13 May 1998), copy in "Asteroids 1998.pdf," NEO History Project collection.

Following this review process, a draft of the paper was posted without fanfare on the University of Pisa website, where it was discovered by Benny J. Peiser, a faculty member at Liverpool John Moores University. On 13 April 1999, Peiser published an account in his widely read CCNET Digest internet forum that chided the professional astronomers, noting that "there is no reason whatsoever why the findings about 1999 AN10 should not be made available to the general public—unless the findings haven't been checked for general accuracy by other NEO researchers." Peiser also speculated that one reason "why the authors may have decided to hide their data could be due to the current NASA guidelines on the reporting of impact probabilities by individual NEOs. After all, NASA is threatening researchers with the withdrawal of funding if they dare to publish such sensitive information in any other form than in a peer reviewed medium." Peiser's speculation was nonsense, but he did demonstrate the problem of public perception whenever Earth impact studies are conducted privately, even if the secrecy lasts only a few days and the impact risk is negligible.[64]

At the August 2000 General Assembly of the IAU in Manchester, England, a NEO Technical Review Committee was approved that could review and verify an Earth impact prediction during a proposed 72-hour hold on any prediction release. This information hold was to be restricted to impact probabilities greater than one in a million for the next 100 years. However, there was an obvious need for transparency, and a few sophisticated amateur orbital specialists, who could also provide impact predictions, would not be bound by the 72-hour restriction suggested by the IAU Committee.

A potential impactor discovered in 2000, 2000 SG344, soon rendered even the 72-hour period largely irrelevant. The object was discovered by David Tholen and Robert Whitely with the University of Hawai'i 2.2-meter telescope on 29 September 2000, and it was quickly identified as a potential threat.[65] Steve Chesley remembers that it was forecast with about a 0.2 percent probability of impact in 2030 and was only a few tens of meters in size. The community went through the formal review process, and while NASA did not make an announcement, the International Astronomical Union did so

---

64. Pre-discovery observations in January 1955 and additional observations made through March 2013 allowed a refinement of this object's orbit, and there are now no impact possibilities for at least the next 100 years.

65. "MPEC 2000-U19 : 2000 SG344," IAU MPC, *https://www.minorplanetcenter.net/iau/ mpec/K00/K00U19.html* (accessed 26 March 2021).

on Friday, 3 November 2000. Yeomans spoke to reporters at JPL that day to explain the discovery and impact predictions.[66] According to Chesley, "Friday night, more data came in, eliminated the whole thing. Meanwhile, the IAU and NASA Headquarters are, I don't know, sailing for the weekend, and the world press is spinning up while the story is dead."[67] Carl Hergenrother had found additional images of the object in the Catalina Sky Survey data archive from May 1999, and those data eliminated the near-term risk. A *Los Angeles Times* reporter pinned Yeomans down on the issue of the disappeared risk the following week. He had followed the rules to the letter and had "no regrets" about the announcement.[68]

The near-real-time, public nature of both NeoDYS and Sentry meant that the idea of waiting for expert consultation no longer made sense. As a result, Ed Weiler, NASA Associate Administrator for the Space Science Enterprise, decided in December 2000 that NASA could quickly release impact predictions once this information was verified and the announcement itself had been vetted by NASA's Public Affairs Office.[69] In practice, Sentry and NEODyS had to agree on the probabilities within a "few tens of percents" in order for an announcement to be made. Otherwise, the scientists supporting them had to dive into the data to figure out why they did not. The fixed period of delay had not worked in scientists' favor; in effect, the switch provided the flexibility to wait for more data if the two systems' forecasts did not agree.

Communicating Impact Hazard

The fundamental problem of how best to communicate the likelihood of asteroid impacts to policy-makers and to the public triggered the development of a couple of risk assessment scales for that purpose. While extremely rare, Earth impacts by large asteroids or comets would have global consequences. Unlike other natural disasters like hurricanes and earthquakes, there

---

66. Usha Lee McFarling, "Scientists Downgrade Asteroid Threat," *Los Angeles Times* (7 November 2000), copy in "Asteroids 2000.pdf," NEO History Project collection.

67. Chesley interview, 9 March 2017; "Much Ado About 2000 SG344 | Science Mission Directorate," *https://science.nasa.gov/science-news/science-at-nasa/2000/ast06nov_2* (accessed 13 June 2017).

68. McFarling, "Scientists Downgrade Asteroid Threat."

69. NEO Program Office notes taken by D. K. Yeomans at NASA Headquarters on 14 December 2000, Yeomans notebook, NEO History Project collection.

have been no large Earth impacts to familiarize the public with such occurrences. Moreover, the topic is provocative and prone to sensationalism. So the very real possibility of these low-probability/high-consequence events is very difficult to communicate to the public. In an effort to provide an easily understood evaluation of a possible threat within the next 100 years, Richard Binzel, at a United Nations conference in 1995, introduced a near-Earth object threat scale that took into consideration the likelihood of an upcoming impact (its impact probability) and the energy of the possible impact in megatons of TNT explosives.[70] Binzel's original index went from 0 (no collision possible), to 1 (extremely improbable), on up to 5 (certain impact on a particular future date).[71]

Subsequently, Binzel's index scale was modified and presented at a June 1999 international conference about NEOs held in Torino (Turin), Italy. The new Torino Scale, as it was renamed, ran from 0 to 10, with colors introduced to represent no threat (0, white), an extremely low threat level (1, green), a threat meriting attention by astronomers (2–4, yellow), threatening (5–7, orange), and certain collision (8–10, red).

The Torino Scale might be appropriate for public communication, but it did not convey the urgency of the event. Would the possible impact occur in several decades or several weeks? To provide a scale more useful to the scientific community for prioritizing objects for additional observations and analysis, Steven Chesley and his colleagues introduced a scale that sought to include the time proximity as well as the impact probability and potential impact energy. In honor of the 2001 international conference on NEOs held in Palermo, Italy, this threat scale for professionals was named the Palermo

---

70. See the glossary for the definitions used in this text. It is worth noting that Binzel and many others refer to his scale as a hazard scale, but as it refers to a specific object and not the entire population of NEOs (the definition of "hazard"), we refer to it as a threat scale in accordance with the Planetary Defense Coordination Office's definitions. These terms are routinely used interchangeably in the literature. Also see Linda Billings, "Words Matter: A Call for Responsible Communication About Asteroid Impact Hazards and Plans for Planetary Defense," *Space Policy* 33 (August 2015): 8–12, *https://doi.org/10.1016/j.spacepol.2015.07.001* (accessed 8 October 2019).

71. R. P. Binzel, "A Near-Earth Object Hazard Index," *Annals of the New York Academy of Sciences* 822, no. 1, *Near-Earth Objects: The United Nations International Conference* (1997): 545–551; R. P. Binzel, "The Torino Impact Hazard Scale," *Planetary and Space Science* 48 (2000): 297–303.

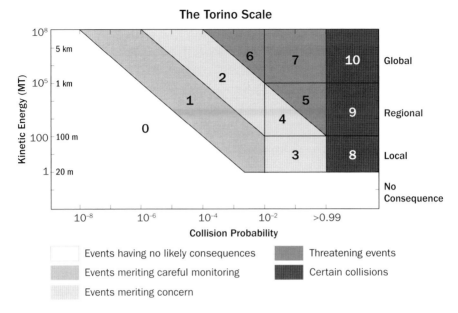

Figure 6-4. The Torino Scale, adopted by the IAU in 1999, is a tool for categorizing potential Earth impact events. An integer scale ranging from 0 to 10 with associated color coding, it is intended primarily to facilitate public communication by the asteroid impact hazard monitoring community. The scale captures the likelihood and consequences of a potential impact event but does not consider the time remaining until the potential impact. More extraordinary events are indicated by a higher Torino Scale value. (Image credit: JPL/Caltech/NASA)

Assessing Asteroid and Comet Impact Hazard Predictions in the 21st Century

No Hazard (White Zone)	0	The likelihood of a collision is zero, or is so low as to be effectively zero. Also applies to small objects such as meteors and bodies that burn up in the atmosphere as well as infrequent meteorite falls that rarely cause damage.
Normal (Green Zone)	1	A routine discovery in which a pass near Earth that poses no unusual level of danger is predicted. Current calculations show the chance of collision is extremely unlikely, with no cause for public attention or public concern. New telescopic observations very likely will lead to reassignment to Level 0.
Meriting Attention by Astronomers (Yellow Zone)	2	A discovery, which may become routine with expanded searches, of an object making a somewhat close but not highly unusual pass near Earth. While meriting attention by astronomers, there is no cause for public attention or public concern as an actual collision is very unlikely. New telescopic observations very likely will lead to reassignment to Level 0.

(continued on following page)

Meriting Attention by Astronomers (Yellow Zone)	3	A close encounter, meriting attention by astronomers. Current calculations give a 1% or greater chance of collision capable of localized destruction. Most likely, new telescopic observations will lead to reassignment to Level 0. Attention by public and by public officials is merited if the encounter is less than a decade away.
	4	A close encounter, meriting attention by astronomers. Current calculations give a 1% or greater chance of collision capable of regional devastation. Most likely, new telescopic observations will lead to reassignment to Level 0. Attention by public and by public officials is merited if the encounter is less than a decade away.
Threatening (Orange Zone)	5	A close encounter posing a serious, but still uncertain, threat of regional devastation. Critical attention by astronomers is needed to determine conclusively whether a collision will occur. If the encounter is less than a decade away, governmental contingency planning may be warranted.
	6	A close encounter by a large object posing a serious but still uncertain threat of a global catastrophe. Critical attention by astronomers is needed to determine conclusively whether a collision will occur. If the encounter is less than three decades away, governmental contingency planning may be warranted.
	7	A very close encounter by a large object, which if occurring this century, poses an unprecedented but still uncertain threat of a global catastrophe. For such a threat in this century, international contingency planning is warranted, especially to determine urgently and conclusively whether a collision will occur.
Certain Collisions (Red Zone)	8	A collision is certain, capable of causing localized destruction for an impact over land or possibly a tsunami if close offshore. Such events occur on average between once per 50 years and once per several thousand years.
	9	A collision is certain, capable of causing unprecedented regional devastation for a land impact or the threat of a major tsunami for an ocean impact. Such events occur on average between once per 10,000 years and once per 100,000 years.
	10	A collision is certain, capable of causing global climatic catastrophe that may threaten the future of civilization as we know it, whether impacting land or ocean. Such events occur on average once per 100,000 years, or less often.

For more information, see D. Morrison, C. R. Chapman, D. Steel, and R. P. Binzel, "Impacts and the Public: Communicating the Nature of the Impact Hazard," in *Mitigation of Hazardous Comets and Asteroids*, ed. M. J. S. Belton, T. H. Morgan, N. H. Samarasinha, and D. K. Yeomans (Cambridge, U.K.: Cambridge University Press, 2004).

Scale.[72] The Palermo Scale is logarithmic, so that a value of −2 indicates that the detected potential impact event is only 1 percent as likely as an impact from a random object of the same size or larger before the potential impact date in question (i.e., Palermo scale = 0) and a Palermo scale of +2 indicates an event that is 100 times more likely than this same random event (the background level). Palermo scale values less than −2 reflect events for which there are no likely consequences; values between −2 and 0 merit careful monitoring; and higher values merit an increasing level of concern. A Palermo scale value as high as 0 is very rare. While there have been hundreds of objects for which the Torino scale value was initially above zero, the vast majority of these cases were quite temporary, with Torino scale values reset to zero once additional follow-up observations were available to refine the object's orbit, thus reducing its future position uncertainties and eliminating possible future impacts. Both Torino and Palermo scale ratings are published on JPL's Sentry and the European Space Agency's (ESA's) NEODyS web pages, providing an essentially real-time estimation of the threat posed by potentially hazardous asteroids.[73]

Gene Shoemaker just missed an era in which rigorous quantification of asteroid risk could be accomplished. He had died in a car accident in July 1997 in Alice Springs, Australia, while studying craters. Carolyn was injured but survived.[74] In 1999, the Lunar Prospector spacecraft carried some of his ashes to the Moon, and the Near Earth Asteroid Rendezvous mission that touched asteroid 433 Eros in February 2001 was renamed in his honor as well. Eleanor Helin retired from JPL in 2002 and passed away in 2009. In 2014, the Palomar Observatory developed an exhibit in her honor around the disused 18-inch Schmidt that she and the Shoemakers had used for many years.[75]

By the time Helin's NEAT project closed down in 2007, it had discovered 442 near-Earth asteroids; her previous film-based PCAS survey had found 65

---

72. S. R. Chesley, P. W. Chodas, A. Milani, G. B. Valsecchi, and D. K. Yeomans, "Quantifying the Risk Posed by Potential Earth Impacts," *Icarus* 159 (2002): 423–432.

73. Comprehensive, up-to-date lists of objects on the JPL Sentry risk page and the European NEODyS risk page are available at *https://cneos.jpl.nasa.gov/sentry/* (accessed 15 June 2017) and *https://newton.spacedys.com/neodys/* (accessed 13 April 2021).

74. David H. Levy, *Shoemaker by Levy* (Princeton, NJ: Princeton University Press, 2000), pp. 260–262.

75. See *http://www.astro.caltech.edu/palomar/visitor/visitorcenter/helinCommemorative/* (accessed 5 July 2017).

over a considerably longer timeframe. Rapid change in detector and computer technology had made possible a great expansion in knowledge of near-Earth objects, while plummeting costs made feasible supporting several automated asteroid surveys on very small (for NASA) budgets.

While the efficient discovery and tracking of NEOs are important steps toward quantifying their risk to Earth, they are only part of the required activities for understanding their potential hazards. In order to understand the magnitude of potential impact hazards, and to inform possible deflection options, it is also necessary to characterize NEO physical properties, including their sizes, densities, shapes, structures, and compositions. In the next chapter, we turn to the characterization of near-Earth objects.

CHAPTER 7

STUDYING THE RUBBLE
OF THE SOLAR SYSTEM

Introduction

Since the 1990s, comets and asteroids have been studied using increasingly sophisticated ground-based and space-based techniques. Albedo, size, and shape measurements—both direct and indirect—have been carried out for hundreds of objects, and radar observations have been used to determine asteroid shapes and sizes with a resolution exceeded only by a few close-up, spacecraft-based observations.

Much of the research outlined in this chapter concerns main-belt asteroids, rather than near-Earth asteroids. There are two reasons for this. Before 1990, fewer than 150 near-Earth asteroids were known, compared to the thousands of main-belt asteroids available for study. More importantly, as was outlined in chapter 6, near-Earth asteroids dynamically evolve from the main belt, so the physical characterization of main-belt asteroids is directly relevant to the subclass of near-Earth asteroids. By the early 21st century, the simple dichotomy between comets and asteroids that astronomers had made for more than a century was breaking down.

Determining Asteroid and Comet Sizes and Shapes

Most of the observed changes in the reflectance of an asteroid are not intrinsic, but rather due to the viewing geometry. The apparent brightness and polarization of an asteroid's reflected light depend strongly upon its phase angle (Sun-asteroid-observer angle) at the time of the observation. The asteroid's apparent visual brightness, or magnitude, depends upon its size and reflectivity (albedo) but it will be at its brightest when observed at zero phase angle (i.e., at opposition).[1] An asteroid's absolute magnitude (H) is its apparent magnitude at zero phase angle and one au from both the Sun and Earth. This has been determined for all well-observed asteroids.[2] But the same is not true for their albedos. Hence, most asteroid diameter estimates begin with a determination of the asteroid's absolute magnitude, and then a likely albedo is assumed to estimate the diameter—a process that can lead to large uncertainties in their size estimates.

For a small number of asteroids, the diameter can be directly inferred from stellar occultation measurements, whereby observers in diverse locations observe and record the time intervals when the starlight blinks out (or not) as the asteroid passes in front of (or near) the star. With observers located at slightly different latitudes and a knowledge of the asteroid's orbital motion in space, the recorded occultation time intervals can be converted to linear chord measures (or at least partial measures) of the asteroid's two-dimensional shape. This technique was well known as early as 1952, when MIT's Gordon Taylor began issuing predictions, but limited opportunities to observe occultations and a lack of equipment and observers meant progress was very slow. Only three occultation-observing programs were carried out between 1952 and the end of 1974 (one for Juno and two for Pallas). By 1989, there were about 30 asteroids for which some estimates of occultation diameters were available.[3]

Alternately, an object's albedo (and hence size) can be determined from polarization measurements taken at several different phase angles. For a particular asteroid, the ascending slope of the curve found when plotting percentage polarization against phase angle is inversely proportional to its geometric albedo. That is, the darker the albedo, the more quickly the percentage polarization increases with increasing phase angle.[4] The first asteroid polarization

---

1. The geometric albedo is a measure of the asteroid's observed reflected light at zero phase angle when compared to that from an ideal surface (perfect Lambert disk) of the same size and distance that scatters light isotropically.

2. As noted in chapter 1, an astronomical unit (au) is the approximate mean distance between the center of the Sun and the center of Earth. It is set at 149,597,870.7 km.

3. R. L. Millis and D. W. Dunham, "Precise Measurement of Asteroid Sizes and Shapes from Occultations," in *Asteroids II*, ed. R. P. Binzel, T. Gehrels, and M. S. Matthews (Tucson, AZ: University of Arizona Press, 1989), pp. 148–170.

4. The percentage polarization (P) is defined in terms of the intensities of the scattered light polarized along the planes parallel to the scattering plane (the plane containing the observer, Sun, and asteroid), called Ipar, and along the plane perpendicular to it

measurements were made photographically at Meudon, France, in 1934 by Bernard Lyot, who provided rough polarization curves for Ceres and Vesta. More accurate photoelectric polarization measurements were provided by S. Provin at the U.S. Naval Observatory in 1955 for Ceres, Pallas, and Iris, followed by a published polarization curve for Icarus by Tom Gehrels in 1970 and a curve for asteroid Flora published by Joseph Veverka at Harvard in 1971.[5] By 2015, there were about 350 asteroids for which polarization curves were available.[6] As useful as this technique can be, however, the vast majority of asteroid size estimates today are made using infrared or radar observations.

Water vapor in Earth's atmosphere absorbs much of the infrared spectrum, and ground-based infrared observers often limit their measurements to the 1.0- to 2.5-micron region for this reason. However, ground-based telescopes that have been optimized for infrared observing can provide data well beyond this range if specific opaque absorbing regions of the atmosphere are avoided. Often, ground-based and space-based infrared telescopic cameras and spectrographs are cooled cryogenically to improve sensitivity.

Turning infrared observations into albedo and size estimates involves using a thermal model that solves the energy balance relationship between incoming solar radiation and outgoing reflected sunlight and thermal reradiation. The thermal model generates an effective spherical diameter and albedo that will simultaneously match the observed reflected sunlight and the asteroid's thermal re-emission. The object's effective spherical diameter is determined directly, and this diameter, along with the asteroid's determined absolute magnitude in the visible region, then provides the object's geometric albedo.[7]

(Iper). Then P(%) = (Iper – Ipar)/(Iper + Ipar). P(%) is then plotted against the phase angle to determine the slope of the ascending branch of this curve, which is a function of the asteroid's albedo. I. A. Belskaya et al., "Asteroid Polarimetry," in *Asteroids IV*, ed. P. Michel, F. E. DeMeo, and W. F. Bottke, Jr. (Tucson, AZ: University of Arizona Press, 2015), pp. 151–163.

5. The Lyot, Provin, Gehrels, and Veverka polarization curves are reproduced in A. Dollfus, "Physical Studies of Asteroids by Polarization of the Light," in *Physical Studies of Minor Planets*, ed. T. Gehrels (Washington, DC: NASA SP-267, 1971), pp. 95–116.

6. I. Belskaya et al., "Asteroid Polarimetry," in *Asteroids IV*, ed. P. Michel, F. E. DeMeo, and W. F. Bottke, Jr. (Tucson, AZ: University of Arizona Press, 2015), pp. 151–163.

7. An asteroid's diameter in kilometers (D), its absolute visual magnitude (H), and its albedo (Pv) are related by the expression D = 1329 × $10^{-H/5}$ $Pv^{-1/2}$.

Modern thermal models also enable estimates for an object's resistance to temperature change—that is, its thermal inertia. Think about your last beach experience. The granular sand can get very hot under the Sun, but it cools down quickly after sunset. On the other hand, solid boulders near the beach, with their higher thermal inertia, warm up and cool down far more slowly. The standard thermal model (STM) assumes that an object is smooth and spherical, has no thermal inertia, and has been observed at small phase angles. This STM assumes that the object rotates very slowly, with no thermal emission on its dark hemisphere. These assumptions are often met for relatively large main-belt asteroids, but it is often not so for the small, irregularly shaped near-Earth asteroids that cannot always be observed at small phase angles.

David A. Allen, using the University of Minnesota's 30-inch O'Brien telescope, was the first to use infrared observations and a simplified thermal model to estimate an asteroid's diameter in 1970, when he determined that the diameter of Vesta was 573 kilometers—within 10 percent of the mean diameter determined by the Dawn spacecraft in 2011 (i.e., 525 kilometers). This size was significantly larger than the generally accepted measure of Vesta at the time, about 390 kilometers, which had been derived by E. E. Barnard via measuring the asteroid's width with a micrometer back in 1894–95.[8]

In 1973, David Morrison, then of the Institute for Astronomy in Hawai'i, set out to observe every known asteroid in opposition after February of that year with a magnitude brighter than 11.5 (and therefore several kilometers in size) with the 2.2-meter telescope on Maunakea. He observed 33 (of 36 available), finding geometric albedos ranging from 0.03 (324 Bamberga and two others) to 0.22 (Vesta), and diameters from about 553 kilometers (Vesta) to 5 kilometers (887 Alinda).[9] Earth's Moon, for comparison, has an average geometric albedo of 0.12 and was considered a relatively dark object among the solar system's small bodies. In his analysis, Morrison commented that his sample of asteroids showed no correlation between size and albedo.[10] This was an important finding because large-scale surveys of asteroids, like Helin and

---

8. David A. Allen, "Infrared Diameter of Vesta," *Nature* 227, no. 5254 (11 July 1970): 158–159, doi:10.1038/227158a0; on Barnard's measurement, see Audouin Dollfus, "Diameter Measurements of Asteroids," in *Physical Studies of the Minor Planets*, ed. T. Gehrels (Washington, DC: NASA SP-267, 1971), pp. 25–32.

9. D. Morrison, "Radiometric Diameters and Albedos of 40 Asteroids," *Astrophysical Journal* 194 (1 November 1974): 203–212, doi:10.1086/153236.

10. Ibid.

Shoemaker's PCAS, would continue to be done based only on visual magnitudes, simply because magnitudes could be measured by optical telescopes relatively easily. A bias in albedo toward larger or smaller asteroids would thus require corrections to the size distributions inferred from magnitudes by these surveys.

In 1977, Morrison gathered together information on 186 large main-belt asteroids for which diameters had been computed from infrared observations as well as from polarimetry. Morrison found the distribution of geometric albedos to be bimodal, with peaks at about 0.035 and 0.15, corresponding to the C (carbonaceous) and S (silicaceous) spectral types respectively.[11]

These techniques were almost entirely carried out on large main-belt asteroids, but since astronomers considered the main belt to be the source of planet-crossing asteroids (via gravitational perturbations), they were directly relevant to estimating the sizes of near-Earth asteroids, too. Morrison's 1973 work included only one NEO (433 Eros), an Amor-class asteroid, which he estimated at about 24 kilometers in diameter. Many years later, he commented that "when I was doing the observations at Maunakea, there simply weren't NEOs available. At least I didn't know about them. There were darn few, and very few that would be bright enough to observe even with the Infrared Telescope Facility"[12]—a facility yet to be built in 1973.

The standard thermal model, which assumed no thermal inertia (i.e., slow rotators), was quite successful for the large main-belt asteroids that were the targets of most of the early infrared observations. But it was not useful in all cases. In 1974, to avoid an unrealistic standard thermal model assumption that an asteroid radiates all of its heat isotropically on the day side, Morrison and Terry J. Jones introduced what has become known as a beaming parameter (η, eta), which effectively adjusts the asteroid temperature at the subsolar point.[13] Larry Lebofsky of the Planetary Science Institute and colleagues developed a nonstandard thermal model for asteroids with thermal inertia

11. David Morrison, "Asteroid Sizes and Albedos," *Icarus* 31, no. 2 (1 June 1977): 185–220, doi:10.1016/0019-1035(77)90034-3.

12. David Morrison, interview by Yeomans and Conway, 17 May 2016, transcript in NASA History Division HRC.

13. Because of multiple reflections on a rough surface, the subsolar point can be warmer than it would be otherwise. T. J. Jones and D. Morrison, "Recalibration of the Photometric/Radiometric Method of Determining Asteroid Sizes," *Astronomical Journal* 79 (1974): 892–895.

(e.g., fast rotators) in 1978 to fit the observations of the C-type near-Earth asteroid 1580 Betulia. Lebofsky's team also provided a refined standard thermal model in 1986, wherein they calibrated their thermal model using accurate diameters of Ceres and Pallas determined by stellar occultations.[14] At that point, thermal models existed for the two end member cases: objects without and with thermal inertia (e.g., slow and fast rotators, respectively). In 1990, John Spencer of the University of Hawai'i Institute for Astronomy published a thermal model for a rough-surfaced asteroid, thus removing the assumption of a smooth surface required by the earlier models and showing that the beaming parameter depended upon surface roughness.[15] Published in 1998, the more sophisticated Near-Earth Asteroid Thermal Model (NEATM), developed by Alan W. Harris of DLR, could be used to provide diameters and albedo as well as information on the asteroid's thermal inertia and roughness.[16] It is still being used as of this writing.

A large majority of all ground-based, near-IR spectroscopic observations of near-Earth objects have been carried out using the SpeX spectrograph at the NASA Infrared Telescope Facility (IRTF).[17] The IRTF is a 3-meter telescope, optimized for near-IR observations, located at an altitude of 13,600 feet near the summit of Maunakea on the Big Island of Hawai'i and operated by the University of Hawai'i for NASA. It was originally built to provide infrared observations in support of NASA's Voyager missions to the outer planets and went into service in 1979. Approximately 50 percent of the observing time is used for NASA mission support and solar system science—including asteroids.[18]

Providing spectral information over the wavelength region from 0.8 to 5.5 microns, the SpeX spectrograph became operational in 2000, the same year that Schelte "Bobby" Bus began working at the IRTF. Speaking of the

---

14. L. A. Lebofsky et al., "Visual and Radiometric Photometry of 1580 Betulia," *Icarus* 35 (1978): 336–343; L. A. Lebofsky et al., "A Refined Thermal Model for Asteroids Based on Observations of 1 Ceres and 2 Pallas," *Icarus* 68 (1986): 239–251.

15. John Spencer, "A Rough-Surface Thermophysical Model of Airless Planets," *Icarus* 83 (1990): 27–38.

16. A. W. Harris, "A Thermal Model for Near-Earth Asteroids," *Icarus* 131 (1998): 291–301.

17. S. J. Bus, interview by Yeomans and Conway, 25 January 2016, transcript in NASA History Division HRC.

18. S. J. Bus et al., "The NASA Infrared Telescope Facility," 2009, *https://arxiv.org/pdf/0911.0132.pdf* (accessed 3 May 2017).

Bus-DeMeo Taxonomy Key

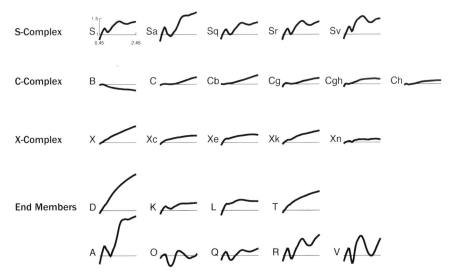

Figure 7-1. In this illustration representing the Bus-DeMeo asteroid spectral classification system (taxonomy), each individual classification (e.g., S, Sa, B, C, etc.) has a representative spectrum whereby the spectral reflectance value is given as a function of wavelength from 0.45 to 2.45 microns. The reflectance has been transformed (normalized) so that at 0.55 microns, the unitless reflectance value is defined as 1.0. For a more complete explanation, see *http://smass.mit.edu/busdemeoclass.html*.

IRTF, Bus noted that "[e]ven though it's an old telescope, it's built like a battleship. I mean, it's a very heavy structure. It has survived the years well [and]…advancements in technology and instrumentation have really kept us on the cutting edge."[19]

The SpeX instrument enabled extending asteroid taxonomy into the near-infrared. David Tholen's taxonomy, which he developed while at the University of Hawai'i, had been based upon visual measurements in eight colors. It was refined by Bus and then extended into the near-IR as a result of the work of Francesca DeMeo of MIT and colleagues.[20] Bus and DeMeo were both graduate students of MIT Professor Richard "Rick" Binzel, who insisted that any new asteroid taxonomy had to be an understandable extension to

19. Bus interview, 25 January 2016.
20. S. J. Bus and R. P. Binzel, "Phase II of the Small Main-Belt Asteroid Spectroscopic Survey: A Feature-Based Taxonomy," *Icarus* 158 (2002): 146–177; F. E. DeMeo et al., "An Extension of the Bus Asteroid Taxonomy into the Near Infrared," *Icarus* 202 (2009): 160–180.

existing taxonomies. Binzel noted that "each time we wanted to build on what was previous. We didn't want to reinvent it. We just wanted to augment it and keep it consistent, because if you change it every time, then everyone is lost." He also emphasized that asteroid taxonomy was a method of classifying spectra and not a method of identifying the minerals in asteroids. Binzel joked with his graduate students that "I'm going to tattoo on your forehead in reverse, so that when you look into the mirror, every morning, you will see that "taxonomy is not mineralogy."[21]

Observing in the infrared region of asteroid spectra allows the water of hydration and water bands near 3 microns to be identified.[22] In dark asteroids, this absorption feature is often the only one present in reflectance spectra. A mineral is said to be hydrated if it contains the OH (hydroxyl) radical or, less likely, water (H_2O) itself. Larry Lebofsky had first identified a hydration band in a spectrum of Ceres in 1978. Since then, hydrated and anhydrous (dry) asteroids in almost every taxonomic class have been identified.[23] In 2010, Andrew Rivkin and Joshua Emery, as well as Humberto Campins and colleagues, identified a unique spectral band near 3.1 microns that they attributed to water ice frost and organics on the surface of outer main-belt asteroid 24 Themis.[24] Interestingly, the C-type Themis family of asteroids, which was likely formed as the result of a collisional breakup of a single large body, has a few so-called active asteroids, like 7968 Elst-Pizarro

---

21. R. Binzel, interview by Yeomans and Conway, 17 October 2016, transcript in NASA History Division HRC.

22. A correlated spectral feature near 0.7 micron usually indicates that the 3-micron feature is present, but the absence of this feature does not necessarily indicate the absence of the 3-micron feature. F. Vilas and M. J. Gaffey, "Phyllosilicate Absorption Features in Main-Belt and Outer-Belt Asteroid Reflectance Spectra," *Science* 246 (1989): 790–792.

23. A. S. Rivkin et al., "Hydrated Minerals on Asteroids: The Astronomical Record," in *Asteroids III,* ed. W. F. Bottke, Jr., A. Cellino, P. Paolicchi, and R. P. Binzel (Tucson, AZ: University of Arizona Press, 2002), pp. 235–253.

24. Andrew S. Rivkin and Joshua P. Emery, "Detection of Ice and Organics on an Asteroidal Surface," *Nature* 464 (2010): 1322–1323; H. Campins et al., "Water Ice and Organics on the Surface of the Asteroid 24 Themis," *Nature* 464 (2010): 1320–1321.

and 118401 LINEAR[25]—perhaps active due to water vaporization carrying dust off their surfaces.[26]

The first space-based IR telescope to observe minor planets was the Infrared Astronomical Satellite (IRAS), which was launched into an Earth polar orbit on 25 January 1983 and operated for 10 months. IRAS was the product of a partnership between the Netherlands, the United Kingdom, and the United States, and it was designed to provide an unbiased infrared survey of the entire sky. It utilized a 60-centimeter telescope and observed 96 percent of the sky in four IR wavelengths (12, 25, 60, and 100 microns). It observed about 2,200 asteroids and enabled determinations for their diameters and albedos. IRAS discovered four asteroids (including 3200 Phaethon) and five comets, including 1983 H1/IRAS-Araki-Alcock, which holds the record for the closest comet-Earth approach in the 20th century (0.031 au on 11 May 1983).[27] To archive and distribute the IRAS data and a variety of subsequent datasets, a new facility called the Infrared Processing and Analysis Center (IPAC) was opened in 1986 at Caltech with an operations budget provided by NASA.[28]

Space-based IR observations of asteroids have also been enabled by several other missions, including the Mid-Course Space Experiment (MSX) that was operated under the Ballistic Missile Defense Organization (BMDO) during 1995–98 and the Japanese AKARI telescope that was operational during 2006–07 and 2008–11.[29] AKARI provided information on asteroid 25143 Itokawa and 162173 Ryugu, the target bodies for the Japanese Hayabusa

25. Active asteroids 7968 Elst-Pizarro and 118401 LINEAR are also designated as Comets 133P/Elst-Pizarro and 176P/LINEAR.

26. D. Jewitt, Henry Hsieh, and Jessica Agarwal, "The Active Asteroids," in *Asteroids IV*, ed. P. Michel, F. E. DeMeo, and W. F. Bottke, Jr. (Tucson, AZ: University of Arizona Press, 2015), pp. 221–241.

27. E. F. Tedesco et al., "The Supplemental IRAS Minor Planet Survey," *Astronomical Journal* 123 (2002): 1056–1085; A. Mainzer, F. Usui, and D. E. Trilling, "Space-Based Thermal Infrared Studies of Asteroids," in *Asteroids IV*, ed. P. Michel, F. E. DeMeo, and W. F. Bottke, Jr. (Tucson, AZ: University of Arizona Press, 2015), pp. 89–106.

28. Peter J. Westwick, *Into the Black: JPL and the American Space Program, 1976–1994* (New Haven, CT: Yale University Press, 2007).

29. E. F. Tedesco et al., "The Midcourse Space Experiment Infrared Minor Planet Survey," *Astronomical Journal* 124 (2002): 583–591.

Table 7-1. Comet and asteroid mission timelines.

Target Body	Mission Name	Flyby or Rendezvous Date	Encounter Dist. (km)	Best Pixel Scale (m)	Max. Target Extent (km)
Comets					
21P/Giacobini-Zinner	ICE	11 Sept. 1985	7,800	[a]	—
1P/Halley	VEGA 1	6 Mar. 1986	8,890	[a]	—
	Suisei	8 Mar. 1986	150,000	[a]	—
	VEGA 2	9 Mar. 1986	8,030	[a]	—
	Sakigake	11 Mar. 1986	7 million	[a]	—
	Giotto	14 Mar. 1986	596	45	15
26P/Grigg-Skjellerup	Giotto	10 Jul. 1992	200	[b]	—
19P/Borrelly	DS1	22 Sept. 2001	2,171	47	8.0
81P/Wild 2	Stardust	2 Jan. 2004	234 & ER	15	5.5
9P/Tempel 1	Deep Impact	4 Jul. 2005	500	1	8.0
103P/Hartley 2	EPOXI[c]	4 Nov. 2010	694	4	2.3
9P/Tempel 1	Stardust-NExT	15 Feb. 2011	178	11	8.0
67P/C-G	Rosetta	12 Nov. 2014	R&L	0.002	4.1
Asteroids					
951 Gaspra	Galileo	29 Oct. 1991	1,600[d]	54	18.2
243 Ida/Dactyl	Galileo	28 Aug. 1993	2,391	31	59.8/1.6
253 Mathilde	NEAR	27 Jun. 1997	1,212	160	66
9969 Braille	DS1	29 Jul. 1999	26[e]	120	2
433 Eros	NEAR	12 Feb. 2000	R&L	0.01	34
5535 Annefrank	Stardust	2 Nov. 2002	3,079	185	6.6
2867 Steins	Rosetta	5 Sept. 2008	803	80	6.7
21 Lutetia	Rosetta	10 Jul. 2010	3,162	60	121
25143 Itokawa	Hayabusa	12 Sept. 2005	R&L&ER	0.006	0.535
4179 Toutatis	Chang'e-2	13 Dec. 2012	3.2[f]	10	4.75
4 Vesta	Dawn	Jul. 2011	R	20	573
1 Ceres	Dawn	Mar. 2015	R	35	965
162173 Ryugu	Hayabusa2	Jun. 2018	R,L&ER[g]	0.005	0.87
101955 Bennu	OSIRIS-REx	Dec. 2018	R,L&ER[h]	0.004	0.49

Key: **R** = Rendezvous; **L** = Landing; **ER** = Earth return of sample; **Best image resolution** ~ 3 × Best Pixel Scale; **Target size** = longest dimension; **67P/C-G** = 67P/Churyumov-Gerasimenko. **a.** The Soviet Union sent the VEGA 1 and VEGA 2 to fly by Venus, followed by a flyby of Comet 1P/Halley. At Comet Halley, the camera of the VEGA 1 spacecraft was out of focus and the camera on the VEGA 2 spacecraft provided overexposed images. The ICE, Sakigake, and Suisei spacecraft did not carry cameras. **b.** The Giotto camera was disabled by a dust hit during the Comet Halley flyby. **c.** Once the investigations of Comet Tempel 1 were complete, the mission was renamed EPOXI (Extrasolar Planet Observation and Deep Impact Extended Investigation). **d.** The closest image was taken at 5,300 km. **e.** The closest approach of the Deep Space 1 (DS1) spacecraft was 26 km, but the images were taken from a distance of about 14,000 km. **f.** This Chinese technology test spacecraft was launched into lunar orbit in October 2010, and, after eight months, it began its retargeting to fly past Toutatis. Images were taken from distances of 93 km to 240 km. **g.** Hayabusa2 had a successful Earth return on 6 December 2020. **h.** OSIRIS-REx has a planned Earth return for 24 September 2023.

and Hayabusa2 sample return missions.[30] Using the Spitzer Space Telescope launched in 2003, David Trilling led an observing program aimed at providing diameters and albedos for several hundred near-Earth asteroids. His team found that many near-Earth asteroids smaller than 1 kilometer have high albedos (> 0.35), but this is not true for those objects larger than 1 kilometer. Trilling's team attributed this to the larger objects having been formed via collisions earlier than the small ones and hence subjected to the darkening effects of space weathering over a longer period of time. Others used the Spitzer Space Telescope and the ESA Herschel observatory to characterize asteroid 101955 Bennu, the 2018 target for the Origins, Spectral Interpretation, Resource Identification, Security, and Regolith Explorer (OSIRIS-REx) asteroid sample return mission.[31]

By far the most space-based IR data for asteroids have been provided by the Wide-field Infrared Survey Explorer (WISE) and Near-Earth Object Wide-field Infrared Survey Explorer (NEOWISE) projects. The WISE mission had been funded by NASA's astrophysics-oriented Explorer program to perform a complete infrared survey of the sky from a polar, Sun-synchronous orbit. Proposed by Edward Wright of the University of California, Los Angeles (UCLA), the telescope was launched on 14 December 2009. Its infrared detectors were cooled by a block of solid hydrogen, and it completed its first survey of the sky in July 2010. After that, depletion of the hydrogen gradually eliminated the usefulness of two of its four infrared channels, and the spacecraft was put in hibernation in February 2011. WISE discovered 135

---

30. T. G. Müller, S. Hasegawa, and F. Usui, "(25143) Itokawa: The Power of Radiometric Techniques for the Investigation of Remote Thermal Observations in the Light of the Hayabusa Rendezvous Results," *Publications of the Astronomical Society of Japan* 66, no. 3 (June 2014): 52 (1–17), doi:10.1093/pasj/psu034; S. Hasegawa et al., "Albedo, Size and Surface Characteristics of Hayabusa-2 Sample-Return Target 162173 1999 JU3 from AKARI and Subaru Observations," *Publications of the Astronomical Society of Japan* 60, no. sp2 (December 2008): S399–S405, doi:10.1093/pasj/60.sp2.S399. "Hayabusa" is the Japanese word for "falcon," and asteroid Itokawa was named in honor of the father of Japanese rocketry, Hideo Itokawa (1912–99).

31. D. E. Trilling et al., "EXPLORENEOs. I. Description and First Results from the Warm Spitzer Near-Earth Object Survey," *Astronomical Journal* 140 (2010): 770–784; D. E. Trilling et al., "NEOSURVEY 1: Initial Results for the Warm Spitzer Exploration Science Survey of Near-Earth Object Properties," *Astronomical Journal* 152, no. 6 (2016): 1–10; J. P. Emery et al., "Thermal Infrared Observations and Thermophysical Characterization of OSIRIS-REx Target Asteroid (101955) Bennu," *Icarus* 234 (2014): 17–35.

near-Earth objects and observed more than 580 before its initial deactivation in February 2011.

Amy Mainzer at JPL, who had developed the fine guidance sensor for the Spitzer infrared space telescope, had proposed to NASA's planetary science program the adaptation of the Panoramic Survey Telescope and Rapid Response System (Pan-STARRS) moving-objects processing software to extract asteroids from the WISE imagery. She and the rest of the WISE science team knew that WISE would see many asteroids. She worked with Spacewatch Principal Investigator Bob McMillan, who was also on the WISE science team, to understand what would have to be done to identify the asteroids captured in WISE's imagery, and with Larry Denneau of the Institute for Astronomy to adapt the unfinished Pan-STARRS software to the task.[32] This add-on effort became known as NEOWISE, and while WISE had been funded by NASA's Astrophysics Division, NEOWISE was funded by the NEO Observations Program of the Planetary Science Division and later by the Planetary Defense Coordination Office after it was formed in 2016.

This initial phase of NEOWISE effectively confirmed that the Spaceguard Survey goal had been met. While there was some concern that the ground-based surveys had produced results—and population models based on them—that were biased by their inability to observe near the Sun, NEOWISE did not have the same set of biases. Instead, since it detected only infrared radiation, the asteroids it discovered tended to be the low-albedo (darker) asteroids that the ground-based optical telescopes were less likely to see.

As of July 2018, NEOWISE had discovered 266 near-Earth asteroids and 28 comets, defined the albedo distribution for near-Earth asteroids, and demonstrated that there are likely far fewer midsized near-Earth asteroids than previously thought.[33] The NEOWISE Principal Investigator, Amy Mainzer, commented, "[T]here are some really dark NEOs in the population, and the interesting thing is that the albedo distribution is roughly the same over a pretty wide span of sizes, so from a few kilometers down to a few hundred

---

32. Amy Mainzer, interview by Conway, 25 June 2017, transcript in NASA History Division HRC; see also Larry Denneau et al., "The Pan-STARRS Moving Object Processing System," *Publications of the Astronomical Society of the Pacific* 125, no. 926 (2013): 357, doi:10.1086/670337.

33. Up-to-date statistics are available at *https://cneos.jpl.nasa.gov/stats/wise.html* (accessed 29 August 2017); A. Mainzer et al., "NEOWISE Observations of Near-Earth Objects: Preliminary Results," *Astrophysical Journal* 743, no. 2 (2011): 156.

meters, they have roughly the same ratio of bright to dark objects."[34] This result underscores Morrison's similar 1974 result for main-belt asteroids.

In 2011, Martin Conners of Athabasca University, Canada, and Christian Veillet of the University of California, Los Angeles, located in NEOWISE data the first known "Trojan" asteroid (2010 TK7) to share the same orbital path around the Sun as Earth.[35] By definition, Trojan asteroids are those that occupy the gravitationally stable points ahead and behind a planet in its orbit. The vast majority of Trojan asteroids currently known share Jupiter's orbit; 2010 TK7 occupies L4, the stable point ahead of Earth. Amy Mainzer has a particular fondness for this Earth Trojan. "I like to think of it as if you have a stream flowing downhill, sometimes you have little eddies behind a rock, and if you stick something there like a leaf, it'll just kind of stay there for a while and keep looping around. Eventually, it'll get perturbed out and go away."[36]

Radar Characterization of Asteroids

Radar techniques have allowed remarkable progress in the characterization of asteroid shapes. In 1982, JPL radar astronomer Ray Jurgens outlined a rough technique that estimated the dimensions of an ellipsoid representation of an asteroid that had been well observed by radar.[37] In 1984, JPL's Steven Ostro provided a method for inverting light curves to provide a rough shape, and four years later, he extended the method even further. For a radar dataset that provided observations of various asteroid spin-pole orientations, Ostro's technique could determine the bounding shape and dimensions of the object when viewed pole-on.[38] The bounding shape, or convex hull, is best imagined as the shape of a rubber band stretched about its equator, so that the asteroid's

---

34. Mainzer interview, 20 June 2017.

35. M. Conners, P. Wiegert, and C. Veillet, "Earth's Trojan Asteroid," *Nature* 475 (2011): 481–483.

36. Mainzer interview, 20 June 2017.

37. R. F. Jurgens, "Radar Backscattering from a Rough Rotating Triaxial Ellipsoid with Applications to the Geodesy of Small Bodies," *Icarus* 49 (1982): 97–108.

38. S. J. Ostro, M. D. Dorogi, and R. Connelly, "Convex-Profile Inversion of Asteroid Lightcurves: Calibration and Application of the Method," *Bulletin of the American Astronomical Society* 16 (1984): 699; S. J. Ostro, R. Connelly, and L. Belkora, "Asteroid Shapes from Radar Echo Spectra: A New Theoretical Approach," *Icarus* 73 (1988): 15–24.

Figure 7-2. Steven Ostro of JPL, left, and Alan W. Harris of the German Aerospace Center (DLR), right, in 1995. (Photo courtesy Donald K. Yeomans)

general shape can be identified while ignoring surface depressions. In 1993, Scott Hudson of Washington State University introduced a technique for the three-dimensional reconstruction of an asteroid well-observed by radar.[39] By employing Hudson's shape modeling techniques, Ostro and colleagues were able to provide shape models for a number of well-observed near-Earth asteroids, and these digital models were often converted into actual plastic asteroid models that became very popular within the community. During the 2001 near-Earth object meeting in Palermo, Italy, Ostro finished his talk on the latest results of radar shape modeling, reached into his backpack, and began tossing small plastic asteroid shape models into the audience—thus creating a chaotic scramble as normally staid scientists rushed forward over seats to retrieve the prizes.[40]

39. S. Hudson, "Three-Dimensional Reconstruction of Asteroids from Radar Data," *Remote Sensing Review* 8 (1993): 195–203.

40. It would be difficult to overestimate the importance of Steve Ostro's work in the use of radar observations to characterize near-Earth objects. He was the driving force for

Of particular interest was the radar shape modeling of Aten asteroid (66391) 1999 KW4, which revealed an equatorial bulge that had likely survived after the asteroid had spun up via the YORP process and shed material that re-accreted into a satellite. The shape of the primary and the presence of the satellite were revealed as a result of Arecibo and Goldstone radar observations taken in May 2001.[41] Ostro and colleagues concluded that this satellite, with a mean diameter of about 500 meters, currently orbits the primary asteroid (mean diameter about 1.5 kilometers) every 17.4 hours. The primary has a rapid rotation period of about 2.8 hours, although earlier in its lifetime it presumably spun a bit faster when shedding material.

As used in 2018, this 3D modeling technique begins with an attempt to fit the radar dataset with an ellipsoid, and if the spin-pole direction is not known in advance, then a trial-and-error grid search is undertaken whereby the pole direction is systematically varied until there is a best fit of the radar observations. The technique can then proceed using more sophisticated shape representations until the rotating shape model allows an accurate representation of the time history of the observed radar frequency bandwidths. To help ensure that the derived shape model evolves toward reality, certain constraints or penalties are assigned if the model begins to evolve into an unrealistic shape. With regard to these constraints, JPL radar scientist Lance Benner notes, "[I]t's very easy to get a model that looks kind of like a sea urchin, that provides a really nice fit to the data, but there's no way that asteroids look like sea urchins."[42] By 2015, the modeling software process could deal with one object of a binary system as well as time-variable spin rates and spin impulses. The former enabled the first YORP detection using radar data by Patrick Taylor of Cornell University and colleagues, while the latter is important for analyzing spin-state changes due to close planetary encounters.[43]

For close Earth-approaching asteroids with strong signals (high signal-to-noise ratio), the resolution in time delay is limited by the rate at which signals

radar observations of near-Earth objects from the mid-1980s until cancer took him far too early in December 2008.

41. S. J. Ostro et al., "Radar Imaging of Binary Near-Earth Asteroid (66391) 1999 KW4," *Science* 314 (2006): 1276–1280.

42. Lance Benner, interview by Yeomans, 23 June 2017, transcript in NASA History Division HRC.

43. P. A. Taylor et al., "Spin Rate of Asteroid (54509) 2000 PH5 Increasing Due to the YORP Effect," *Science* 316 (2007): 274-277.

produced by the transmitter amplifier tubes (klystrons) can be modulated—that is, the rapidity with which the frequency of the transmitted waveform can be varied. At Arecibo, shortly after equipment upgrades in 1999, the highest range resolution was about 15 meters. Five years later, a portable fast sampler data-taking system designed and built by Jean-Luc Margot provided range resolution of up to 7.5 meters.[44] At Goldstone during the interval 1992–2010, the finest resolution was 18.75 meters, but frequency modulation upgrades provided a fivefold improvement to 3.75 meters beginning with observations of 2010 AL30, when this 30-meter object approached Earth to within a third of a lunar distance in January 2010.[45] This increased radar imaging resolution provides finer detail for shape and surface characteristics, which may become important for possible mitigation efforts or resource utilization. In operation since 1963, all Arecibo observations terminated with the final collapse of the 305-meter-diameter antenna in December 2020.[46]

Asteroid Structures

The observed brightness of an asteroid as a function of time (e.g., its light curve) can be used to obtain its rate of rotation as well as some indication of its shape and the orientation of its rotation axis in space.[47] While a single light curve over the relatively short period of time required for one rotation can be used to determine its rotation period, many nights over several years are often needed to obtain the multiple light curves with different viewing geometries that are required for reconstruction of shape and spin-axis orientation. In general, naked-eye and photographic measurement techniques are not precise enough to accurately capture the brightness variations of asteroids with respect to nearby comparison stars. Rather, photoelectric or CCD sensors

---

44. L. A. M. Benner et al., "Radar Observations of Near-Earth and Main-Belt Asteroids," in *Asteroids IV*, ed. P. Michel, F. E. DeMeo, and W. F. Bottke, Jr. (Tucson, AZ: University of Arizona Press, 2015), pp. 165–182.

45. M. A. Slade et al., "First Results of the New Goldstone Delay-Doppler Radar Chirp Imaging System," *Bulletin of the American Astronomical Society* 42 (2010): 1080; Benner et al, "Radar Observations of Near-Earth and Main-Belt Asteroids." With Goldstone transmitting, Arecibo can also acquire images with a resolution of 3.75 meters.

46. Eric Hand, "Arecibo Telescope Collapses, Ending 57-Year Run," *Science*, doi:10.1126/science.abf9573.

47. A brief historical introduction to light curve analysis is given in chapter 3.

are generally required, and CCDs and computer-driven search programs now make light-curve observations and analysis much more efficient than earlier photographic or photoelectric efforts. The first tabulation of asteroid rotation rates, 27 of them, was carried out by Hannes Alfvén in 1964, with many additions provided by Tom Gehrels six years later.[48]

In 1996, JPL's Alan Harris noted that no asteroid yet discovered had a spin period shorter than about 2.2 hours—the spin rate at which centrifugal forces acting upon a patch of a spherical asteroid's surface would equal the asteroid's gravitational attraction, given a reasonable bulk density of 2.7 grams per cubic centimeter (g/cm³). He concluded that this observation might imply a rubble pile structure for most, if not all, asteroids. With no tensile strength, any asteroids rotating more rapidly would fly apart.[49] Petr Pravec of the Czech Republic's Astronomical Institute and JPL's Alan Harris expanded upon this concept in a 2000 paper.[50] The discovery of small asteroids (e.g., 1995 HM and 1998 KY26) spinning faster than the 2.2-hour limit led them to suggest that monolithic asteroids, under stress and with tensile strength, could be abundant among the smaller asteroids. Fast-rotating asteroids in the size range of 200 meters to 10 kilometers tended to have smaller light-curve amplitudes (nearly spherical shapes) as their spin rates increased, which the authors claimed provided further evidence for their rubble-pile structure. That is, any fragment on the end of an elongated body would be the first to be lost to a fast-rotating body. Asteroids larger than about 200 meters were thought to be loosely bound rubble piles with negligible tensile strength to hold them together. This theory thus provided a pleasing and simple categorization into two types of asteroid structures—small monoliths and large rubble piles—but this turned out to be too simplistic.

48. H. Alfvén, "On the Origin of Asteroids," *Icarus* 3 (1964): 52–56; T. Gehrels, "Photometry of Asteroids," in *Surfaces and Interiors of Planets and Satellites*, ed. A. Dollfus (London: Academic Press, 1970), pp. 319–376.

49. A. W. Harris, "The Rotation States of Very Small Asteroids: Evidence for Rubble Pile Structure" (abstract), in *Lunar and Planetary Science XXVII* (Houston: Lunar and Planetary Institute, 1996), pp. 493–494. Derek Richardson had a picturesque definition of a rubble pile. "This structure is literally a pile of rubble, with the organization that you might expect from a bunch of rocks dumped from a truck." D. C. Richardson et al., "Gravitational Aggregates: Evidence and Evolution," in *Asteroids III*, ed. W. F. Bottke, Jr., et al. (Tucson, AZ: University of Arizona, 2002), pp. 501–515.

50. P. Pravec and A. W. Harris, "Fast and Slow Rotation of Asteroids," *Icarus* 148 (2000): 12–20.

University of Washington researcher Keith Holsapple pointed out in 2007 that for asteroids larger than about 10 kilometers, the lack of tensile and cohesive strength makes no difference in the permissible spin rate since the structure of these asteroids is governed by their own gravity—not their strength. Hence these larger asteroids could be either monolithic or rubble piles. Holsapple also pointed out that for small asteroids below a few kilometers in diameter that are dominated by strength considerations, they could rotate considerably faster than the 2.2-hour limit if they had only a modest amount of tensile strength.[51] In 2014, Paul Sánchez and Dan Scheeres, at the University of Colorado, noted that van der Waals surface attractions between the finest material grains in a rubble pile could provide the modest cohesive attraction necessary to hold small, rapidly rotating rubble-pile fragments in place.[52] Thus the neat rubble-pile dichotomy began to fall apart.

Meanwhile, another factor quickly become apparent: whether or not an asteroid is a rubble pile also has to do with its collisional history. There are more than 20 asteroid families—groups of asteroids that share the same orbital and spectral properties and are thought to have originated from the collisional disruption of a larger body. Over long intervals of time, family members evolve on diverging orbital paths so that relatively young families, like the Karin and Themis families, are most easily identified. Several unsuccessful attempts in the 1990s were made to understand the physics of family formation by simulations using complex computer software and hardware assets or by the extrapolation of impact experiments in the laboratory. Simulated catastrophic collisions that could reproduce the observed size distribution of fragments (in particular, the large family members) resulted in simulated ejection velocities of individual fragments that were much too slow for them to overcome their own gravitational attraction. That is, the parent body might be shattered, but not dispersed to form a family. Conversely, if the ejection velocities were high enough for the fragments to escape the parent body after the collision, the resulting fragment size distribution did not match that of observed family members because no large fragment was created.

---

51. K. A. Holsapple, "Spin Limits of Solar System Bodies: From the Small Fast-Rotators to 2003 EL61," *Icarus* 187 (2007): 500–509.

52. P. Sánchez and D. J. Scheeres, "The Strength of Regolith and Rubble Pile Asteroids," *Meteoritics and Planetary Science* 49, no. 5 (2014): 788–811. Van der Waals forces allow a gecko to walk on walls or ceilings.

In the 1990s, computational assets were insufficient to model both the disruption and re-accumulation of the parent body as well as the formation of escaping fragments that could gravitationally re-accumulate and form large family members. In 2001, the French dynamicist Patrick Michel and his colleagues were the first to successfully combine the two processes. They showed that a catastrophic collision of a solid body larger than about 200 meters would result in a realistic family of rubble-pile asteroids with the original solid parent body reduced to a smaller rubble pile itself.[53] These simulations produced both large fragment sizes and high enough ejection velocities to form a dispersed asteroid family. The key was the successful re-accumulation of large family members from smaller fragments that moved at low velocities with respect to one another even as their much higher ejection velocities allowed escape from the original parent body.

Subsequent research by Michel and his colleagues showed that shattered asteroids are more likely to be family parent bodies and that the outcome of a catastrophic collision depends upon the parent body's internal structure, with lower-density parents better able to withstand a collisional disruption. Their simulations could also reproduce the general shape of the asteroid Itokawa, which was the target body for the Japanese rendezvous and sample-return mission Hayabusa.[54]

These computer simulations of collisions also generated numerous asteroids with satellites, and by the time these simulations were possible, asteroid satellites had already been discovered. C. André and Egon von Oppolzer had speculated in 1900 that the light curve of 433 Eros, taken during a close Earth approach in 1900, could be due to an eclipsing binary system.[55] However, NEAR spacecraft observations taken in 2000 found no satellites—only an

53. P. Michel et al., "Collisions and Gravitational Reaccumulation: Forming Asteroid Families and Satellites," *Science* 294 (2001): 1696–1700. Patrick Michel has wondered how he, a guy raised in his parents' 5-star hotel in the showbiz village of Saint-Tropez, ended up being a scientist. Patrick Michel, e-mail correspondence with Don Yeomans, 21 October 2016.

54. P. Michel, W. Benz, and D. C. Richardson, "Disruption of Fragmented Parent Bodies as the Origin of Asteroid Families," *Nature* 421 (2003): 608–611; P. Michel and D. C. Richardson, "Collision and Gravitational Re-accumulation: Possible Formation Mechanism of Asteroid Itokawa," *Astronomy and Astrophysics* 554 (2013): L1 (4 pages).

55. C. André, "Sur le système formé par la planète double (433) Eros," *Astronomishe Nachrichten* 155 (1901): 27–30; Egon von Oppolzer, "Notiz. betr. Planet (433) Eros," *Astronomische Nachrichten* 154 (1901): 297.

elongated single body. In the 1970s, there were a number of spurious reports of asteroid satellites based upon anomalous light curves or blinkouts during stellar occultation observations.[56] The search for satellites continued into the 1980s using asteroid light-curve analysis and radar observations, along with direct telescopic imaging using CCD devices, but, by the end of the 1980s, asteroid satellites were considered nonexistent or at least extremely rare.[57]

The thinking changed dramatically in August 1993 after the Galileo spacecraft, en route to its orbital tour of Jupiter, flew past main-belt asteroid 243 Ida and imaged a small 1.5-kilometer satellite orbiting the 31-kilometer S-type primary. The discovery was made by Galileo Imaging Team member Ann Harch; Dactyl, as it was to be named, showed for the first time that asteroids can have satellites. After the Galileo discovery, several asteroid satellites were identified, mostly using direct ground-based telescopic imaging with adaptive optics, photometric light-curve analysis, or radar observations. The first radar discovery of a near-Earth asteroid satellite was 2000 DP107, which was observed by the Goldstone and Arecibo radars in September and October 2000. The 300-meter-sized satellite revolved about the 800-meter-sized primary with an orbital period of 1.755 days. Using this orbital period plus the semimajor axis of the satellite around the primary, Kepler's third law allowed a mass determination for the primary.[58] In turn, the mass divided by an estimate of the primary's volume—determined from the radar shape model—provided a bulk density estimate of about 1.7 grams per cubic

---

56. A 1979 review by Van Flandern and colleagues gives a complete summary of evidence for asteroid satellites to that time, but none of these suggestions has been shown to be real despite studies with modern techniques. T. C. Van Flandern et al., "Satellites of Asteroids," in *Asteroids*, ed. T. Gehrels (Tucson, AZ: University of Arizona Press, 1979), pp. 443–465.

57. Stuart Weidenschilling et al. and Bill Merline et al. provide summaries of the observations and theory through 1989 and 2002 respectively. S. J. Weidenschilling, P. Paolicchi, and V. Zappalà, "Do Asteroids Have Satellites?" in *Asteroids II*, ed. R. P. Binzel, T. Gehrels, and M. S. Matthews (Tucson, AZ: University of Arizona Press, 1989), pp. 643–658; W. J. Merline et al., "Asteroids Do Have Satellites," in *Asteroids III*, ed. W. F. Bottke, Jr., A. Cellino, P. Paolicchi, and R. P. Binzel (Tucson, AZ: University of Arizona Press, 2002), pp. 289–312.

58. The mean sidereal motion is $n = 2\pi/P$ where P is the orbital period of the satellite. Then, by Kepler's third law, $n^2 a^3 = GM$, where a is the semimajor axis of the satellite, M is the primary's mass and G is the gravitational constant. S. J. Ostro et al., "2000 DP107," *IAU Circular* 7496 (25 September 2000); J. L. Margot et al., "Binary Asteroids in the Near-Earth Object Population," *Science* 296 (2002): 1445–1448.

centimeter, suggesting that the object is an under-dense, porous, rubble-pile aggregate.[59] Thus, for near-Earth objects with radar-observed satellites, it is often possible to constrain interior structure, knowledge that could be vital for dealing with an impact threat or identifying objects that might be suitable for resource utilization.

For near-Earth objects, the majority of asteroid satellite discoveries (roughly 75 percent) occur via radar investigations, with most of the remaining coming from light-curve analyses. Roughly 15 percent of all near-Earth asteroids have a satellite, and at least three have two satellites each.[60]

Asteroid Compositions

As outlined in chapter 3, one of the few solid compositional connections between meteorites of a particular type on Earth and asteroids of a particular spectral class in space are some of the so-called howardite, eucrite, and diogenite (HED) meteorites and the asteroid Vesta. Since the most common meteorite type by far is the ordinary chondrite, one would think that these would be a spectral match with the S-type asteroids, which are the most common asteroid types nearest Earth in the inner main belt. But the spectra of S-type asteroids have significantly redder wavelengths, at odds with the spectra of ordinary chondrite meteorites. This unexpected result is known as the Ordinary Chondrite Paradox.

As early as 1968, Bruce Hapke noted that solar wind particles might alter ("space-weather") the spectral characteristics of the lunar surface. Seven years later, he offered an interpretation that high-speed micrometeoroid impacts could alter the surface of airless bodies by the vapor deposition of tiny iron particles on surface powder grains.[61] These depositions, called nanophase

---

59. The bulk density of liquid water is 1 gram/cubic centimeter and about 3.3 g/cm^3 for ordinary chondrite meteorites.

60. Jean-Luc Margot and colleagues provide a summary of satellite discoveries through 2015. Jean-Luc Margot et al., "Asteroid Systems: Binaries, Triples, and Pairs," in *Asteroids IV*, ed. P. Michel, F. E. DeMeo, and W. F. Bottke, Jr., (Tucson, AZ: University of Arizona Press, 2015), pp. 355–373. Benner interview, 23 June 2017. For a complete list of all asteroid binaries and triple systems, see *http://www.johnstonsarchive. net/astro/asteroidmoons.html* (accessed 30 August 2017).

61. B. Hapke, "Lunar Surface: Composition Inferred from Optical Properties," *Science* 159 (1968): 76–79. B. Hapke, W. Cassidy, and E. Wells, "Effects of Vapor-Phase

iron, would then redden the surface. Despite Hapke's prescient work, almost all planetary scientists in the 1970s, 1980s, and into the 1990s continued to believe that asteroid spectra had to be taken at face value and that most meteorites had to come from relatively few asteroids. This interpretation was likely bolstered in 1984, when near-Earth asteroid 1862 Apollo was identified as the first of what would be called Q-types, with spectra similar to those of ordinary chondrites.[62] But Q-types were relatively rare and small, so there would have to be many undiscovered objects of this type, or else some dynamic process that would preferentially bring them to Earth's neighborhood.

Hints that space weathering alters asteroid spectra came in 1993 with the Galileo spacecraft flyby of asteroids Gaspra and Ida/Dactyl: fresh surfaces exposed by relatively recent impacts showed less reddening than adjacent surfaces.[63] By 2000, Carlé Pieters of Brown University and her colleagues had brought together convincing evidence that micrometeorite and solar wind bombardment on airless bodies (e.g., the Moon) would redden a powdery surface with nanophase iron, thus allowing a compositional match between the common ordinary chondrites and the common S-type asteroids.[64]

The chemical composition measurements of the S-type asteroid 433 Eros by the nearby NEAR-Shoemaker spacecraft in 2000, as well as laboratory analysis of the sample returned from the S-type asteroid Itokawa by the Japanese Hayabusa mission in 2010, solidified the understanding that the common S-type asteroids are indeed the parent bodies of the most common ordinary chondrite meteorites.

Deposition Processes on the Optical, Chemical and Magnetic Properties of the Lunar Regolith," *Moon* 13 (1975): 339–354.

62. L. A. McFadden, M. J. Gaffey, and T. B. McCord, "Mineralogical-Petrochemical Characterization of Near-Earth Asteroids," *Icarus* 59 (1984): 25–40.

63. M. J. S. Belton et al., "The Galileo Encounter with 951 Gaspra: First Pictures of an Asteroid," *Science* 257 (1992): 1647–1652; M. J. S. Belton et al., "First Images of Asteroid 243 Ida," *Science* 265 (1994): 1543–1547.

64. C. M. Pieters et al., "Space Weathering on Airless Bodies: Resolving a Mystery with Lunar Samples," *Meteoritics and Planetary Science* 35 (2000): 1101–1107. Clark Chapman has provided a thorough history of the S-type asteroid paradox in two papers: C. R. Chapman, "S-Type Asteroids, Ordinary Chondrites, and Space Weathering: The Evidence from Galileo's Fly-bys of Gaspra and Ida," *Meteoritics and Planetary Science* 31 (1996): 699–725; C. R. Chapman, "Space Weathering of Asteroid Surfaces," *Annual Review of Earth and Planetary Sciences* 32 (2004): 539–569.

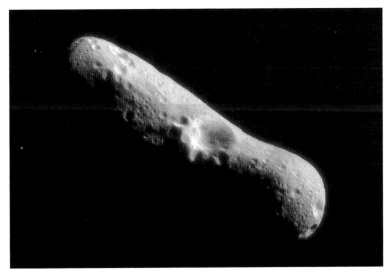

Figure 7-3. 433 Eros was imaged by NEAR on 14 February 2000. The large central crater is about 6 kilometers in diameter. (NASA/JPL/Johns Hopkins University [JHU] Applied Physics Laboratory [APL], image no. PIA02467)

Missions to Comets

As we have already seen with the discussion of the Halley's Comet Armada in chapter 3, much can be learned about comets from spacecraft missions, albeit at much greater cost. The first U.S. imaging mission to a comet was by a technology demonstration mission known as Deep Space 1, whose principal purpose was to prove the usefulness of a low-impulse solar electric propulsion system.[65] The Deep Space 1 flyby of periodic Comet Borrelly revealed a nucleus that appeared to be about half the size of Comet Halley's, but the physical characteristics were similar. It too had collimated gas jets arising from a small fraction of the nucleus, lacked obvious impact craters, and was very dark. JPL's Bonnie Buratti and coworkers determined that its albedo was only 0.029—then the darkest known object in the solar system.[66]

65. Erik M. Conway and Mirella Flores, "Deep Space 1: A Revolution in Space Exploration," *Quest* 14, no. 2 (2007): 41–51.

66. H. U. Keller et al., "In Situ Observations of Cometary Nuclei" in *Comets II*, ed. M. C. Festou, H. U. Keller, and H. A. Weaver (Tucson, AZ: University of Arizona Press, 2004), pp. 211–222; B. J. Buratti et al., "Deep Space 1 photometry of the nucleus of Comet 19P/Borrelly," *Icarus* 167 (2004): 16–29.

Most of NASA's comet and asteroid missions, though, have come from the Agency's Discovery Program, which operates by competition. Scientists at universities and/or NASA Centers propose mission ideas, one or two of which are funded every few years. NEAR-Shoemaker was the first Discovery program mission to a small solar system body. The second was the Stardust spacecraft, which was equipped with particle capture cells. In January 2004, it flew within 500 kilometers of Comet Wild 2, captured more than 10,000 microscopic dust particles (1–300 microns) and returned them to Earth for analysis on 15 January 2006. Donald Brownlee, at the University of Washington, led the Stardust science team.

Prior to the Stardust mission, there was a widely held view that comets were a mixture of ice and noncrystalline interstellar grains that had formed in low-temperature environments.[67] But instead, the Stardust particles larger than a micron turned out to be largely crystalline grains that had been heated in the inner solar nebula to temperatures of 800°C or more before being carried outward to the outer solar nebula by turbulent mixing. Carbon-based, or organic, materials were found, including the discovery of the amino acid glycine, an important building block of life.[68] A few calcium-aluminum inclusion (CAI) fragments, evident in the most primitive meteorites, were found in the Stardust particles, suggesting that these particles formed close to the Sun during the earliest stage of the solar system's development. Apparently, some mixing took place to carry minerals formed near the Sun to the outer solar nebula before Comet Wild 2 formed there.[69]

The nucleus of Comet Wild 2 was roughly spherical, with a diameter of about 5.5 kilometers. It showed smooth areas at the bottom of circular depressions, which were not impact craters but perhaps the result of surface layers

---

67. J. M. Greenburg, "What Are Comets Made Of? A Model Based on Interstellar Dust," in *Comets*, ed. L. L. Wilkening (Tucson, AZ: University of Arizona Press, 1982), pp. 131–163.

68. J. E. Elsila, D. P. Glavin, and J. P. Dworkin, "Cometary Glycine Detected in Samples Returned by Stardust," *Meteoritics and Planetary Science* 44 (2009): 1323–1330. Glycine was also evident in the Rosetta mission target, Comet Churyumov-Gerasimenko.

69. M. E. Zolensky et al., "Mineralogy and Petrology of Comet 81P/Wild 2 Nucleus Samples," *Science* 314 (2006): 1735–1739; D. Brownlee et al., "The Stardust Mission: Analyzing Samples from the Edge of the Solar System," *Annual Review of Earth and Planetary Science* 42 (2014): 179–205.

Figure 7-4. Comet Tempel 1 imaged by Deep Impact's flyby spacecraft after impact on 4 July 2005. The burst of light on the right side of the comet is sunlight reflected from the ejecta thrown up by the impact. (Image credit: NASA/JPL-Caltech/University of Maryland [UMD], image no. PIA02137)

giving way and forming slump depressions or sinkholes. About 20 collimated jets were observed.[70]

In 2005, Stardust was followed by a mission to investigate the nature of Comet Tempel 1's nucleus via a 10.3-kilometer-per-second impact with a 372-kilogram spacecraft, Deep Impact.[71] Michael A'Hearn, a cometary authority at the University of Maryland and Principal Investigator for the mission, proposed a flight plan whereby the impacting spacecraft was released from the larger mother spacecraft 24 hours prior to impact. Analyses of spectra and images taken before, during, and after the impact demonstrated that Comet Tempel 1 was a relatively weak structure and had only a few impact

70. H. U. Keller et al., "In Situ Observations of Cometary Nuclei," n. 65.

71. The comet impact occurred on 4 July 2005—American Independence Day.

craters, an albedo of 0.059, a bulk density of about 0.4 grams per cubic centimeter, an extent of 4.4–8.0 kilometers, and an abundance of jet activity.[72] Some of the jets were collimated, and most produced water vapor (H_2O), but some on the night side were rich in the more volatile carbon dioxide (CO_2). The ratio of cometary production rates of CO_2 compared to H_2O was about 7 percent, and the differing sources of water and carbon dioxide suggested a heterogeneous nucleus.[73] From spectral observations, the University of Maryland's Jessica Sunshine and colleagues detected water ice in the impact ejecta and in three small patches on the surface of the nucleus. However, these surface patches were far too small to explain the total amount of atmospheric water vapor, so most of it must have come from subsurface sources.[74] A'Hearn and colleagues concluded that the top few centimeters of the comet's surface are largely ice-free and the bulk of the H_2O and CO_2 ices are likely within 1 meter of the surface.[75] To account for the observed surface layering and the diverse composition of these layers, Arizona researcher Michael Belton suggested a nucleus model consisting of a pile of randomly stacked layers. These diverse layers were thought to have been produced over time by impacts of comets that originated in differing regions of a non-uniform protoplanetary nebula.[76]

---

72. P. C. Thomas et al., "The Shape, Topography, and Geology of Tempel 1 from Deep Impact Observations," *Icarus* 191 (2007): 51–62; J. E. Richardson, H. J. Melosh, C. M. Lisse, and B. Carcich, "A Ballistic Analysis of the Deep Impact Ejecta Plume: Determining Comet Tempel 1's Gravity, Mass, and Density," *Icarus* 191 (2007): 176–209. Updates for some of Comet Tempel 1's parameters are provided by Veverka et al., "Return to Comet Tempel 1: Overview of Stardust-NExT Results," *Icarus* 222 (2013).

73. L. M. Feaga et al., "Asymmetries in the Distribution of H_2O and CO_2 in the Inner Coma of Comet 9P/Tempel 1 as Observed by Deep Impact," *Icarus* 191 (2007): 134–145.

74. J. M. Sunshine et al., "The Distribution of Water Ice in the Interior of Comet Tempel 1," *Icarus* 191 (2007): 73–83; J. Sunshine et al., "Exposed Water Ice Deposits on the Surface of Comet 9P/Tempel 1" *Science* 311 (2006): 1453–1455.

75. M. F. A'Hearn et al., "Deep Impact and Sample Return," *Earth, Planets and Space* 60 (2008): 61–66.

76. M. Belton et al., "The Internal Structure of Jupiter Family Cometary Nuclei from Deep Impact Observations: The "TALPS" or "Layered Pile" Model," *Icarus* 187 (2007): 332–344. This model, which postulated the formation of layered cometary nuclei by the successive collisions of diverse, smaller, primordial comets, was nicknamed the TALPS model (TALPS spelled backward is SPLAT).

After the Comet Tempel 1 encounter, the Deep Impact flyby spacecraft was retargeted to fly within 700 kilometers of Comet Hartley 2 on 4 November 2010. The mission was then renamed the Extrasolar Planet Observation and Deep Impact Extended Investigation (EPOXI). Comet Hartley 2's longest extent is about 2.3 kilometers with a bi-lobed shape. A'Hearn described the shape as a cross between a bowling pin and a pickle. The smallest of the cometary spacecraft targets, Comet Hartley 2 was hyperactive, with water vapor sublimating from the smooth waist region and CO_2-rich jets dragging out water ice chunks, which then sublimated above the smaller end of the nucleus.

Space-based infrared observations of Hartley 2 by the ESA's Herschel Space Telescope provided the surprising result that this comet had a deuterium-to-hydrogen (D/H) ratio similar to that in Earth's oceans, as well as some meteorites and asteroids—suggesting that short-period comets may have contributed much of the water to the early Earth.[77] However, the six other comets for which these measurements were made had D/H ratios about twice that of Earth's oceans, suggesting that primitive asteroids, rather than comets, probably provided much of Earth's water. The idea that comets might only be a minor contributor to Earth's water was underscored once the D/H ratio for the Rosetta mission target, Comet Churyumov-Gerasimenko, was found to be more than three times the value for Earth's oceans.[78]

After the successful flyby of Comet Wild 2 by the Stardust spacecraft in 2004 and the return of the dust samples to Earth two years later, the Stardust mission was renamed Stardust-NExT and retargeted to fly past Comet Tempel 1 in mid-February 2011. Compared to the earlier Deep Impact mission images, the additional Stardust-NExT images of Comet Tempel 1 revealed very few noticeable changes in surface features. The crater formed by the collision of the Deep Impact probe into the porous cometary surface in

---

77. The Herschel spacecraft observations of Hartley 2 determined the D/H ratio from the ratio of heavy water to regular water. In heavy water, one of the two hydrogen atoms has been replaced by the hydrogen isotope deuterium. P. Hartogh et al., "Ocean-like Water in the Jupiter Family Comet 103P/Hartley 2," *Nature* 478 (2011): 218–220.

78. M. Fulle et al., "Unexpected and Significant Findings in Comet 67P/Churyumov-Gerasimenko: An Interdisciplinary View," *Monthly Notices of the Royal Astronomical Society* 462 (2016): S2–S8.

Figure 7-5. Comet Churyumov-Gerasimenko imaged by ESA's Rosetta spacecraft from a distance of about 68 kilometers on 14 March 2015. (Image credit: ESA/Rosetta/NAVCAM, image no. PIA19687)

2005 was barely recognizable as a 50-meter-wide depression in the Stardust-NExT images five and a half years later.[79]

The European Space Agency's Rosetta spacecraft, after a 10-year flight, which included two asteroid flybys and three gravity-assists from Earth and another from Mars, arrived at Comet Churyumov-Gerasimenko in August 2014. It then began an intensive two-year study.[80] The Philae lander was released from the mother spacecraft on 12 November 2014, but due to the failure of two devices meant to secure the lander to the surface, it bounced

---

79. NExT stands for New Exploration of Tempel 1. Veverka et al., "Return to Comet Tempel 1: Overview of Stardust-NExT Results," *Icarus* 222 (2013): 424–435.

80. Fulle et al., "Unexpected and Significant Findings."

twice and landed in a crevasse, thus limiting its usefulness.[81] Nevertheless, acoustic signals with different wavelengths were generated by a hammering device on Philae, and these signals were then picked up by sensors in the lander's three legs. The longer-wavelength signals penetrated deeper into the comet's surface layers, and modeling efforts determined that an upper crust some 10 to 50 centimeters thick overlay a much softer interior—a rugged and stiff upper crust over a much less stiff, softer interior.[82]

Rosetta's 21 science instrument packages, two-year study period, and close-up imaging garnered a number of important discoveries. Rosetta provided the first bulk density determination of a cometary nucleus made directly from mass and volume estimates. The determined value of 0.53 grams per cubic centimeter was half that of water. The porosity was determined to be about 70 percent, which implied that the nucleus had far more empty pore space than solid material. The shape of the nucleus was a bizarre double-lobed object 4.1 kilometers at its longest extent that appeared to have been formed by two smaller cometary bodies.[83] It was frequently likened to the shape of a giant "rubber ducky" bathroom toy.

During the three months around perihelion, some 34 dust jets were noted, a third of them collimated. These jets appeared at cometary noon or sunrise—perhaps when deeper volatile pockets had a chance to warm up, or when volatiles closer to the surface first experienced sunlight.[84] Wide and shallow pits (some active) were suggested to be sinkholes, and very noticeable surface changes were evident, with some roundish features growing at a rate of a few centimeters per hour—presumably a result of the vaporization of water ice just below the surface.[85] The mass of the dust leaving the comet was four

---

81. The names Rosetta and Philae came from the ancient Egyptian Rosetta stone slab and Philae obelisk that helped decipher Egyptian hieroglyphs.

82. M. Knapmeyer, H.-H. Fischer, J. Knollenberg, K. J. Seidensticker, K. Thiel, W. Arnold, C. Faber, and D. Möhlmann, "Structure and Elastic Parameters of the Near Surface of Abydos Site on Comet 67P/Churyumov-Gerasimenko, as Obtained by SESAME/CASSE Listening to the MUPUS Insertion Phase," *Icarus* 310 (2018): 165–193.

83. Fulle et al., "Unexpected and Significant Findings."

84. J.-B. Vincent et al., "Large Heterogeneities in Comet 67P as Revealed by Active Pits from Sinkhole Collapse," *Nature* 523 (2015): 63–68; J.-B. Vincent et al., "Summer Fireworks on Comet 67P," *Monthly Notices of the Royal Astronomical Society* 462 (2016): S184–S194.

85. O. Groussin et al., "Temporal Morphological Changes in the Imhotep Region of Comet 67P/Churyumov-Gerasimenko," *Astronomy and Astrophysics* 583 (2015): A36.

times greater than the mass of the gas departing, and copious amounts of the very volatile molecular oxygen, nitrogen, and argon were taken to mean that Comet Churyumov-Gerasimenko, once formed, had never experienced high temperatures.[86]

The missions to comets substantially revised the understanding of the cometary nuclei from monolithic, bright, dirty ice balls to under-dense, dark, icy dirtballs. When spacecraft began visiting asteroids, there would also be some surprises and revisions to the understanding of these objects.

Missions to Asteroids

While not nearly as obvious and showy as comets, there are far more asteroids in Earth's neighborhood than comets. For every comet that comes close to Earth, there are more than one hundred asteroids that do the same.[87] In terms of planetary defense, asteroids are the primary concern, and because many of them are believed to be very primitive remnants of the early solar system formation process, they offer clues to the thermal and physical conditions under which the planets, including Earth, formed.

By the late 20th century, Earth-based observations of asteroids had established that they had diverse shapes, rotation rates, and compositions, but there were a number of questions that could be answered only with space-based observations and sample returns. As is always the case for successful flight projects, there were complete surprises that raised new questions and interpretations. Among these were the discovery of the first (of many) asteroid satellites and the realization that some C-type asteroids were very-low-density, highly porous objects.

Launched in 1989, the Galileo spacecraft took six years to reach Jupiter. Before its 1995 rendezvous with Jupiter, the spacecraft conducted scientific investigations of asteroids Gaspra (1991) and Ida (1993), as well as the 1994 collision of Comet Shoemaker-Levy 9 with Jupiter. From its peanut-like shape, paucity of large craters, and plethora of small craters, Gaspra was theorized by Richard Greenberg and colleagues to have a rubble-pile structure,

---

86. Fulle et al., "Unexpected and Significant Findings."

87. G. Stokes et al., *Study To Determine the Feasibility of Extending the Search for Near-Earth Objects to Smaller Limiting Diameters: Report of the Near-Earth Object Science Definition Team* (NASA, 22 August 2003), *https://cneos.jpl.nasa.gov/doc/neo_report2003.html* (accessed 14 September 2021).

in which impact-induced seismic shaking had erased much of the cratered surface upon which small, fresh craters have since formed. On the other hand, Clark Chapman suggested that Gaspra could be a hard, stony-iron, mono-lithic structure upon which even powerful impacts could form only relatively small craters.[88] Since the flyby distance of 1,600 kilometers was too distant to affect the spacecraft's trajectory, and hence determine the asteroid's mass, no bulk density determination was possible to distinguish between the two very different hypotheses.

More comprehensive data were collected during the 1993 flyby of asteroid Ida because it was larger and its faster rotation rate brought more of the sur-face into view. The surprising discovery of Dactyl also allowed a rough mass determination of Ida via Kepler's third law, yielding a bulk density determina-tion of 2.6 grams per cubic centimeter.[89] Given this low value—far less than the 3.3 grams per cubic centimeter of ordinary chondritic material—porosity was likely an important contributor to Ida's interior volume. As was the case for Gaspra, photometric observations suggested a surface regolith rather than bare rock. Ida and Dactyl have similar compositions, and Chapman posited that Ida likely formed as an independent, perhaps rubble-pile, body in the catastrophic disruption of the parent body of the Koronis family, of which it is a member.[90]

The first of the successful, low-cost, science-focused NASA Discovery missions, the Near Earth Asteroid Rendezvous (NEAR) mission, flew past the C-type asteroid 253 Mathilde on its way to its rendezvous with the S-type, near-Earth asteroid 433 Eros on 14 February 2000.[91] At Mathilde,

88. R. Greenberg et al., "Collisional History of Gaspra," *Icarus* 107 (1994): 84–97; C. R. Chapman, "Gaspra and Ida: Implications of Spacecraft Reconnaissance for NEO Issues," in *Near-Earth Objects: The United Nations International Conference*, vol. 822, ed. J. L. Remo (New York: Annals of the New York Academy of Sciences, 1997), pp. 227–235.

89. R. J. Sullivan et al., "Asteroid Geology from Galileo and NEAR Shoemaker Data," in *Asteroids III*, ed. W. F. Bottke, Jr., A. Cellino, P. Paolicchi, and R. P. Binzel (Tucson, AZ: University of Arizona Press, 2002), pp. 331–350.

90. Clark Chapman, "Cratering on Asteroids from Galileo and NEAR Shoemaker," in *Asteroids III*, ed. W. F. Bottke, Jr., A. Cellino, P. Paolicchi, and R. P. Binzel (Tucson, AZ: University of Arizona Press, 2002), pp. 315–330.

91. NASA's Discovery Program, founded in 1992, is a series of low-cost, highly focused scientific space missions. To date, the Discovery missions have been NEAR; Mars Pathfinder; Lunar Prospector; Stardust; Genesis; COmet Nucleus TOUR

only the Multi-Spectral Imager was turned on to save power. Even so, the NEAR-Shoemaker spacecraft observations of Mathilde revolutionized thinking about asteroidal cratering, completely overturning some pre-encounter assumptions. It was commonly thought that impactors large enough to form huge craters would be too large to avoid catastrophic disruption and that such impacts would resurface the object via seismic shaking and ejecta blanketing.[92] Mathilde observations revealed at least four huge craters with dimensions as large as or larger than Mathilde's radius. Mathilde had not only survived each impact, but the shapes of previous craters were also preserved, and surface ejecta was absent. The flyby was close enough for a mass determination, thus providing a bulk density estimate of only 1.3 grams per cubic centimeter—much lower than expected. Yet Mathilde had clearly survived a pounding by large impactors.[93] On the day of the encounter (27 June 1997), when the NEAR science team met at the Johns Hopkins Applied Physics Lab in Maryland, Don Yeomans remembers that it was Gene Shoemaker, with characteristic genius, who first linked the two surprises, noting that the reason Mathilde survived the large impactors could well be due to its extreme porosity. Two years later, the Boeing Company researcher Kevin Housen and colleagues reported upon laboratory impact experiments showing that porous targets suffer compaction without generating ejecta.[94] Much like a bullet fired into a porous pile of sand, Mathilde had simply swallowed the impactors! One mystery remains: Mathilde's slow rotation period of 17.4 days is a surprise since impacts were expected to have spun it up.

(CONTOUR (failed launch); MErcury Surface, Space ENvironment, GEochemistry, and Ranging (MESSENGER); Deep Impact; Dawn; Kepler; and Gravity Recovery and Interior Laboratory (GRAIL). NEAR Mission Manager Robert Farquhar was proud of the fact that the spacecraft began orbiting Eros (named for the Greek god of love) on Valentine's Day. He also had two brass plaques surreptitiously affixed to the interior of the spacecraft to honor his late first wife, Bonnie, and his second wife, Irina. Robert Farquhar, *Fifty Years on the Space Frontier: Halo Orbits, Comets, Asteroids and More* (Denver: Outskirts Press, 2011).

92. Chapman, "Cratering on Asteroids from Galileo and NEAR Shoemaker."

93. D. K. Yeomans et al., "Estimating the Mass of Asteroid 253 Mathilde from Tracking Data During the NEAR Flyby," *Science* 278 (1997): 2106–2109; J. Veverka et al., "NEAR Encounter with Asteroid 253 Mathilde: Overview," *Icarus* 140 (1999): 3–16.

94. K. R. Housen, K. A. Holsapple, and M. E. Voss, "Compaction as the Origin of the Unusual Craters on the Asteroid Mathilde," *Nature* 402 (1999): 155–157.

Upon entering Eros orbit on 14 February 2000, the NEAR spacecraft was renamed NEAR-Shoemaker, in honor of Gene Shoemaker, who had passed away in a tragic automobile accident in Australia just three weeks after the Mathilde flyby.[95] The spacecraft remained in orbit about Eros for a year before softly setting down on the surface on 12 February 2001. The density of Eros was determined to be a uniform 2.67 grams per cubic centimeter—similar to that of Ida—implying a porosity of less than 30 percent. Its measured gravity field was consistent with a uniformly dense body; this, in conjunction with the presence of lengthy surface grooves and ridges, suggested that Eros was a consolidated body, albeit one with a fractured substrate.[96] The Near Infrared Spectrometer, along with the X-ray and Gamma Ray Spectrometers, provided data consistent with Eros having an ordinary chondritic composition, but the link was not definitive.[97] Another surprise was the complete lack of a magnetic field for Eros. Most meteorites, including chondrites that originate from S-type asteroids like Eros, are much more magnetized than Eros itself.[98] This absence has yet to be explained.

Overcoming several in-flight difficulties, the Japanese Space Agency's Hayabusa spacecraft successfully carried out a rendezvous with asteroid Itokawa in September 2005 and brought nearly 1,600 tiny surface samples back to Earth five years later. Hayabusa's misadventures included a hydrazine fuel leak, the failure of two of the three reaction wheels that kept the spacecraft properly oriented, a micro-rover lost to space, the failure of the surface sample collection device, and a dead battery. Fortunately, ingenious engineering fixes were implemented along the way and some dust samples (kicked up by the failed sample collector) made it into the collection device and were brought back to Earth.

Itokawa's surface was rougher and more boulder-rich than that of Eros, with a shape resembling a sea otter's head and body, a bulk density of 1.9 grams per cubic centimeter, and an estimated porosity of 41 percent—somewhat higher

95. Although only the spacecraft was renamed NEAR-Shoemaker, this name is often applied to the mission as well.

96. J. Veverka et al., "NEAR at Eros: Imaging and Spectral Results," *Science* 289 (2000): 2088–2097.

97. A. F. Cheng, "Near Earth Asteroid Rendezvous: Mission Summary," in *Asteroids III*, ed. W. F. Bottke, Jr., A. Cellino, P. Paolicchi, and R. P. Binzel (Tucson, AZ: University of Arizona Press, 2002), pp. 351–366.

98. Ibid.

than that of a pile of sand. The principal axes of the "head" and "body" have different orientations, suggesting that they were once separate bodies that combined after a slow collision into a rough rubble-pile structure. The paucity of craters is attributed to impact-induced resurfacing, and its rubble-pile nature can explain its low bulk density, high porosity, boulder-rich appearance, and bi-lobed shape.[99] Infrared and x-ray fluorescence spectroscopy experiments determined that Itokawa was probably chondritic in composition.[100]

Of the nearly 1,600 tiny dust particles that were brought back for study, two-thirds were silicate mineral grains made up of only olivine, only pyroxene, or only feldspar. The remaining (mostly silicate) grains were polymineralic.[101] The samples were a match to ordinary chondritic material, offering a final confirmation that the common S-type asteroids are indeed the source of the most common meteorites, the ordinary chondrites.

The Hayabusa sample analyses also shed light on the likely thermal history and formation of Itokawa. When the Itokawa parent body condensed from the solar nebula, some of its constituent atoms were short-lived radioactive isotopes like aluminum-26 (^{26}Al). As the proto-asteroid accumulated more material, including more ^{26}Al, it continued to heat up as these radioactive atoms decayed.[102] The asteroid likely reached a temperature of about 800°C before it began a long-term cooling phase as the ^{26}Al was depleted. As the asteroid temperature would have increased with depth, an original parent body of at least 20 kilometers would have been required to generate some of Itokawa's silicate grains.[103] This slow-cooking process formed the crystalline silicate particles, like olivine, that are ubiquitous in asteroids and cometary dust. Tomoki Nakamura noted that, based upon the Hayabusa dust particle analysis, "...the Itokawa parent S-class asteroid was originally much larger, experienced intense thermal metamorphism, and was then catastrophically

99. A. Fujiwara et al., "The Rubble-Pile Asteroid Itokawa as Observed by Hayabusa," *Science* 312 (2006): 1330–1334.

100. M. Abe et al., "Near-Infrared Spectral Results of Asteroid Itokawa from the Hayabusa Spacecraft," *Science* 312 (2006): 1334–1338; T. Okada et al., "X-Ray Fluorescence Spectrometry of Asteroid Itokawa by Hayabusa," *Science* 312 (2006): 1338–1341.

101. T. Nakamura et al., "Itokawa Dust Particles: A Direct Link Between S-Type Asteroids and Ordinary Chondrites," *Science* 333 (2011): 1113–1116.

102. R. E. Grimm and H. Y. McSween, "Heliocentric Zoning of the Asteroid Belt by Aluminum-26 Heating," *Science* 259 (1993): 653–655.

103. Nakamura et al., "Itokawa Dust Particles."

disaggregated by one or more impacts into many small pieces, some of which re-accreted into the present greatly diminished, rubble-pile asteroid…."[104]

Scientists finally got a close-up look at the largest main-belt asteroid, Vesta, in mid-2011. The Dawn spacecraft effected a rendezvous in July, stayed for nearly 14 months, and then continued on for a multiyear visit with the dwarf planet Ceres, beginning in March 2015. This lengthy activity was enabled by a xenon ion drive propulsion system that had been tested during the earlier Deep Space 1 mission.

The Dawn observations of Vesta pointed toward a differentiated body (core-mantle-crust) with triaxial outer dimensions of 573 by 557 by 446 kilometers. They confirmed earlier Hubble Space Telescope observations of a giant impact basin in the southern hemisphere, named Rheasilvia, and also confirmed that a group of igneous-processed, achondritic meteorites (the so-called HEDs) originated from this impact event.[105] The current silicate surface crust of Vesta was formed as a result of past impact events, but its formation early in the solar system's history required that it have pervasive heating and separation into layers (differentiation), with the heating being provided by the decay of short-lived radionuclides like aluminum-26.[106] Gravity and shape models suggest that it has an iron-nickel core about 110 kilometers in radius, a larger mantle region, and an upper crust with an average depth of about 22 kilometers.[107]

---

104. Ibid.

105. C. T. Russell, H. Y. McSween, R. Jaumann, and C. A. Raymond, "The Dawn Mission to Vesta and Ceres," in *Asteroids IV,* ed. P. Michel, F. E. DeMeo, and W. F. Bottke, Jr., (Tucson, AZ: University of Arizona Press, 2015), pp. 419–432; P. C. Thomas et al., "Impact Excavation on Asteroid 4 Vesta: Hubble Space Telescope Results," *Science* 277 (1997): 1492–1495.

106. Grimm and McSween, "Heliocentric Zoning of the Asteroid Belt by Aluminum-26 Heating." ^{26}Al decay likely enabled hydrated silicates in some asteroids since, in the presence of liquid water, heating allows aqueous alteration to convert the normally dry silicates olivine and pyroxene to hydrated silicates. The current scientific consensus is that ^{26}Al heating drove water out of S-types in the inner main belt and enabled aqueous alteration and hydration for C-types in the mid-belt region, but was insufficient to do the same for the more distant D-type Trojan asteroids. The first mention of ^{26}Al heating in the early solar system was by Harold Urey in 1955. H. C. Urey, "The Cosmic Abundances of Potassium, Uranium and Thorium and the Heat Balances of the Earth, the Moon and Mars," *Proceedings of the National Academy of Sciences* 41 (1955): 127–144.

107. Russell et al., "The Dawn Mission to Vesta and Ceres."

While Dawn saw a rocky surface, dry and without activity, at Vesta, the same could not be said for Ceres.

Dawn's observations at Ceres revealed a surface composed mostly of layered ammoniated silicate minerals (phyllosilicates)—and it was as dark as fresh asphalt. There is evidence that significant subsurface water ice is responsible for flat crater floors containing pits, material flows, localized sublimation regions, and a 4-kilometer mountain that appears to be the result of a relatively recent cryovolcanic ice extrusion event. The bulk density of Ceres and gravity field suggest a silicate core with a rocky-ice mantle that has abundant water ice. Several small, square-kilometer-sized patches of water ice were spotted on the surface, along with a transient water-vapor atmosphere that had also been detected in 2012 by the Herschel Space Telescope.[108] Enigmatic bright spots noted in a 92-kilometer crater called Occator may have formed when an impact caused the briny subsurface water to erupt through the surface and sublimate, thus leaving bright carbonates and other salt assemblage deposits behind.[109]

In the late 20th century, the simple paradigm of comets as bright conglomerates of ices and interstellar dust grains gave way to the more complex paradigm of comets as very dark, fragile, icy dirtballs that formed from outer solar system ices and inner solar system dust. This apparent paradox pointed strongly to mixing in the early solar system. Some asteroids suggest early solar system mixing, which can explain the formation of primitive, crystalline dust and the melting that must have taken place in the inner solar system.[110] Some asteroids show evidence of surface or subsurface ices and outgassing, and some comets that once showed evidence of cometary outgassing no longer do so.[111]

---

108. C. T. Russell et al., "Dawn Arrives at Ceres: Exploration of a Small, Volatile-Rich World," *Science* 353 (2016): 1008–1010; J. Ph. Combe et al., "Detection of Local H_2O Exposed at the Surface of Ceres," *Science* 353 (6303), (2016): aaf3010-1/6; M. Küppers et al., "Localized Sources of Water Vapour on the Dwarf Planet (1) Ceres," *Nature* 505 (2014): 525–527.

109. M. C. DeSanctis et al., "Bright Carbonate Deposits as Evidence of Aqueous Alteration on (1) Ceres," *Nature* 536 (2016): 54–57.

110. Early solar system mixing would naturally follow from the large-scale migrations of the major planets involved with the development of the early solar system. A. Morbidelli, D. P. O'Brien, D. A. Minton, and W.F. Bottke, Jr., "The Dynamical Evolution of the Asteroid Belt," in *Asteroids IV*, ed. P. Michel, F. E. DeMeo, and W. F. Bottke, Jr., (Tucson, AZ: University of Arizona Press, 2015), pp. 493–507.

111. Jewitt et al., "The Active Asteroids," *Asteroids IV*, pp. 221–242.

Rather than separate categories for comets and asteroids, it now seems appropriate to think of a single group of small solar system bodies that run the gamut from fragile active cometary fluff balls to coherent icy dirtballs with subsurface ices, then from inactive and loosely bound rubble piles to coherent yet fractured rocky bodies, and finally to solid slabs of nickel-iron. These small bodies of the solar system have an extraordinary diversity of sizes, shapes, structures, densities, porosities, and compositions.

The only meaningful distinction that remains between comets and asteroids is whether or not they show activity. This activity is most often caused by the vaporization of volatile ices (mostly water ice), which controls the surface characteristics of comets, while impact events control the surface characteristics of asteroids. In the early 21st century, ground-based and space-based observations of comets and asteroids showed that they could no longer be considered as separate classifications of whirling rocks and dirty snowballs that formed in place within our solar system.

CHAPTER 8

IMPACTS AS NATURAL HAZARDS

Introduction

If the late-20th- and early-21st-century research on small solar system bodies undermined the old distinction between "asteroids" and "comets," the first decade of the 21st century witnessed changing perceptions of the risk from near-Earth objects within the scientific community. One event in particular—the discovery of 2004 MN4, a potentially hazardous 340-meter asteroid, generated public interest and led to rising demands for increased attention to what NASA's Lindley Johnson had labeled "planetary defense." In the 2005 NASA budget authorization, Congress again asked for a study of NEO survey capabilities and of space missions to track and deflect hazardous asteroids.[1] This "Near-Earth Object Survey and Deflection Analysis of Alternatives" report was delivered in 2007, though new funds did not actually arrive for the NEO program until 2010.[2] By then, the active NEO surveys had

---

1. National Aeronautics and Space Administration Authorization Act of 2005, PL 109-155, *https://www.congress.gov/bill/109th-congress/senate-bill/1281/text* (accessed 3 June 2019).

2. There are two versions of this document. The published version was "NASA, Near-Earth Object Survey and Deflection Analysis of Alternatives: Report to Congress," March 2007, *https://www.nasa.gov/pdf/171331main_NEO_report_march07.pdf* (accessed 14 September 2021). It was heavily redacted from the original committee draft, and the draft was later released as well: Office of Program Analysis and Evaluation, "2006 Near-Earth Object Survey and Deflection Study (DRAFT)," NASA, 28 December 2006, "2006-NearEarthObjectSurveyAndDeflectionStudy-NASA.pdf," NEO History Project collection. Both versions are also available at *https://cneos.jpl.nasa.gov/doc/neo_report2007.html* (accessed 14 September 2021).

nearly achieved the Spaceguard goal of finding 90 percent of the 1-kilometer and larger near-Earth asteroids and were pursuing a new goal, detection of 90 percent of near-Earth asteroids 140 or more meters in diameter, with two important new assets, neither primarily funded by the Near-Earth Objects Observations Program.

In October 2002, astronauts Edward T. Lu and Russell L. "Rusty" Schweickart joined with planetary scientists Piet Hut and Clark Chapman to form the B612 Foundation. Named for the asteroid home in Antoine de Saint-Exupéry's *The Little Prince*, the nonprofit organization evolved out of an informal planetary defense workshop at Johnson Space Center late the previous year.[3] Their discussion centered on the use of low-thrust propulsion as a means to divert near-Earth objects. The previous year had seen the NASA Near-Earth Objects Observations Program's NEO discovery effort pass the halfway mark toward completion of the Spaceguard Survey goal, but little progress on deflection had been made since the hazards workshop a decade before. The B612 Foundation's activism over the NEO issue helped raise further public awareness, but it also collided with some of NASA's plans.

Expansion of the NEO Survey Goal

Having passed, at least approximately, the halfway point in reaching the Spaceguard Survey goal in 2000, NASA's planetary astronomy program manager Thomas H. Morgan established a panel to examine questions related to what, if anything, should be done about the myriad smaller near-Earth objects that exist. While the impact hazard of the 1-kilometer and larger asteroid population was quickly coming into clear view, objects smaller than a kilometer in diameter were far more common and could reasonably be expected to be devastating at a regional scale. How should that hazard be addressed?

Morgan invited Grant Stokes, of the MIT Lincoln Laboratories and leader of the LINEAR NEO survey, to chair a Science Definition Team and Yeomans of JPL to act as vice-chair. Eight others were invited to join the nine-month study, including population model specialists, Principal Investigators from some of the NEO surveys, sensor specialists, and representatives of the Department of Defense. Morgan's charter for the group required cost

---

3. "B612 History," 22 March 2004, *https://web.archive.org/web/20040322172509/http://www. b612foundation.org/about/history.html*; retrieved from Archive.org on 28 August 2017.

estimates for the programs he expected them to propose, as well as a cost-benefit analysis, and the resulting report became the 2003 NEO Science Definition Team Report.[4]

An important factor in deciding to empanel a new study was advocacy for the construction of a new large telescope, known as the Large Synoptic Survey Telescope (LSST).[5] Originally named the Dark Matter Telescope, this was to be an 8-meter-aperture telescope with a very wide field of view, designed specifically to image the entire visible sky several times a month. Other subdisciplines beyond near-Earth object astronomy had discovered value in repeat imaging of the sky—"time domain astronomy" was one increasingly common label—including those interested in the universe's hidden mass of dark matter. But the LSST was to be very expensive, and its advocates had begun to try to recruit support from within the NEO community too. NASA's Associate Administrator for Space Science, astronomer Edward J. Weiler, had already been approached informally about supporting it and was not about to do so without analysis.[6]

At their first meeting in September 2002, the Science Definition Team (SDT) discussed a variety of preparatory issues, including how to address the subject of risk. Simon "Pete" Worden, then representing the Air Force's Office of Scientific Research, framed the debate as a question of "what to worry about more, the very rare huge mass extinction event or the rather frequent Tunguska type event."[7] The Spaceguard Survey goal was designed to characterize the risk of the extinction-scale event but contributed little to understanding the risk of smaller events. The 1-meter-class telescopes that were performing the current surveys were too small to adequately capture the population of 100-meter-scale objects. Yet, as Alan W. Harris pointed out, most smaller-scale events fall in uninhabited regions of Earth's surface,

---

4.   Near-Earth Object Science Definition Team, "Study To Determine the Feasibility of Extending the Search for Near-Earth Objects to Smaller Limiting Diameters," 22 August 2003, *https://www.nasa.gov/sites/default/files/atoms/files/pdco-neoreport030825.pdf* (accessed 28 April 2021).

5.   The LSST was later renamed the Rubin Observatory in honor of astronomer Vera C. Rubin.

6.   "SDT History Book," tab "2nd Meeting," 17 October 2002. Document courtesy of Lindley Johnson, copy in SDT June 2003 documents, "Smaller Limiting Diameters History Book.pdf," NEO History Project collection.

7.   Quoted from "SDT History Book," tab "1st Meeting," 5 September 2002.

killing no one.[8] But if a Tunguska-scale object were to explode over a major city instead of empty wilderness, it could kill millions of people and destroy tens or even hundreds of billions of dollars' worth of infrastructure and other kinds of property. Since the team's charter required a cost-benefit analysis, the group had to identify the key risk that they would address, figure out how to place a cost on that risk, and compare it to the cost of a search system designed to reduce the risk. As Yeomans put it during the discussion, "we need to predict damage cost of such an impact, to compare with the benefit of doing the search."[9]

This task went beyond the population models that Harris and others had been making. Risk assessment and cost-benefit analysis meant having to assess what damage would be done by which kinds of impacts—a problem that was not well understood. At the second meeting, discussion focused on how to figure this out. Pete Worden commented that nuclear weapons tests were not necessarily very informative since they had not been designed to address a key question: what size of asteroid should be the threshold for the new goal? Not having been carried out to "almost but not quite" cause damage to the surface, nuclear weapons tests did little to define a lower limit to destructive potential. There had also not been much study of the possibility that impacts at sea could cause tsunamis. William Bottke of the Southwest Research Institute was aware of a 1968 report that indicated that nuclear weapons were not particularly effective at initiating tsunamis, but that was it. Harris commented at this meeting that this kind of natural-hazard analysis for cosmic impacts was mostly a hobby, as no one funded it.[10]

The Science Definition Team considered the tsunami issue to be the least well understood and decided to hold a special workshop to help the team develop a better understanding of it. Asteroid impacts were unlike the great earthquakes that generally produce tsunamis because they would generate shorter-wavelength waves, but these would still have longer wavelengths than the other very common kind of waves, those driven by winds. So there were no direct, and valid, analogies available. Two very divergent opinions on the matter emerged. One view was that the relatively short wavelengths of impact-generated waves would cause them to break far out on continental shelves and

---

8. Harris retired from JPL in 2002 but continued working via the Planetary Science Institute.
9. "SDT History Book," tab "1st Meeting," 5 September 2002.
10. "SDT History Book," tab "2nd Meeting," 17 October 2002.

dissipate most of their energy offshore, so there would be little inundation of the coasts and thus little damage. Another analysis had found the opposite—giant waves that, as JPL's Steven Chesley remembered, would roll up the Appalachian mountains. (This scenario tended to show up in asteroid-impact movies.) A third analysis had found a fairly significant effect for impactors of 200–300 meters in size, based on analysis of the energy remaining in the water after dissipation of some of the energy in breaking on continental shelves.[11] "The water has to go somewhere," Chesley commented later.[12] In the opinion of the group, however, that analysis had overestimated the frequency of impacts by quite a bit, so Chesley worked with the study's lead author, Steven Ward of the University of California, Santa Cruz, to publish a new analysis using more representative asteroid population numbers.[13]

One ground rule the team adopted was to consider only the subset of asteroids that were potentially hazardous, meaning that their orbits had a small chance of intersecting Earth's in the next few hundred years. These were known as PHAs, "Potentially Hazardous Asteroids." Only about 20 percent of the known NEOs met this definition.[14] Stokes put it bluntly in one presentation: "PHAs represent the collision danger—the rest are chaff," he said.[15] They also concluded early on that comets represented much less hazard than previously thought. Yeomans's analysis showed that long-period comets represented only about a percent of the overall impact risk since, in near-Earth space, asteroids outnumber comets by more than 100 to 1.

When the SDT members had resolved all of their questions about risk, they found that the hazard due to PHAs effectively broke into four size domains. Asteroids of less than 50 meters were no hazard, as in most cases they would explode too high in the atmosphere to produce damage at the

---

11. Steven N. Ward and Erik Asphaug, "Asteroid Impact Tsunami: A Probabilistic Hazard Assessment," *Icarus* 145, no. 1 (1 May 2000): 64–78, doi:10.1006/icar.1999.6336.

12. Stephen Chesley, interview by Conway, September 2017.

13. Steven R. Chesley and Steven N. Ward, "A Quantitative Assessment of the Human and Economic Hazard from Impact-Generated Tsunami," *Natural Hazards* 38, no. 3 (July 2006): 355–374, doi:10.1007/s11069-005-1921y.

14. Technically, they defined "potentially hazardous" as having a Minimum Orbital Intersection Distance (MOID) less than 0.05 astronomical units from Earth's orbit. Often, the definition of a PHA includes the restriction that it be absolute magnitude 22 or brighter—or about 140 meters or larger in diameter.

15. NEO SDT, "Study To Determine the Feasibility of Extending the Search for NEOs to Smaller Limiting Diameters," p. 12.

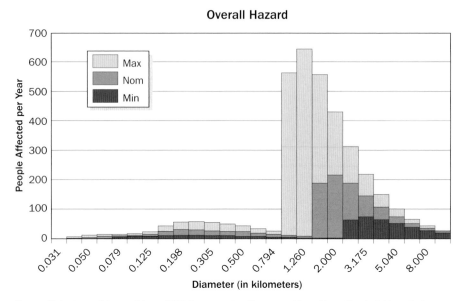

Figure 8-1. Overall hazard from NEO impacts by diameter. From Near-Earth Object Science Definition Team, "Study To Determine the Feasibility of Extending the Search for Near-Earth Objects to Smaller Limiting Diameters," 22 August 2003, p. 34.

surface. Between 40 meters and 150 meters, the principal hazard came from airbursts over inhabited areas. Between 200 meters and 1 kilometer in diameter, the tsunami risk was dominant, because most of Earth's surface is water. Asteroids below 200 meters would probably not create tsunamis because the energy from their explosions in midair would not transmit much energy into the oceans, while the 1-kilometer and larger asteroids would make the tsunami largely irrelevant given all the other damage. The largest asteroids, as Figure 8-1 shows, also produced the largest hazard, even though they were extraordinarily rare events, occurring on scales of millions of years.

The group had chosen the metrics of deaths and damage per year in order to quantitatively compare the asteroid risk to other natural hazards. They also needed metrics like this for the cost-benefit analysis they were required to prepare by their charter, so while it may seem odd to present annualized potential deaths from a hazard that has apparently never killed anyone, this kind of analysis was necessary.[16]

---

16. A new analysis of the surviving Tunguska evidence concludes that at least three people were killed by the blast. See Peter Jenniskens et al., "Tunguska Eyewitness Accounts,

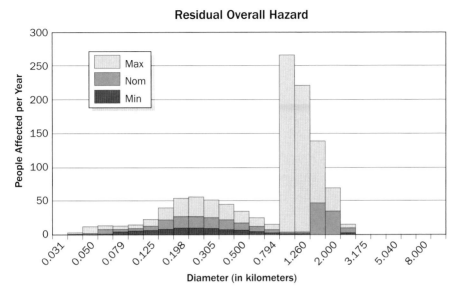

Figure 8-2. Remaining hazard after completion of the "Spaceguard Survey" goal, expected in 2008. Note that the vertical scale of this chart is one-half that of figure 8-1, reflecting an overall reduction in remaining hazard. From NEO SDT, "Study To Determine the Feasibility of Extending the Search for Near-Earth Objects to Smaller Limiting Diameters," p. 35.

The team also quantified what they called the "residual hazard," the hazard that would remain after 2008, the deadline for meeting the Spaceguard Survey requirement. Finding 90 percent of the 1-kilometer or larger asteroids would reduce the overall unknown hazard substantially, as a comparison of Figures 8-2 and 8-1 suggests, since the overall hazard depended heavily on the largest, least-frequent impacts. But it did little to reduce the uncertainty in risk from smaller impactors that were rarely detected by the existing NEO surveys, so that risk grew in comparison. They calculated that the risk remaining after achievement of the Spaceguard Survey goal was about 293 casualties per year (down from 1,250 casualties per year prior to the Spaceguard Survey). Converted to dollars, the cost of casualties was $1.6 million per person and $98,000 in infrastructure per person; for the (much wealthier)

Injuries, and Casualties," *Icarus*, "Tunguska" 327 (15 July 2019): 4–18, *https://doi. org/10.1016/j.icarus.2019.01.001* (accessed 28 October 2019).

United States, it amounted to $6.96 million per person and $734,000 per person, respectively.[17]

Stokes's Science Definition Team also had to conceptualize a set of new NEO search systems whose projected costs could be compared to the "benefit" of reducing these potential damage numbers. Stokes's own LINEAR survey, which at the time was the most successful at finding NEOs by far, served as one point of comparison for this effort. LINEAR used five images per night of each patch of sky (compared to Spacewatch's three) to discern asteroids and did much of its own discovery follow-up. Both these things effectively reduced the amount of sky it could cover each night. During one meeting, Stokes was adamant about the need for whatever systems they conceived of to do most of its own follow-up observing despite the resulting reduction in coverage. A survey system that was efficient at discovering the several hundred thousand smaller NEOs they expected to find would overwhelm any third-party attempts to "keep up" with the new objects posted on the Minor Planet Center's confirmation page every day, so the SDT had to accept a great deal of self–follow-up in any survey, even if it seemed to slow progress. Panel member Spahr, then at the MPC, was similarly adamant at their fifth meeting about the need to have at least three detection tracks in three weeks of a new object in order to develop an adequate orbit for it, with two of those three in the first two nights.[18]

There was also some pressure to develop space-based options for the study. NASA was unlikely to build major new ground facilities, and this was, after all, a study for NASA. But at that moment, NEO discovery was entirely the domain of ground facilities, and that was likely to continue for at least several more years. So the SDT pursued analyses of ground- and space-based options, as well as a couple of possible hybrids.

For their space-based discussions, the team initially started with both infrared and optical systems but quickly dropped discussion of infrared systems. Their mandate was to evolve a system that could be operational in the 2008/2009 timeframe, and they did not believe that the infrared focal plane technology that existed in 2002 would be adequate. Suitable thermal infrared

---

17. NEO SDT, "Study to Determine the Feasibility of Extending the Search for Near-Earth Objects to Smaller Limiting Diameters," p. 110.

18. "SDT History Book," tab "5th Meeting," 11–12 March 2003, NEO History Project collection.

detectors either had to be cooled cryogenically, limiting the lifespan of the telescope to that of the cryogen, or required active cooling devices that were in development but not yet space-qualified. The technology for optical space-borne telescopes was more readily available.[19]

That decision then led the SDT into a discussion of orbits. One possibility was a Venus-trailing orbit that would look back toward Earth. Looking outward toward Earth meant that it would not have to deal with the Sun blocking its view of portions of the Earth-approaching asteroid population each day, so it would be ideal for the purpose of cataloging all the PHAs they were concerned with. But that orbit's distance from Earth would reduce the amount of imagery that the telescope could return, and the spacecraft would also be out of communication with Earth for weeks at a time, when it and Earth were on opposite sides of the Sun. It would be the worst option in terms of providing early warning of a near-term impact. It would also be the most expensive of the space-based options.[20]

The other options were low-Earth orbit and two of the gravitationally quasi-stable "Lagrange" points: L1, which lies between Earth and the Sun, and L2, which is on the far side of Earth on the Sun-Earth line. The SDT concluded that the L2 position would be superior for the general NEO discovery effort because the telescope could observe NEOs at their brightest near opposition and its only obstructed view was toward the Sun

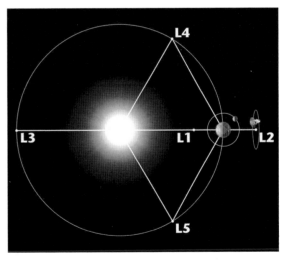

Figure 8-3. Earth-Sun Lagrange points. Numerous solar science missions operate at L1, while NASA's James Webb Space Telescope orbits L2. "Trojan" asteroids, like 2010 TK7, occupy L4 and L5. (Image credit: NASA/WMAP Science Team)

19. "SDT History Book," tab "2nd Meeting," 17 October 2002; NEO SDT, "Study To Determine the Feasibility of Extending the Search for Near-Earth Objects to Smaller Limiting Diameters," p. 43.

20. NEO SDT, "Study To Determine the Feasibility of Extending the Search for Near-Earth Objects to Smaller Limiting Diameters," pp. 47–48.

(from L2, the Sun and Earth always appear close together, so the telescope would have only one "keep out zone"). So they decided to carry into their cost-benefit analysis only the low-Earth orbit (LEO), L2, and Venus-trailing mission concepts, though the SDT also analyzed several variants of these, with different telescope apertures.[21]

For dedicated ground-based NEO survey telescopes, the SDT explored a variety of different apertures from 1 meter to 8 meters, different CCD pixel sizes, different locations, and different numbers of telescopes, via a Lincoln Laboratory simulation tool called the Fast Resident Object Surveillance Simulation Tool (FROSST). Much of that modeling work was done by Jenifer Evans of the MIT Lincoln Laboratory, who was also a member of the SDT. There was quite a bit of discussion of search strategies and of the issue of location—the Tucson, Arizona, area or Maunakea, Hawai'i—which turned out strongly in favor of Maunakea. The "seeing," or stability of the atmosphere that a telescope would look through, and weather on Maunakea were so much better that using Hawai'i represented a halving of telescope aperture— a 2-meter telescope on Maunakea could perform as well as a 4-meter telescope in Arizona. This situation affected the cost profoundly; Hawai'i was a more expensive location in which to build observatories, but the smaller telescope would still be less expensive overall.[22]

The Science Definition Team's risk analysis led them to the conclusion that the new goal for NEO hazard reduction should be to "construct a search system that is capable of retiring 90% *of the risk* posed by collisions with asteroids whose diameters are less than one kilometer" (emphasis added).[23] Integrating the unknown hazard presented by the various object sizes gave them the result: 90 percent of the remaining hazard would be eliminated by cataloging 90 percent of the objects more than 140 meters in diameter.

Once they had determined their goal, the SDT evaluated the performance of the myriad of potential survey systems. Here, the time allotted for the survey was a powerful factor that favored the space-based options. Only the Venus-trailing space telescopes could actually meet the 10-year goal, and they were deemed the most expensive options. While none of the ground systems could meet a 10-year deadline, several could possibly succeed given

---

21. Ibid., pp. 47–48.
22. Ibid., p. 80.
23. Ibid., quoted from p. 112.

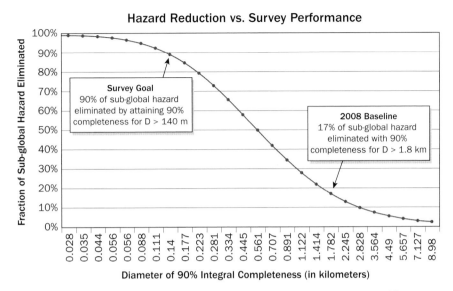

Figure 8-4. Hazard reduction from sub-global (less than 1 kilometer in diameter) impactors as a function of survey completeness. On the right, the 2008 Baseline case reflects completion of the "Spaceguard Survey" goal. The new goal is specified on the left. (NEO SDT, "Study To Determine the Feasibility of Extending the Search for Near-Earth Objects to Smaller Limiting Diameters," p. 113)

20 years. Costs ranged from $236 million to $397 million (in 2003 dollars), for the least expensive ground system (a pair of 4-meter telescopes, which might take 20 years to complete the survey) and a 2-meter Venus-trailing telescope, which might do it in 8 years.[24] All of their candidate systems would pay for themselves by reducing the unknown hazard within the first two years of their survey operation, primarily by retiring the remaining risk of a large asteroid impact.

Grant Stokes briefed the team's findings to the new NASA Near-Earth Objects Observations Program executive, Lindley N. Johnson, who had succeeded Tom Morgan as program executive in June 2003. Johnson was then a recently retired U.S. Air Force lieutenant colonel who had been involved with Air Force space surveillance for most of his career. He had also, with Worden, been an advocate of making the near-Earth object hazard one of the U.S. Air Force's missions. He had prepared a paper on asteroid threat mitigation

---

24. Ibid., pp. 115–117.

Figure 8-5. Lindley N. Johnson, left, and Eleanor Helin, right, in June 1996. Helin named an asteroid she had discovered (1989 CJ1) "5905 Johnson" to honor his support for NEO astronomy while a serving U.S. Air Force officer. (Image courtesy of Lindley Johnson)

for the Air University's "Spacecast 2020" meeting in 1994 (see chapter 5).[25] While he and Worden had not been very successful in convincing Air Force senior officers to adopt what Johnson had named "planetary defense" as a core Air Force mission, they had been able to get the various collaborative efforts already discussed going and funded. Johnson already knew Stokes from the Ground-based Electro-Optical Deep-Space Surveillance system upgrade efforts that the Air Force had funded in the 1990s for its space surveillance mission. While Johnson was new to NASA, he was not new to the subject or its community.

The 2004 MN4 Story

A year after the SDT study was completed, another NEO discovery generated a lot of press attention and challenged the NEO community's preparations for dealing with potential impacts. Originally designated as 2004 MN4 and now

---

25. Lindley N. Johnson, "Preparing for Planetary Defense: Detection and Interception of Asteroids on Collision Course with Earth," Air University, 9–10 November 1994, *http://commons.erau.edu/space-congress-proceedings/proceedings-1995-32nd/april-25-1995/18/* (accessed 3 June 2019), copy in NEO History Project collection.

known as 99942 Apophis, it was first definitively spotted on 19 June 2004 by Roy Tucker of the University of Arizona and David Tholen and Fabricio Bernardi of the University of Hawai'i. As Tholen tells the story, he had been advocating looking for near-Earth asteroids near the Sun during the 2 hours after sunset and before sunrise because the surveys being carried out at the time avoided this region. The telescopes in use at the time would not find much there, partly because the asteroids would not be fully illuminated by the Sun (they would appear as small crescents, if our eyes could resolve them at all), and partly because whatever telescope was being used would have to be looking near the horizon, where there is more air between the telescope and space, and therefore more scattering of the available light. The seeing, in other words, would not be as good as if the same telescope were looking at the point in the sky directly opposite the Sun (which is called opposition). Because the surveys had all been focused on opposition and the simulated asteroid population models were built and checked against the survey data, the population models were biased toward asteroids with most of their orbit further from the Sun than Earth. But Tholen could not think of any reason why there would not be asteroids sunward of Earth—they would just be harder to find. "That doesn't appear physical to me," he said in 2017. "You'd have to assume that the Earth is sweeping up *everything* in order to prevent anything from propagating inward, and that just seems so unphysical to me."[26]

In 2002, NASA's Thomas Morgan had funded a three-year proposal of Tholen's to do some observing with a new 8k CCD camera that was to be built for the University of Hawai'i 2.2-meter telescope on Maunakea. But there were delays in finishing the camera, so eventually Tholen got time on the 8.2-meter Subaru telescope. He and Bernardi found a fast-moving object on the second of the two nights they had on that telescope, 17 June (universal time). Since they could not follow up on it right away from the Subaru telescope but had scheduled time on Kitt Peak's 90-inch telescope, they contacted Roy Tucker at the University of Arizona and asked him to look for it. They then flew to Arizona too. Tucker found an object within the observing field that was moving at the right speed, according to specifications that Tholen had e-mailed him, but not quite in the right spot. After a second night of observations, they sent the data to the Minor Planet Center on the 20th.

26. David Tholen, interview by Conway, 20 June 2017, transcript in NASA History Division HRC.

MPC's calculations from their first handful of observations suggested that it was more likely to be a Mars-crossing asteroid than a near-Earth object. They did not have time on the telescope for the next two nights, and then Kitt Peak had two days of rain, so they ran out of time to gather more observations before the object effectively went out of sight behind the Sun.[27]

The Siding Spring Observatory in Australia, still part of the Catalina Sky Survey at the time, picked it up again on 18 December 2004. After it went up on the MPC's NEO confirmation page, two other observatories reported it on the 19th, and another on the 20th. The MPC also linked these new observations back to the June observations at Kitt Peak, so the object had a six-month observational arc and got its initial designation as 2004 MN4.

The object 2004 MN4 triggered a great deal of effort over the Christmas holidays because the University of Pisa's NEODyS system and JPL's Sentry system both gave it an impact probability of 1 in 233 during a close approach on 13 April 2029, earning it a Torino scale rating of 2.[28] Right away, Chesley at JPL worried that the observations Tucker and Tholen had made at Kitt Peak might be problematic, and he had Tholen recheck them. They proved to have clock errors, which Tholen finally resolved on 23 December. Other observers scrambled to make new observations and search for so-called "precovery" observations, observations that had been previously made but in which 2004 MN4 had not been recognized, in order to extend the observational arc. On the 24th, Robert McNaught at the Siding Spring Observatory provided a number of new observations. These briefly made the impact probability even worse. Lindley Johnson had a Christmas Eve teleconference with Yeomans about whether they should make any kind of announcement, and they decided against it. The potential impact was decades in the future, and radar observations could be made in a month, so they decided to wait for those observations

---

27. Tholen interview, 20 June 2017; Steven R. Chesley, "Potential Impact Detection for Near-Earth Asteroids: The Case of 99942 Apophis (2004 MN4)," *Proceedings of the International Astronomical Union* 1, no. S229 (August 2005): 215–228, doi:10.1017/S1743921305006769; María Eugenia Sansaturio and Oscar Arratia, "Apophis: The Story Behind the Scenes," *Earth, Moon, and Planets* 102, no. 1–4 (June 2008): 425–434, doi:10.1007/s11038-007-9165-3.

28. Donald K. Yeomans et al., "Near-Earth Asteroid 2004 MN4 Reaches Highest Score to Date on Hazard Scale," 23 December 2004, *https://cneos.jpl.nasa.gov/news/news146.html* (accessed 11 September 2017). The Torino scale is explained in chapter 6.

to be taken and processed.[29] But on 27 December, the University of Arizona's Spacewatch group located it on images taken on 15 March 2004.[30] When those observations were taken into account, the probability of impact for 2029 disappeared. Subsequent returns, however, in 2036 and 2052, still held non-zero impact probabilities.

The asteroid's trajectory brought it close enough to Earth in January 2005 for measurements to be attempted using the Arecibo radar, and a group of JPL radar astronomers flew to Puerto Rico in late January. They obtained detections on three nights, and these confirmed that the 13 April 2029 close approach would miss Earth, though only barely.[31] The predicted orbit had it passing at a distance of 6 Earth radii—about the altitude at which geosynchronous communications satellites operate.

The initial flurry of activity around 2004 MN4 initially went largely unnoticed by the American media due to the enormous Sumatra-Andaman earthquake and subsequent tsunami that occurred on 26 December.[32] Other news headlines notwithstanding, the asteroid did not stay out of the news for very long, partly because there were still very close flybys of Earth projected. The 2029 flyby would alter the asteroid's trajectory, converting it from an "Aten"-class asteroid into an "Apollo"-class asteroid. The shift in its orbit would also make the orbit more uncertain, so impacts during close approaches in 2036 and beyond were harder to rule out.

As the discoverers, Tucker and Tholen had the opportunity to suggest a name for the asteroid, which received the permanent MPC catalog number 99942. There was a tradition that Aten-class asteroids be named for Egyptian gods and Apollo-class asteroids be named for Greek gods; they chose the name Apep for the asteroid, an Egyptian god of the underworld who was known to the Greeks as Apophis. The Minor Planet Center awarded 2004

---

29. Johnson comments on manuscript draft, copy in NEO History Project collection.

30. Chesley interview; Sansaturio and Arratia, "Apophis: The Story Behind the Scenes," pp. 425–434.

31. Jon D. Giorgini et al., "Predicting the Earth Encounters of (99942) Apophis," *Icarus* 193 (2008): 1–19.

32. California Institute of Technology, "What Happened During the 2004 Sumatra Earthquake," *http://www.tectonics.caltech.edu/outreach/highlights/sumatra/what.html* (accessed 29 April 2021); Thorne Lay et al., "The Great Sumatra-Andaman Earthquake of 26 December 2004," *Science* 308, no. 5725 (20 May 2005): 1127–1133, doi:10.1126/science.1112250.

MN4 the name 99942 Apophis in July 2005.[33] Apophis also happened to be the name of a principal villain in a television show called *Stargate SG-1*; indeed, in one 2002 episode, the godlike villain Apophis launches an asteroid at Earth to destroy it.[34] The radar observations from 2005 put 99942 Apophis in the 300-meter size class, making it not quite a planet-killer, but still enormously destructive should its orbit change enough during the 2029 encounter to convert it into an impactor.

Former astronaut Russell "Rusty" Schweickart, who was chair of the B612 Foundation and also active in the Association of Space Explorers, led the way in calling for an effort to better understand the asteroid's orbit. In May 2005, Schweickart presented an analysis at a National Space Society meeting in Washington, DC, that called for placing a radio transponder on Apophis to improve knowledge of its orbit. There would be future opportunities to obtain more radar data on the asteroid, but he did not believe that they would improve the orbit knowledge enough to rule out the 2036 impact possibility. He also contended that, because of the energies involved in shifting the orbit after the 2029 encounter, any deflection attempt would have to be started long before the 2029 encounter—so a decision to actually make a deflection attempt would have to be made around 2014.[35] Schweickart wanted the tracking mission started right away so that it could provide years of data in advance of the 2014 decision that might be necessary.

The crux of the matter was this: in order to receive just the right gravitational kick to come back around and strike Earth in 2036, Apophis had to pass through a very small "window" of space during its 2029 encounter—only a few hundred yards wide. Relatively speaking, anyway, it would not take very much energy to move the asteroid away from that window (which became known as a "keyhole" in the asteroid community, explained in chapter 6). But trying to move the asteroid away later would take far more energy.

Schweickart also wrote a letter to the NASA Administrator, Michael Griffin, advocating this approach, and Lindley Johnson sent it to Chesley

---

33. Tholen interview; Sansaturio and Arratia, "Apophis: The Story Behind the Scenes," pp. 425–434.

34. "Fail Safe," SGCommand, *http://stargate.wikia.com/wiki/Fail_Safe* (accessed 29 April 2021).

35. Russell L. Schweickart, "A Call to (Considered) Action," presented at the National Space Society International Space Development Conference, 20 May 2005, Washington, DC.

at JPL for analysis. The issue Chesley had to answer was really one of time: what decisions would have to be made, and when? Apophis's orbit around the Sun was slightly larger than Earth's (with a somewhat longer period), but not greatly so. In the years before the 2029 encounter, it would cross Earth's orbit several more times. During some of those orbits, it would be invisible from Earth; during others, it would be visible to telescopes but not to radar; in still others, it would be close enough to make radar measurements. Chesley predicted that the Arecibo radar would be able to observe Apophis in May 2006, February and July 2013, and October 2020; even without telescopic observations (for which there were more opportunities) or the transponder mission, these would reduce the uncertainty of the orbit solutions for 2029 considerably.[36]

Chesley also evaluated what the best timing would be for a transponder mission, if one were to be necessary. In an important sense, knowledge of a planetary body's orbit decays if the body is not observed frequently. Partly, that is simply due to unavoidable errors in the observations (no instrument is perfect), but it is also because there are small forces that cannot be measured easily that affect the motions of the bodies. In the case of asteroids, one important small force is the Yarkovsky effect, explained in chapter 6. From one orbit to the next, this effect makes little difference, but across many orbits, it could have a large effect. Therefore, if the Apophis tracking mission were initiated too early—perhaps launching in 2008 and ending a year or two later—it would not actually provide the uncertainty reduction Schweickart expected.

Chesley modeled how the uncertainty in knowledge of 99942 Apophis's orbit would evolve, both with and without the expected future radar data, in order to figure out what point in time would be optimal for launching a one-year tracking mission. He found that the tracking mission should reach the asteroid in 2019. Since it would take four years to develop and launch such a mission, a decision to do it did not have to be made until after the 2013 Arecibo radar observations.[37] That made the 2013 radar observations rather

36. Chesley interview; Steven R. Chesley, "Potential Impact Detection for Near-Earth Asteroids: The Case of 99942 Apophis (2004 MN4)," *Proceedings of the International Astronomical Union* 1, no. S229 (August 2005): 215–228, doi:10.1017/S1743921305006769.

37. Steven R. Chesley, "Potential Impact Detection for Near-Earth Asteroids: The Case of 99942 Apophis (2004 MN4)," *Proceedings of the International Astronomical Union* 1, no. S229 (August 2005): 215–228, doi:10.1017/S1743921305006769.

important, of course, but it meant that nothing extraordinary had to be done before then.

That answer was received with some satisfaction back at NASA Headquarters, which at the time was immersed in planning for a new space exploration policy known as the Vision for Space Exploration (VSE). While it had received presidential endorsement in 2004, the VSE faced an uphill funding battle in Congress and was never fully funded.[38] A short-term, high-public-profile, and potentially expensive asteroid-tracking mission to Apophis that would divert attention from the Agency leadership's preferred goal was not a desirable outcome.

But the possibility that 99942 Apophis might become a bigger problem was not lost on Congress. The NEO program's chief patron in Congress had been George E. Brown, a California Congressman who had written the Spaceguard Survey goal into NASA's authorization language in 1995 and passed away in 1999. It gained a new patron in Representative Dana Rohrabacher, who in mid-2004 introduced a George E. Brown Near-Earth Object Survey Act as a kind of memorial that would require NASA to establish a program to find all the 100-meter or larger near-Earth asteroids by 2020. The bill died in committee in 2004, but he resubmitted it in March 2005, citing 99942 Apophis specifically in his prepared statement, and this time it passed.[39]

As incorporated into the 2005 NASA Authorization Act later that year, the Brown Act required NASA to provide Congress with an analysis of how best to carry out this next-generation near-Earth object survey, including space- and ground-based alternatives, and an analysis of methods to deflect potentially hazardous discoveries. At NASA's request, the target asteroid size had been changed from 100 meters to 140 meters, to better conform to the Science Definition Team findings. They would have one year to provide the

---

38. The VSE's challenges are discussed in Government Accountability Office (GAO), "Constellation Program Cost and Schedule Will Remain Uncertain Until a Sound Business Case Is Established," GAO-09-844, August 2009. Also see Glen R. Asner and Stephen J. Garber, *Origins of 21st-Century Space Travel: A History of NASA's Decadal Planning Team and the Vision for Space Exploration, 1999–2004* (Washington, DC: NASA SP-2019-4415, 2019).

39. George E. Brown, Jr. Near-Earth Object Survey Act, H. Rept. 109-158, 109th Cong., 1st sess., 27 June 2005, *https://www.congress.gov/congressional-report/109th-congress/house-report/158/1* (accessed 11 August 2017).

study, and each year for five years, the NASA Administrator had to provide a progress report to Congress.[40]

The 2005 NASA Authorization Act also made "detecting, tracking, cataloguing, and characterizing near-Earth asteroids and comets" a statutory responsibility by amending the Agency's founding law, the National Aeronautics and Space Act of 1958.[41] The amendment specified that the purpose of doing this was to provide "warning and mitigation of the potential hazard of such near-Earth objects to the Earth." Cosmic hazard had finally entered the terrain of federal law.

The congressionally mandated study due in 2006 was run by NASA's Office of Program Analysis and Evaluation (OPAE) and drew on many of the same people the earlier study had.[42] But it focused more on the employment of already-planned wide-field survey observatories, under the idea that telescopes already in the design phase were likely to be completed earlier than entirely new designs. One new facility considered in the OPAE study was the Large Synoptic Survey Telescope (LSST) that had helped motivate the 2003 Science Definition Team study; another, the Panoramic Survey Telescope and Rapid Response System (Pan-STARRS), was under development by the University of Hawai'i's Institute for Astronomy. These were quite different designs, though they shared a common purpose: each intended to do wide-field, repeat imaging of the entire accessible sky. Cosmologist Nick Kaiser of the Institute for Astronomy, who led the initial design of Pan-STARRS, explained later that "what we were arguing was, there was a better way to do wide-field imaging, which wasn't to build a single big telescope, but to build an array of small telescopes."[43] While LSST was to be an 8-meter-aperture, single-mirror telescope designed for a 10-degree field of view and a gigapixel-scale camera, Pan-STARRS was to be a network of 1.8-meter telescopes with large cameras. Neither was intended primarily for asteroid hunting, though.

---

40. NASA Authorization Act, Pub. L. No. PL 109-155 (2005), *https://www.gpo.gov/fdsys/pkg/PLAW-109publ155/pdf/PLAW-109publ155.pdf* (accessed 30 April 2021).

41. Ibid.

42. NASA Office of Program Analysis and Evaluation, "2006 Near-Earth Object Survey and Deflection Study (DRAFT)," 28 December 2006, *https://cneos.jpl.nasa.gov/doc/neo_report2007.html* (accessed 5 June 2019). This website contains links to both versions of this study.

43. Nick Kaiser, interview by Conway, 20 June 2017, transcript in NASA History Division HRC.

Both LSST and Pan-STARRS were directed primarily at cosmological questions, and in particular the hunt for dark matter. But their advocates were always quick to point out that their data would contribute to many other areas in astronomy, including asteroid surveys.

In Congress, Pan-STARRS had an ally in Senator Daniel Inouye of Hawai'i, and this gave it an edge over LSST. Inouye, first elected to the Senate in 1963, was able to fund the first Pan-STARRS telescope's construction via an earmark in the Air Force budget. Pan-STARRS 1 was being built on an existing site on Haleakalā, Maui, as a functioning prototype for the operational array of four telescopes, which were to be built on Maunakea. As the congressionally mandated study was being set up, Pan-STARRS 1's team expected to have the telescope in operation in 2007. But at this stage of its planning, Pan-STARRS 1's team intended most of the telescope's time to be allocated to imaging the entire visible sky 10 to 20 times in five color filters (known as the 3π Steradian Survey).[44] Only about 10 percent of the telescope's time was to be dedicated specifically to NEOs.

The OPAE study, which Johnson named the Analysis of Alternatives study (or AOA), had two main thrusts. The first was its review of survey alternatives, which came to somewhat different conclusions than had Grant Stokes's 2003 Science Definition Team. If only ground systems were considered, only one option for reaching the 2020 goal existed: building a copy of LSST dedicated only to the NEO hunt, and sharing the planned Pan-STARRS 4 (as the full quad-telescope system was known) and LSST itself. This had a life-cycle cost of $820 million. If the deadline were extended to 2024, two other ground-based options would exist: in one, an eight-telescope version of Pan-STARRS plus sharing of Pan-STARRS 4 and LSST could do it for a cost of $560 million. In the other, a dedicated "LSST-like" observatory could do it, but for $870 million.[45]

In considering space-based survey options, the AOA study emphasized infrared telescopes over optical, reversing the earlier SDT decision. This decision was made partly because of newly available technology, but also because they interpreted the congressional language as requiring characterization of

44. K. C. Chambers, E. A. Magnier, N. Metcalfe, H. A. Flewelling, M. E. Huber, C. Z. Waters, L. Denneau, et al., "The Pan-STARRS1 Surveys," ArXiv E-Prints, December 2016, arXiv:1612.05560 (accessed 5 June 2019).

45. NASA OPAE, "2006 Near-Earth Object Survey and Deflection Study (DRAFT)," summarized from p. 12.

NEOs, not just discovery, and the infrared telescopes enabled much better size, and hence mass, estimates. Since a factor of 2 error in size meant up to a factor of 8 error in mass, getting more accurate mass estimates earlier would make a big difference in planning an asteroid deflection effort, should one be deemed necessary. So, while the space-based infrared telescopes would all cost more than the ground-based options and had higher technical and cost risks, they were rated highly. Here the telescopes in Venus-like orbits dominated the analysis because they would eliminate the observational bias produced by Earth-based telescopes' inability to survey efficiently near the Sun.[46]

The second thrust of the AOA study concerned the deflection of asteroid threats. Clearly motivated by 99942 Apophis, this section emphasized the ability to respond to relatively near-term threats. The team analyzed various scenarios involving the use of nuclear and conventional explosives, kinetic impactors, and what they called "slow push" techniques—primarily a gravity tractor and a "space tug." Other techniques that did not necessarily involve sending spacecraft to the asteroid in question were mentioned, such as focusing lasers or mirrors on it to impart an impulse, but they were not analyzed. The key issue for deflecting an asteroid lies in the amount of velocity change that must be imposed to change an asteroid's orbit enough to miss Earth, and for a given launch mass, nuclear devices could produce the most change. They also carried the most risk of fragmenting the body, which would complicate the mitigation problem enormously. Nevertheless, nuclear options dominated the analysis, as they had during the 1992 deflection study involving the national laboratories.

That emphasis on nuclear options made the resulting document controversial, and NASA leadership did not help matters when they initially refused to release the full study. It was deemed too detailed for a congressional response, and instead of the full 275-page study, Congress was sent a 28-page summary that included what amounted to a dismissal of the entire enterprise: "NASA recommends that the program continue as currently planned, and we will also take advantage of opportunities using potential-dual-use telescopes and spacecraft—and partner with other agencies as feasible—to attempt to

---

46. NASA OPAE, "2006 Near-Earth Object Survey and Deflection Study (DRAFT)," summarized from pp. 12–13.

achieve the legislated goal within 15 years. However, due to current budget constraints, NASA cannot initiate a new program at this time."[47]

NASA officials quickly changed their minds about not releasing the full report, perhaps because it was already circulating in an electronic draft anyway. But rather than finishing the report, they provided the unfinished draft with several pages of errata appended to it. These covered some, though not all, of the draft's problems. The tactic did not save them from loud public criticism.

Rusty Schweickart and Clark Chapman, via the B612 Foundation, both wrote detailed technical criticisms, released them to the public, and sent them to the NASA Administrator. Schweickart's critique focused on the choice of example deflection cases. The study team had chosen Apophis and another few-hundred-meter-class asteroid, (144898) 2004 VD17, which had a potential impact in 2102. These were unrepresentative of the actual threat, he contended, which would largely come from smaller asteroids that did not require the high impulse of nuclear devices. He also thought that there were many more "keyholes" that a high-impulse blast might shove the asteroid into, avoiding an imminent impact threat only to increase the odds of a later one.[48] In essence, he was arguing that nuclear devices were too crude a tool to use for deflection. More finesse was in order, and the slow-push techniques were more appropriate.

Clark Chapman's critique focused on the nuclear issue too, though from a different perspective. The report's authors had, in his opinion, chosen the nuclear option first and then worked from there toward a conclusion that this option required the least knowledge about the specific asteroid to be deflected. That view he considered absurd; one would still need to know mass, whether the asteroid was solid or a "rubble pile," whether it was rapidly tumbling or not, etc. The core flaw of the document, though, was its assumption that maximal impulse was best.

---

47. NASA, "Near-Earth Object Survey and Deflection Analysis of Alternatives: Report to Congress," March 2007, p. 4, *https://cneos.jpl.nasa.gov/doc/neo_report2007.html* (accessed 5 June 2019).

48. Russell L. Schweickart, "Technical Critique of NASA's Report to Congress and Associated of '2006 Near-Earth Object Survey and Deflection Study: Final Report' Published 28 Dec. 2006," *https://b612foundation.org/b612-response-to-nasas-2007-neo-report-to-congress-nasas-neo-report-to-congress-stirred-considerable-controversy-due-to-both-its-rejection-of-congresss-request-for-a-recom/* (bottom of page; accessed 14 September 2017), copy in NEO History Project collection.

The appropriate figure-of-merit is to evaluate the fraction of expected deflections required during the next century which can be satisfied by a deflection system that is (a) sufficient (with appropriate margin), (b) most precise and controllable (so we know what we are doing and what we have done, such as **not** placing the NEO into a keyhole), and (c) most gentle (so that the NEO, if a rubble pile or other loose assemblage, will not come apart unpredictably).[49]

He had many other critiques. For example, the study had focused on using the already-canceled Jupiter Icy Moons Orbiter nuclear-electric propulsion system as its basis for the gravity tractor and space tug mission variants, and not the successfully flown solar-electric Deep Space 1 mission, for example.[50] This biased its cost and technological readiness evaluations. But his core critique was that too large an impulse was likely to be as bad as too little. Asteroid deflection needed to be tuned to the specific asteroid, not treated in a one-size-fits-all fashion.

Schweickart and Chapman met with some of the report's authors in NASA's Office of Program Analysis and Evaluation on 18 June 2007. They did not come away satisfied with the answers they heard there, and they also were not shy about telling reporters about their unhappy experience. Leonard David featured their criticisms in a 2007 article for *Ad Astra*, for example.[51]

The analysis run by Yeomans's NEO Program Office at JPL in response to the letter that Schweickart had first sent to Administrator Griffin back in June 2005 had also concluded that the specific case of Apophis did not require a nuclear solution. The NASA Discovery Program's 2005 Deep Impact mission, which had collided a 370-kilogram impactor with Comet Tempel 1, had

---

49. Clark R. Chapman, "Critique of '2006 Near-Earth Object Survey and Deflection Study: Final Report' Published 28 December 2006 by NASA HQ. Program Analysis & Evaluation Office," *https://b612foundation.org/b612-response-to-nasas-2007-neo-report-to-congress-nasas-neo-report-to-congress-stirred-considerable-controversy-due-to-both-its-rejection-of-congresss-request-for-a-recom/* (bottom of page; accessed 14 September 2017), copy in NEO History Project collection.

50. On the Jupiter Icy Moons Orbiter, see, for example, Leonard David, "NASA's Prometheus: Fire, Smoke and Mirrors," Space.com, *https://www.space.com/929-nasas-prometheus-fire-smoke-mirrors.html* (accessed 5 June 2019).

51. Leonard David, "Fair Warning, Deadly Response: The Asteroid Threat," *Ad Astra* 19, no. 3 (2007), *https://space.nss.org/fair-warning-deadly-response-the-asteroid-threat/* (accessed 14 September 2021).

already demonstrated the ability to autonomously manage a very-high-speed impact with a celestial body.[52] While the comet was too massive to allow the far smaller impactor to make an obvious impact-induced change in its orbit, the mission had resolved the most technically difficult part of such a mission. Because moving 99942 Apophis outside the 600-meter "keyhole" in 2029 would not require changing the asteroid's velocity by very much as long as it was done a few years earlier, a 1,000-kilogram impactor could do it in 2024 or 2026.[53]

Asteroid 99942 Apophis reappeared in late 2012 as expected, and radar observations were acquired early the next year by teams of investigators at both Goldstone and Arecibo. Analysis of their data confirmed that on 13 April 2029, it would approach to within 5 Earth radii of Earth's surface and would have its orbit altered from an Aten-class to an Apollo-class asteroid.[54] The principal uncertainty in predicting its future orbit was due to the Yarkovsky effect, which the radar teams had been unable to measure. In 2013, a separate group of researchers calculated the probability that the Yarkovsky effect might push the asteroid into one of the gravitational "keyholes" that could exist, concluding that there was slightly more than a one-in-a-million chance of entering one that could lead to a 2068 impact. Radar observations in 2021 might allow determination of the Yarkovsky effect on Apophis, offering the possibility that the (unlikely) 2068 impact could be ruled out then.[55]

For the small community of researchers interested in pursuing a next-generation NEO survey, nothing much came of the five years of advocacy that began with the SDT effort in 2002. While Congress had approved a new goal, the discovery of 90 percent of the 140-meter and larger asteroids by 2020, NASA leaders had refused to propose a program to meet that goal, and

---

52. M. F. A'Hearn, "Deep Impact: Excavating Comet Tempel 1," *Science* 310 (2005), doi:10.1126/science.1118923. See chapter 7 for more detail on Deep Impact.

53. D. K. Yeomans et al., "Briefing on NASA's Near-Earth Object Program Office, Oct. 2005," courtesy of D. Yeomans, copy in NEO History Project collection.

54. Marina Brozović et al., "Goldstone and Arecibo Radar Observations of (99942) Apophis in 2012–2013," *Icarus* 300 (January 2018): 115–128, *https://doi.org/10.1016/j. icarus.2017.08.032.*

55. D. Farnocchia et al., "Yarkovsky-Driven Impact Risk Analysis for Asteroid (99942) Apophis," *Icarus* 224, no. 1 (May 2013): 192–200, *https://doi.org/10.1016/j. icarus.2013.02.020.*

Congress did not force the issue. Asked about the official reticence to answer Congress's demands years later, Lindley Johnson commented:

> [T]here's been a continual kind of a chicken-and-egg thing going on between the Agency, the Administration, and Congress about funding for all of this. Congress has taken the position that, "We've requested you do this, so you should submit to us a budget of what it takes to do it." And over on this side, the Agency side, it's been, "Well, you know, we have so many priorities and a limited budget to do them in, that this just hasn't risen high enough on our priority list to push out something that's already in the budget." And given we have to stay below a top level as given to us by the Administration, OMB, we just can't fit it in the budget constraints that we, the Agency, have to abide by.[56]

The NEO program budget remained constant through 2010 at about $4 million per year, enough to continue some of the existing surveys, but not to initiate a new observatory. The appearance of Apophis in David Tholen's "sweet spot" of previously overlooked sky did not change this reality, despite bringing new attention to cosmic hazards.

Next-Generation Surveys

During the first decades of the 21st century, two observatories built for other purposes became dedicated to the hunt for NEOs. One of these observatories, the WISE mission repurposed as NEOWISE, we discussed in chapter 7. The other repurposed observatory, Pan-STARRS, was in a sense the progenitor of the one new NEO observatory built after 2010, the Asteroid Terrestrial-impact Last Alert System, or ATLAS. Both were built by the University of Hawai'i Institute for Astronomy (IfA).

Pan-STARRS 1 began its initial observing program in mid-2010. To finance its scientific operations, Nick Kaiser and Kenneth Chambers had assembled a science consortium around a series of key projects that were to be carried out during the telescope's first three years of operations. Two of these, the 3π Steradian Survey and the Medium Deep Survey, made up the

---

56. Lindley Johnson, interview by Yeomans and Conway, 29 April 2016, copy in NASA History Division HRC.

lion's share of the observing time. Their first attempts to extract NEOs from the telescope's imagery failed because they relied only on pairs of images, which resulted in high numbers of false positives. LINEAR used five images for that reason, and as did NEOWISE. In September 2010, they switched to four images (the number used by the Catalina Sky Survey) and were immediately successful at finding NEOs, though not particularly efficient at it at first.[57] IfA astronomer Richard Wainscoat, who led the asteroid portion of the survey, explained that their initial observing plan had been to search near the Sun, the largely unobserved region in which Tucker and Tholen had found Apophis, but that did not work out. Wainscoat commented in 2017 that "the 'sweet spots' turned out to be difficult, because the trade-winds blow out of the direction of the northeast, which is where the sweet spot is going to be in the morning."[58] The winds hit Haleakalā crater and create turbulence that hinders seeing, in addition to the other known problems of observing near the Sun. As the other surveys already had, Pan-STARRS shifted to mostly opposition surveys to increase its discovery rate.

Pan-STARRS also suffered from another thorny problem: the lack of adequate follow-up observations, particularly for objects in the southern sky. Pan-STARRS observations were largely automated, but the data mining of the images was not quite real-time; without a human in the processing loop, it could not do same-night follow-up of new discoveries, as the Catalina Sky Survey did. Waiting until the next night meant that some objects were never spotted again; Tholen was funded by the NEO Observations Program to perform follow-up for Pan-STARRS and others using the Canada-France-Hawai'i telescope on Maunakea, but as the Pan-STARRS's discovery rate increased, that effort was overwhelmed. By 2017, there were millions of detections in the Minor Planet Center's "Isolated Tracklet File," which contained

---

57. Richard Wainscoat et al., "The Pan-STARRS Search for Near Earth Objects," *Proceedings of the International Astronomical Union* 10, no. S318 (August 2015): 293–298, doi:10.1017/S1743921315009187.

58. Richard Wainscoat, interview by Conway, 20 June 2017. The "sweet spots" are locations in space, near the ecliptic and 90–120 degrees from opposition, where the largest hazardous objects would preferentially be found. See S. Chesley and T. Spahr, "Earth Impactors: Orbital Characteristics and Warning Times," in *Mitigation of Hazardous Comets and Asteroids*, ed. M. J. S. Belton, T. H. Morgan, N. Samarasinha, and D. K. Yeomans (Cambridge: Cambridge University Press, 2004), pp. 22–37.

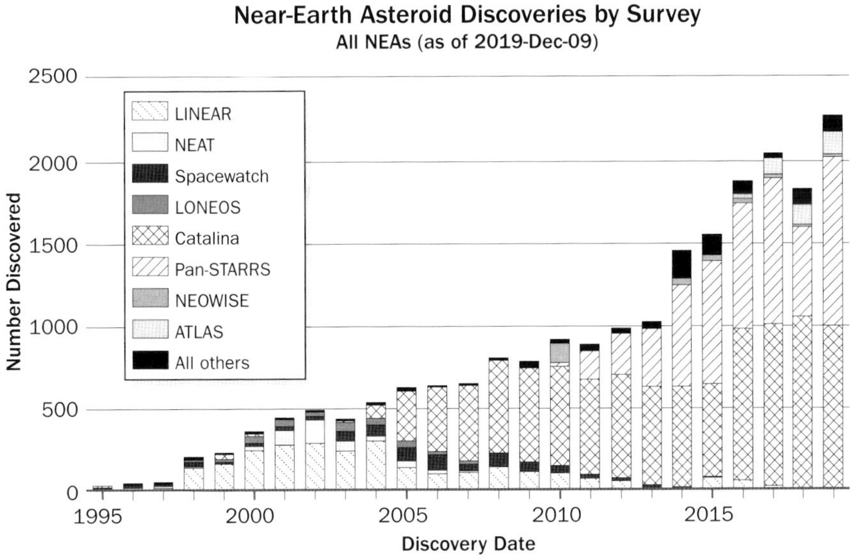

Figure 8-6. NEO discovery statistics as of December 2019. (Image from the Center for Near-Earth Objects Studies, *http://cneos.jpl.nasa.gov*)

all of the putative "asteroids" that had not received enough (or any) follow-up observations to generate an orbit.[59]

In 2014, Pan-STARRS 1 became mostly dedicated to the near-Earth objects mission, and Pan-STARRS 2 joined it when it became operational. Pan-STARRS 2 was ultimately built on the same site as Pan-STARRS 1 on Haleakalā, Maui, not on Maunakea as originally planned, and not to the same design. As Figure 8-6 indicates, Pan-STARRS began to rival the Catalina Sky Survey once it became dedicated to NEO discovery.

John Tonry, who was part of the IfA team developing the Pan-STARRS cameras, proposed a very different kind of observatory in 2009. Instead of trying to find all of a particular size of asteroid in pursuit of a survey goal, his proposed "Asteroid Terrestrial Impact Last Alert System (ATLAS)" was designed to find asteroids of any size during their "death plunge," the final days or weeks before impact. "We suggest that the best mitigation strategy

---

59. On the tracklets issue, see Robert J. Weryk, Richard J. Wainscoat, and Gareth Williams, "Isolated Tracklet Linking," in *American Astronomical Society Division for Planetary Sciences Meeting Abstracts*, vol. 49, no. 103.02, 2017, *http://adsabs.harvard. edu/abs/2017DPS....4910302W* (accessed 3 May 2021).

in the near term is simply to move people out of the way," he wrote.[60] With a few days' warning for a likely impact site, most people, and quite a bit of mobile property, could be moved out of harm's way. An observatory, or a network of observatories, that made fast scans of the entire sky nightly (he hoped for twice per night) could provide a week's warning of a 50-meter asteroid approach and three weeks' warning for a 140-meter asteroid. These rapid scans would capture only relatively bright (and therefore only very large or very close) asteroids, so they would not contribute much to the Brown Act survey goal. But such a system would provide early warning at relatively small cost. The telescopes would be small, 0.4 meters or so, and fully automated.

Tonry explained later that he was trying to figure out "how best to build the optimal survey system per unit dollar." He was less interested in the asteroid task per se than in the larger astronomical goal of transient events, which, because of the Pan-STARRS program's relatively slow pace, he thought was not being well-served. ATLAS would not have Pan-STARRS's depth, but it could survey the sky much more rapidly. Tonry approached the Keck Foundation for funding initially but lost out to another project, and he proposed to NASA's NEO Observations program three times before finally getting the idea funded in September 2012.[61] The expected cost was $5 million over five years. An even smaller "Pathfinder" telescope was the first to be built as a proof of concept and software testbed, at the National Oceanic and Atmospheric Administration's atmospheric research observatory on Mauna Loa, Hawai'i. It began operating in 2013. The first full-sized telescope, built at a disused U.S. Air Force site on Haleakalā, achieved first light in May 2015; the second replaced the Pathfinder telescope on Mauna Loa and began operating in February 2017.[62] On the project's blog—a tool that did not exist when Tom Gehrels had begun advocating for an electronic asteroid-hunting telescope in the 1970s—Tonry commented in March 2017 that a human only

---

60. John L. Tonry, "An Early Warning System for Asteroid Impact," *Publications of the Astronomical Society of the Pacific* 123, no. 899 (January 2011): 58–73.

61. John Tonry, interview by Conway, 20 June 2017, transcript in NASA History Division HRC.

62. "ATLAS Update 10," 30 December 2013, ATLAS website, *http://fallingstar.com/ua20131230.php* (accessed 3 May 2021); Ari Heinze, "ATLAS Telescope 2 Installed on Mauna Loa," ATLAS website, 6 February 2017, *https://fallingstar.com/ua20170315.php* (accessed 14 September 2021); "ATLAS Update #14," ATLAS website, 30 March 2015, *https://fallingstar.com/ua20150330.php* (accessed 14 September 2021).

had to visit the Haleakalā site once a month for maintenance. Everything else had been automated, even weather observations, and the project's website allowed anyone to see what the two telescopes saw in real time.[63]

Taking on the rather expensive Pan-STARRS operation, NEOWISE, and Tonry's small ATLAS became possible for NASA's Near-Earth Objects Observations Program after a policy shift in 2010. But the policy shift was not due to another close asteroid flyby. Instead, it was linked to NASA's human spaceflight program.

The Astronaut Imperative

The new Obama administration entered office in January 2009 faced with the largest financial crisis in the United States since the 1929 stock market collapse, and NASA itself faced soaring costs for the prior administration's Vision for Space Exploration policy and its attendant development program, known as Constellation.[64] The White House placed the Constellation program under review by an independent panel led by retired Lockheed Martin chairman Norman Augustine. The Augustine Commission ultimately reported that NASA would need a budget increase of several billion dollars per year to succeed.[65] Both the administration and Congress sought to cut the NASA budget, though, and did. The NASA budget shrank from $18.7 billion in fiscal year 2010 to $16.8 billion in fiscal year 2013, before Congress began to restore it. By then, the Administration had canceled the Constellation program and instead focused NASA's human spaceflight mission on a near-term goal of sending astronauts to a near-Earth asteroid.

President Obama announced the new goal at a 15 April 2010 event at Kennedy Space Center and embedded it in the new U.S. National Space Policy

63. "ATLAS Update #18," March 2017, ATLAS website, *http://fallingstar.com/ua20170315. php* (accessed 3 May 2021).

64. On the Constellation program, see Government Accountability Office, "NASA Constellation Program Cost and Schedule Will Remain Uncertain Until a Sound Business Case Is Established," GAO-09-883, 26 August 2009. For its origins, see Glen R. Asner and Stephen J. Garber, *Origins of 21st Century Space Travel: A History of NASA's Decadal Planning Team and the Vision for Space Exploration, 1999–2004* (Washington, DC: NASA SP-2019-4415, 2019).

65. Review of U.S. Human Spaceflight Plans Committee, "Seeking a Human Spaceflight Program Worthy of a Great Nation," NASA, Washington DC, 2009, *https://www. nasa.gov/pdf/396093main_HSF_Cmte_FinalReport.pdf* (accessed 28 September 2017).

in June.[66] While the idea never generated much interest in Congress, it helped change the fiscal landscape for NASA's Near-Earth Objects Observations Program. If one wished to send astronauts to a near-Earth asteroid, one had to find a NEO that was close enough to be reached and large enough to be landed on, but not spinning too rapidly. And that meant spending more money on NEO discovery and characterization despite the overall shrink-age of the Agency's budget. The NASA NEO Observations Program's budget went from $3.7 million in fiscal year 2009 to $40.5 million in fiscal year 2014. These budget increases allowed Johnson to take over the funding of Pan-STARRS after 2014 and WISE's reactivation as NEOWISE in 2011, as well as funding some technology development efforts at JPL in support of a proposal by Amy Mainzer to develop a mission optimized for NEO discovery and characterization called NEOCam. The Near-Earth Objects Observations Program also funded expanded efforts to characterize a number of NEAs that might be suitable for exploration, including one (101955 Bennu) that became the target of the United States' first asteroid sample-return mission, the Origins, Spectral Interpretation, Resource Identification, Security, and Regolith Explorer mission, or (more mercifully), OSIRIS-REx.[67]

No space telescope dedicated to completing the Brown survey goals came from the expanded budget, though. In 2011, the B612 Foundation struck a deal with Ball Aerospace to develop an infrared, Venus-trailing solar-orbit telescope for this purpose, based on the spacecraft bus developed for the Kepler exoplanet mission. They also signed a Space Act Agreement in 2012 with NASA for the use of the Deep Space Network for communications and for participation in technical reviews.[68] This action effectively forestalled any NASA effort along the same lines, since the Agency senior management would hardly advocate for a duplicative effort. But the B612 Foundation proved

---

66. Barack Obama, "Remarks by the President on Space Exploration in the 21st Century," John F. Kennedy Space Center, Merritt Island, Florida, 15 April 2010, *https://www.nasa.gov/news/media/trans/obama_ksc_trans.html* (accessed 23 September 2017); "National Space Policy of the United States of America," 28 June 2010, *https://www.nasa.gov/sites/default/files/national_space_policy_6-28-10.pdf* (accessed 23 September 2017).

67. See *https://www.asteroidmission.org/objectives/osiris-rex-acronym/* (accessed 3 May 2021); see also OSIRIS-REx press kit, August 2016, *https://www.nasa.gov/sites/default/files/atoms/files/osiris-rex_press_kit.pdf* (accessed 3 May 2021).

68. Edward T. Lu et al., "The B612 Foundation Sentinel Space Telescope," *New Space* 1, no. 1 (March 2013): 42–45, doi:10.1089/space.2013.1500.

Figure 8-7. NEOCam detector development. The cylindrical object in the center is a cryogenic chamber used for cooling the detectors. Left to right, Judy Pipher of the University of Rochester, Amy Mainzer of the University of Arizona, and Mark McKelvey of NASA Ames Research Center. (Image courtesy of Amy Mainzer)

unable to raise even a small fraction of the funding necessary to build their Sentinel mission, as they named it, let alone launch and operate it. In 2015, NASA terminated the Space Act Agreement, as the Foundation had not begun the telescope's development phase in 2014, as required by the agreement.[69]

JPL's Mainzer submitted the NEOCam proposal to NASA's Discovery program competition again in 2015, but it was not one of the two selectees announced in 2017.[70] Instead, the NEOCam project was funded for further

---

69. "NASA Terminates Space Act Agreement with B612 Foundation for Sentinel Spacecraft," *https://spacepolicyonline.com/news/nasa-terminates-space-act-agreement-with-b612-foundation-for-sentinel-spacecraft/* (accessed 28 August 2017); "B612 Presses Ahead with Asteroid Mission Despite Setbacks," *http://spacenews.com/b612-presses-ahead-with-asteroid-mission-despite-setbacks/* (accessed 23 September 2017).

70. NEOCam was submitted to the Discovery program three times, in 2006, 2010, and 2015. See *https://neocam.ipac.caltech.edu/page/mission* (accessed 10 June 2019); "NASA Selects Two Missions to Explore the Early Solar System," 4 January 2017, *https://www.jpl.nasa.gov/news/news.php?feature=6713* (accessed 10 June 2019).

development of the telescope's enabling technology, its infrared detectors. The Discovery program's selection committee had not seen NEO discovery as high-priority science, a consequence of it having become a matter of policy.[71]

During the 2000s, the possibility of cosmic impact came to be seen as a natural hazard, like earthquakes or hurricanes, and scientists began to treat impacts in a similar fashion. They deployed a narrative of risk. These other hazards are managed, to a degree at least, via cost-benefit analysis, and as we have seen in this chapter, asteroid scientists began to deploy cost-benefit analysis as part of their effort to justify increased funding for NEO discovery and characterization. This risk narrative was not the only narrative available, though. The 2010 decision to send astronauts to a near-Earth asteroid made the narrative of exploration available. Perhaps asteroids could be stepping-stones for human expansion into the solar system.

71. Committee on Near-Earth Object Observations in the Infrared and Visible Wavelengths, "Finding Hazardous Asteroids Using Infrared and Visible Wavelength Telescopes," July 2019 (pre-publication draft), p. S-5.

CHAPTER 9

ASTEROIDS AS STEPPINGSTONES AND RESOURCES

On 15 April 1962, a cartoon entitled "Asteroid Arrester" appeared in newspapers across the United States. An installment of Arthur Radebaugh's syndicated strip *Closer Than We Think*, the one-panel comic depicts a spacecraft latching onto an asteroid and firing retrorockets to divert its course. "This would make it possible to study the origins of the solar system, possibly increase our store of minerals and even learn about the beginnings of life," the caption claims.[1] These different potential goals illustrate the overlapping interpretations of asteroids—and near-Earth objects in particular—that emerged throughout the latter half of the 20th century and into the 21st, intersecting with the narrative of risk discussed in the previous chapters.

The strong emphasis on NEOs as dangerous threats to humanity spurred attempts on both civilian and military sides to formulate an adequate mitigation strategy in the event of an impending Earth impact. These studies in turn drew attention to two new objectives: human exploration of asteroids and exploitation of their natural resources. Both ideas were rooted in science fiction, an early example being Garett P. Serviss's 1898 story "Edison's Conquest of Mars," an unofficial sequel to *The War of the Worlds*, in which daring Earth inventors and scientists (including Thomas Edison himself and Lord Kelvin, among others) come upon an asteroid made of pure gold being mined by a colony of Martians.[2]

---

1.   A. Radebaugh, "Asteroid Arrester," *Closer Than We Think*, *Chicago Tribune* (15 April 1962).

2.   Garrett Putnam Serviss, "Edison's Conquest of Mars," 1898, *https://www.gutenberg. org/files/19141/19141-h/19141-h.htm* (accessed 7 May 2021).

The scarcity of mineral resources on Mars was consistent with a notion of planetary evolution that was emerging in early-20th-century planetary science. Percival Lowell, for example, wrote in 1906 about one phase of the process he believed to be the reason behind his "canals": "Study of the several planets of our solar system, notably the Earth, Moon, and Mars, reveals tolerably legibly an interesting phase of a planet's career, which apparently must happen to all such bodies, and evidently has happened or is happening to these three: the transition of its surface from a terraqueous to a purely terrestrial condition."[3] If Mars was in transition from a vibrant Earth-like planet to a lifeless Moon-like one, then this evolutionary path would eventually take its course on our own planet. As Robert Markley traces in his book *Dying Planet: Mars in Science and the Imagination*, this theme of loss and dwindling resources played out in both science fiction and environmental concerns of the day. These anxieties persisted and would later be rekindled by the 1972 "Limits to Growth" report by the Club of Rome, which sparked a wave of modern interest in the prospect of space colonization as an alternative to terrestrial disaster.[4] In contrast to the specter of resource depletion at home, the seeming abundance of valuable space resources allowed for a ready comparison between the cold reaches of space and the American frontier, where new resources could always be found. By the middle of the century, the archetype of the asteroid miner as explorer and frontiersman was well established in the realm of science fiction, and in the decades that followed, a surge of enthusiasm for space settlement and resource utilization pushed to bring the sci-fi trope of conquering the final frontier into reality. "Without a frontier to grow in," wrote space advocate Robert Zubrin in 1994, "not only American society, but the entire global civilization based upon Western enlightenment values of humanism, reason, science and progress will die."[5]

Thus, neither asteroid exploration nor resource extraction was a new idea, but both gained currency decades later when framed as the opportunity to develop multi-use technologies in support of planetary defense. These emerging objectives, like the defense strategies that elevated them to prominence,

---

3.   P. Lowell, *Mars and Its Canals* (London: The Macmillan Company, Ltd., 1906).

4.   "Limits to Growth," Club of Rome report, *https://www.clubofrome.org/report/the-limits-to-growth/*.

5.   R. Zubrin, "The Significance of the Martian Frontier," *Ad Astra* (September/October 1994), *https://space.nss.org/the-significance-of-the-martian-frontier-by-robert-zubrin/* (accessed 14 September 2021).

also introduced new legal challenges. As we saw in chapter 5, deflection strategies involving nuclear weapons proposed at the 1992 Interception Workshop stirred controversy among the scientific contingent of the meeting. Such methods also raised legal and ethical concerns that, even as of this writing, have yet to be clarified in international space policy. Similarly, the exploitation of mineral and water resources contained within asteroids is not without gray areas of its own. Asteroid mining may have been first conceived as a natural extension of American Manifest Destiny, but the private enterprises that aim to pursue it are subject to the laws of the 21st century, not the 19th.

NEOs as Resource Repositories and Exploration Steppingstones

On 10 May 1962—just one week after Radebaugh's "Asteroid Arrester" comic was published—U.S. Vice President and chairman of the National Space Council Lyndon B. Johnson gave a speech at the World's Fair in Chicago. "Someday we will be able to bring an asteroid containing billions of dollars['] worth of critically needed metals close to Earth to provide a vast source of mineral wealth to our factories," the newspapers reported.[6] Although exploring an asteroid, much less retrieving one, was not one of NASA's primary goals in the first decade of the Agency's existence, the prospect of visiting and potentially utilizing the small bodies of the solar system with robotic or crewed missions was not entirely beyond consideration. A 1961 study by Goddard Space Flight Center scientist Su-Shu Huang considered how close an asteroid would need to pass to be captured into Earth orbit, concluding that such a feat could be achieved with one of the handful of near-Earth asteroids that had been discovered by that time. While Huang also briefly mentions the possibility of ejecting the asteroid's own material to influence its velocity (the basic concept behind the mass driver), Huang's motivation for exploring asteroid retrieval was the technical accomplishment itself. He

---

6. Matt Novak, "Asteroid Mining's Peculiar Past," *BBC Future*, 18 November 2014, *http://www.bbc.com/future/story/20130129-asteroid-minings-peculiar-past* (accessed 12 June 2019); e.g., "Use of 'Captured' Asteroid for Mineral Supply Studied," *Express and News* (San Antonio) (7 November 1964): 9.

concludes, "This proposed project would require a major effort but, if realized, it would be a lasting mark of human achievement."[7]

NASA considered asteroid exploration on various other occasions throughout the 1960s and 1970s, at first focusing largely on the asteroid belt but gradually including Earth-approaching asteroids as well. A 1964 document on future mission planning proposed a long-term goal of sending robotic probes to the asteroid belt by the end of the next decade, and the Office of Manned Space Flight's 1969 "Five Year Plan" considered plans to send crewed missions to asteroids.[8] Pioneer 10 became the first spacecraft to cross the asteroid belt in 1973, demonstrating for future missions that it could be safely done.[9]

In 1975 and 1977, two Space Settlement workshops were held at NASA Ames Research Center to explore various technological challenges associated with the colonization of space, including the utilization of resources gleaned from the Moon and asteroids.[10] Both workshops were directed by Gerard K. O'Neill, the Princeton University physicist and space activist who first proposed the mass driver in 1974.[11] Inspired in part by the Club of Rome's 1972 pessimistic "Limits to Growth" report, O'Neill became a leading figure in a new push toward space colonization in the face of the predicted depletion

---

7. Su-Shu Huang, "Velocity Modification for Earth Capture of an Astronomical Body in the Solar System," NASA Technical Note D-1140, December, 1961, *https://ntrs.nasa. gov/archive/nasa/casi.ntrs.nasa.gov/19980227403.pdf* (accessed 12 June 2019).

8. Michelle K. Dailey, "The Long and Storied Path to Human Asteroid Exploration," 16 April 2013, *https://www.nasa.gov/topics/history/features/asteroids.html* (accessed 12 June 2019).

9. Richard O. Fimmel, William Swindell, and Eric Burgess, *Pioneer Odyssey* (Washington, DC: NASA SP-349, 1977), *https://history.nasa.gov/SP-349/sp349.htm* (accessed 12 June 2019): "But before sophisticated missions to the outer planets could be planned, at least one spacecraft had to penetrate and survive passage through the asteroid belt" (chap. 2).

10. Final reports of both workshops: *Space Settlements: A Design Study* (Washington, DC: NASA SP-413, 1977), *http://large.stanford.edu/courses/2016/ph240/martelaro2/ docs/nasa-sp-413.pdf,* and *Space Resources and Space Settlements* (Washington, DC: NASA SP-428, 1979), *https://space.nss.org/settlement/nasa/spaceres/index.html* (accessed 14 September 2021).

11. Gerard K. O'Neill, "The Colonization of Space," *Physics Today* 27, no. 9 (September 1974): 32–40.

of resources on Earth.[12] The 1975 settlement study, which was cosponsored by Stanford University and the American Society for Engineering Education (ASEE), focused on the requirements to build a sustainable colony near a source of substantial natural resources. While the report acknowledged that some asteroids had been discovered to pass by Earth well within the orbit of Mars, the Moon was selected as the study's target location because its properties were better known. Even so, the asteroid resources available for use were estimated to be practically limitless: "the total quantity of materials within only a few known large asteroids is enough to permit building space colonies with a total land area many thousands of times that of the Earth."[13]

The final report of the 1977 Space Settlement workshop focused even more on the potential advantages of asteroids—especially Earth-approaching ones—to furnish the necessary materials to sustain a space colony. Several papers detailed strategies to divert an object using various types of mass driver, or "asteroid retriever." One paper, led by former astronaut Brian O'Leary, concluded that "the asteroid-retrieval option is competitive with the retrieval of lunar materials for space manufacturing, while a carbonaceous object would provide a distinctive advantage over the Earth as a source of consumables and raw materials for biomass."[14] In order to boost the inventory of near-Earth objects available for mining, O'Leary called for an augmented Earth- and space-based telescopic search and follow-up program, accompanied by robotic precursor missions to rendezvous and land on one or more NEOs by the mid-1980s.

A companion paper coauthored by O'Leary with the University of Hawai'i's Michael Gaffey and JPL's Eleanor Helin, "An Assessment of Near-Earth Asteroid Resources," drew on chemical analyses of meteorites as well as reflectance spectra of Apollo–Amor group asteroids to estimate that NEOs might be rich sources of volatile material (including water, carbon, and carbon compounds) and metals (such as nickel-iron). With only 40 objects then identified, however, they emphasized the strong need to collect more observations,

---

12. W. P. McCray, *The Visioneers: How a Group of Elite Scientists Pursued Space Colonies, Nanotechnologies, and a Limitless Future* (Princeton, NJ: Princeton University Press, 2012).

13. Richard D. Johnson and Charles Holbrow, eds., *Space Settlements: A Design Study* (Washington, DC: NASA SP-413, 1977), p. 53.

14. John Billingham, William Gilbreath, and Brian O'Leary, eds., *Space Resources and Space Settlements* (Washington, DC: NASA SP-428, 1979), pp. 173, 187.

concluding that a mission to survey one or more of these objects, as well as support for detection and characterization programs, should be undertaken to support any large-scale future operations in space, "since asteroidal bodies appear to be the least expensive source of certain needed raw materials."[15]

In the Moon Treaty Hearings before Congress on 29 May 1980, several questions were devoted to the development of extraterrestrial resources—not only on the Moon but contained within asteroids and other small bodies as well. NASA Administrator Robert A. Frosch made it clear that, while both the observational capability to better characterize these objects and the technology to make any present resources available would require decades to achieve, the value of such an endeavor to the U.S. space program was apparent. Citing potential stores of carbon, iron, nickel, cobalt, and chromium, as well as precious metals such as platinum, osmium, rhodium, rhenium, and iridium, Frosch noted that NASA had recently devoted new attention to understanding the necessary steps toward making cost-effective use of space materials. "NASA has the responsibility to create and develop the space technologies and plan the possible missions needed to make effective use of these resources," he testified. Among the potential uses listed was asteroid retrieval to Earth.[16]

In 1984, President Ronald Reagan issued an executive order establishing a committee to evaluate the nation's space program. The National Commission on Space, led by former NASA Administrator Thomas O. Paine, included physicist Luis Alvarez, astronaut Neil Armstrong, and space advocate Gerard K. O'Neill and put out a 1986 advocacy document entitled "Pioneering the Space Frontier." This report drew on colonialist tropes of the settlement of North America to promote a new mission for the 21st century: "To lead the exploration and development of the space frontier, advancing science, technology, and enterprise, and building institutions and systems that make accessible vast new resources and support human settlements beyond Earth orbit,

---

15. Ibid., p. 191, "An Assessment of Near-Earth Asteroid Resources."

16. "The Moon Treaty: Hearings Before the Subcommittee on Science, Technology, and Space of the Committee on Commerce, Science, and Transportation, United States Senate, on Agreement Governing the Activities of States on the Moon and Other Celestial Bodies," 96th Cong., 2nd sess. (29 and 31 July 1980) (Frosch testimony), pp. 36–46.

from the highlands of the Moon to the plains of Mars."[17] Among the many objectives outlined, the Commission included prospecting of Earth-crossing asteroids as "particularly promising for exploration and resource utilization."[18]

A 1984 workshop cosponsored by the California Space Institute, ASEE, and NASA's Lyndon B. Johnson Space Center considered the best available avenues to fulfill that responsibility, resulting in a Space Resources report published in 1992.[19] A candidate plan proposed sending robotic probes to near-Earth asteroids beginning around 2005, with mining operations beginning around 2015 to bring water and metals back to geosynchronous orbit to fuel activities on the Moon.[20] Although the new space initiative announced by President George H. W. Bush on 20 July 1989 and NASA's official response to it focused solely on the Moon and Mars as exploration milestones, the Space Resources report highlighted the role of near-Earth asteroid resources to achieve those ambitious goals, concluding that "near-Earth resources can indeed foster the growth of human activities in space."[21]

As we have seen throughout the previous chapters, increasing numbers of detected near-Earth objects, together with the 1980 Alvarez hypothesis and other new insights into the role that impacts may have played in our planet's history, led to the reevaluation of these bodies as hazards to life on Earth. As that risk narrative unfolded, discussions of NEOs as potential sources of natural resources began to be framed as a useful side effect in service of planetary defense. Whereas the 1977 Space Settlement study proposed ways for improvements to the nascent NEO search programs to benefit asteroid resource characterization, the mitigation of risk argument began to appear in the studies of the 1990s. In the 1992 Interception Workshop, for example, the "Astrodynamics of Interception" section considered three scenarios for an impending impact, two with a long lead time and one with only a year or less. In the first case, in which an Earth-crossing asteroid with a well-determined orbit is projected to intercept Earth more than one orbit in advance, the

---

17. "Pioneering the Space Frontier," An Exciting Vision of Our Next Fifty Years in Space," in *The Report of the National Commission on Space* (1986), *https://history.nasa.gov/painerep/begin.html* (accessed 12 June 2019).

18. Ibid., p. 65.

19. M. F. McKay, D. S. McKay, and M. B. Duke, *Space Resources: Energy, Power, and Transport*, vols. 1–4 (Washington, DC: NASA SP-509), 1992.

20. Ibid., pp. 3–4.

21. Ibid., p. vi.

report notes that programs addressing the threat "will have beneficial spin-offs for other NASA programs and for science because the long warning times permit detailed scientific explorations and investigations of extraterrestrial resources."[22] In the same workshop, a solar sail idea was also proposed as an opportunity to both deflect an incoming near-Earth asteroid and safely retrieve it, potentially using the asteroid material in a space-based fabrication scheme to construct the sails themselves.[23]

Perhaps the most obvious technology proposed for asteroid deflection with additional prospecting and exploration benefits was the mass driver concept. The idea that valuable resources could be mined in the process of producing the necessary reaction mass to drive propulsion had long been considered by O'Neill, O'Leary, and other participants in the Space Settlements workshops, and the concept was mentioned again in the 1992 Interception Workshop.[24] Considered a "medium-term innovative technology (less than 20 years)," a crewed mission was proposed to bring mining equipment to the approaching NEO. Reaction mass for deflection would come from spent fuel tanks from launch, followed by tailings from the mining operation installed by the crew. Although the study participants foresaw many challenges to this approach (not least of which was the "physical control of a large object, including management of angular and linear momentum"[25]), the multiple objectives of such a scheme made it attractive to some, as long as resource extraction was framed as a fortuitous spinoff of deflection. As the report phrased it, "One benefit is that this system could return large quantities of asteroidal resources to cislunar space."[26]

This emphasis on resource mining as an added bonus in the pursuit of threat mitigation was reiterated throughout the 1990s and into the new century, sometimes accompanied by citations of scientific study as a second spinoff advantage. The 1995 Planetary Defense Conference (PDC) report presented the prospect of a working planetary defense system as "an enormous

22. Gregory H. Canavan, Johndale C. Solem, and John D. G. Rather, eds., *Proceedings of the Near-Earth Object Interception Workshop* (Los Alamos, NM: Los Alamos National Laboratory, February 1993), p. 86.

23. Ibid., p. 231.

24. E.g., O'Leary et al., "Retrieval of Asteroidal Materials," in *Space Resources and Space Settlements* (Washington, DC: NASA SP-428, 1979), pp. 173–189.

25. Ibid., p. 230.

26. Ibid.

bargain in terms of the mitigated risk as well as the advance of science and the exploitation of space resources."[27] The 2003 NRC Decadal Survey in Planetary Science was much more focused on the scientific questions that could be answered by further study of NEOs. Nevertheless, the survey also recognized both the "potential mitigation of hazards to Earth that arrive from space, and provision of knowledge about space resources that are available for utilization" as a significant driver of solar system exploration.[28] At the Planetary Defense Conference in 2004, George Friedman of the University of Southern California was the lead author on a paper entitled "Mass Drivers for Planetary Defense," which made the case that establishing profitable asteroid mining ventures would naturally serve the interests of planetary defense because "it will be an easy operation to apply enough delta V with a mass driver to change a certain collision to a clean miss, given a few years' lead time."[29] By leveraging the ability to deflect an asteroid in its orbit while carrying out mining activities, the paper concluded, this technique could generate "potentially enormous" returns.[30]

The latter half of the 2000s was marked by a recession and a shrinking budget for NASA, which had been struggling to meet its ambitious space initiative goals for some time.[31] Its centerpiece human exploration program, known as Constellation, was also overrunning its budget. In 2009, the Obama administration had created a blue-ribbon review committee led by Norman Augustine to recommend what to do about the Constellation Program. Augustine's committee found that NASA's human exploration program could not be carried out with the funding level Congress was willing to provide: "In fact," they wrote, "the Committee finds that no plan compatible

27. J. Nuckolls, *Proceedings of the Planetary Defense Workshop*, Lawrence Livermore National Laboratory, California, 22–26 May 1995 (Lawrence Livermore National Laboratory, CA: No. CONF—9505266), p. 527.

28. National Research Council, *New Frontiers in the Solar System: An Integrated Exploration Strategy* (Washington, DC: National Academies Press, 2003), *https://books.google.com/books?hl=en&lr=&id=6RN9pbFefvcC&*.

29. George Friedman et al., "Mass Drivers for Planetary Defense" (presented at the 2004 Planetary Defense Conference: "Protecting Earth from Asteroids," AIAA SPACE Forum, Orange County, CA, 23–26 February 2004), p. 9, *https://arc.aiaa.org/doi/abs/10.2514/6.2004-1450* (accessed 12 June 2019).

30. Ibid., p. 10.

31. See chapter 8.

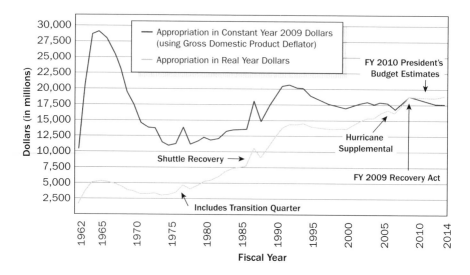

Figure 9-1. NASA historical budget and projection from 2009 to 2014. (Review of U.S. Human Spaceflight Plans Committee, *Seeking a Human Spaceflight Program Worthy of a Great Nation* [October 2009])

with the FY 2010 budget profile permits human exploration to continue in any meaningful way."[32]

Within this atmosphere, another potential advantage to pursuing a near-Earth asteroid mission began to gain attention: the possibility of a convenient exploration steppingstone. In late 2006, the Advanced Projects Office within NASA's Constellation Program had sponsored a study to determine whether sending the Orion Crew Exploration Vehicle (CEV) to a near-Earth asteroid using the Ares family of launch vehicles (then under development within the Constellation Program) would be a feasible mission to undertake. Led by David Korsmeyer, the study included representatives from across NASA, and its key findings were later summarized in a 2009 paper.[33] In addition to the scientific objective of better understanding the structure and composition of near-Earth asteroids, the study found ample other reasons to support such a crewed mission. These included "more practical applications such as

---

32. Review of U.S. Human Spaceflight Plans Committee, *Seeking a Human Spaceflight Program Worthy of a Great Nation* (October 2009), p. 16, *https://www.nasa.gov/pdf/396093main_HSF_Cmte_FinalReport.pdf* (accessed 12 June 2019).

33. P. A. Abell et al., "Scientific Exploration of Near-Earth Objects Via the Orion Crew Exploration Vehicle," *Meteoritics & Planetary Science* 44, no. 12 (December 2009): 1825–1836.

resource extraction and utilization (e.g., water, precious metals, volatiles, etc.) and NEO hazard mitigation (e.g., determining material properties, internal structures, macro-porosities, etc.),"[34] as well as the opportunity to gather technical and engineering data on spacecraft operations and to test sample collection techniques. Moreover, one of the goals of the Constellation system was flexibility; ideally, it would be capable of accommodating missions to a variety of destinations. A crewed mission to a near-Earth asteroid would demonstrate critical technologies for future, less "destination-driven" space exploration, marking an important milestone toward this goal.[35]

The Augustine committee had incorporated human exploration of a near-Earth asteroid into one of its recommended future, and less expensive, pathways for NASA to follow, and this idea became a cornerstone of President Obama's speech at Kennedy Space Center on 15 April 2010. "We'll start by sending astronauts to an asteroid for the first time in history," he said, but that would only be a beginning. "By the mid-2030s, I believe we can send humans to orbit Mars and return them safely to Earth. And a landing on Mars will follow."[36] The White House had already canceled the Constellation Program in February, intending to replace it with full commercialization of launch services and new technology investments aimed at making a future commercially procured heavy-lift vehicle more affordable. Congress ultimately did not allow this, and in 2011 it reached an agreement with the White House to partly fund commercialization as well as financing a new heavy-lift, government-owned vehicle named the Space Launch System, or SLS.[37]

34. Ibid., p. 1835.

35. Ibid., p. 1826.

36. "President Barack Obama on Space Exploration in the 21st Century," 15 April 2010, *https://www.nasa.gov/news/media/trans/obama_ksc_trans.html*.

37. Jason Davis, "Space in Transition: How Obama's White House charted a new course for NASA," 22 August 2016, *http://www.planetary.org/blogs/jason-davis/2016/20160822-horizon-goal-part-3.html* (accessed 10 May 2021); Jason Davis, "To Mars, with a Monster Rocket: How Politicians and Engineers Created NASA's Space Launch System," 3 October 2016, *http://www.planetary.org/blogs/jason-davis/2016/20161003-horizon-goal-part-4.html* (accessed 12 June 2019). Also see Departments of Commerce and Justice, and Science, and Related Agencies Appropriations Bill, 2011, 11th Cong., 2nd sess., Report 111-229 (22 July 2010), pp. 122–123, *https://www.congress.gov/congressional-report/111th-congress/senate-report/229/1?overview=closed* (accessed 11 June 2019).

The asteroid mission remained a notional goal throughout this period of turmoil. Two key issues emerged during this time. One was simply that very few potential asteroid targets were known. A short study of possible candidates that might be listed in the Minor Planet Center and JPL small bodies databases was performed to make a list of potential candidates, and very quickly the Small Bodies Assessment Group (SBAG) began advocating for an expanded asteroid search program.[38] More seriously for the human mission potential, though, was the simple fact that going to an asteroid meant spending months in space, and that meant that a long-duration habitat needed to be designed. But nothing like that fit into NASA's future budget profile.

Soon after the President's announcement in 2010, a study organized by electric propulsion engineer John Brophy at NASA's Jet Propulsion Laboratory (JPL) was initiated to consider the feasibility of capturing a small near-Earth asteroid and bringing it back to the International Space Station (ISS) using high-power solar-electric propulsion (SEP), a near-term technology then under development.[39] While the study found several challenges, its primary conclusion was that "no show stoppers were identified for the approach that would return an entire 10,000-kg asteroid to the ISS in a mission that could be launched by the end of this decade."[40]

Louis D. Friedman of the Planetary Society then entered the picture. An intern at the Society, Marco Tantardini, was very interested in asteroid science and became an advocate; they contacted Martin Lo, a trajectory designer at JPL, who invited Brophy to help run a follow-on study of asteroid retrieval at

---

38. P. A. Abell, B. W. Barbee, R. G. Mink, D. R. Adamo, C. M. Alberding, D. D. Mazanek, N. Johnson, et al., "The Near-Earth Object Human Space Flight Accessible Targets Study (NHATS) List of Near-Earth Asteroids: Identifying Potential Targets for Future Exploration," n.d., 3. The Center for Near Earth Object Studies (CNEOS) page maintains an up-to-date list: *https://cneos.jpl.nasa.gov/nhats/* (accessed 11 May 2021). On SBAG, see Small Bodies Assessment Group Findings, 9 August 2010, *https://www.lpi.usra.edu/sbag/findings/index.shtml#sbag3* (accessed 23 October 2018).

39. John R. Brophy et al., "300 kW Solar Electric Propulsion System Configuration for Human Exploration of Near-Earth Asteroids," AIAA paper 2011-5514 (presented at the 47th AIAA/ASME/SAE/ASEE Joint Propulsion Conference and Exhibit, San Diego, CA, 31 July 2011).

40. J. R. Brophy et al., "Asteroid Return Mission Feasibility Study," AIAA paper 5565 (2011), p. 1, *https://trs.jpl.nasa.gov/bitstream/handle/2014/43897/11-2709_A1b.pdf* (accessed 12 June 2019).

the Keck Institute for Space Studies.[41] Two workshops, one taking place in September 2011 and the other following in February 2012, were jointly sponsored by the Keck Institute for Space Studies (KISS) and JPL to consider a slightly different scenario: the retrieval of a near-Earth asteroid into a high lunar orbit by a robotic spacecraft, followed by a crewed visit to the asteroid. This much shorter mission would not require the development of a long-duration habitat. This study, summarized in a final report released in April 2012, showcased a shift in the defense-centered rhetoric of previous decades.[42] Rather than resource development and exploration goals offered as a spinoff of planetary defense, the KISS asteroid retrieval mission touted fully multi-use technologies, available for human exploration, resource exploitation, and hazard mitigation. The crux of it would be the development of a very large solar-electric propulsion module to move the asteroid.[43] In addition to meeting the President's goal of sending astronauts to an asteroid by 2025, it would offer a cost-effective opportunity to gain valuable human operational experience in space that could translate to much longer missions to more distant NEOs. It would also represent a new kind of partnership between robotic and crewed missions: one partner to retrieve the asteroid and the other to explore and process its materials. In addition, having a near-Earth asteroid parked in lunar orbit would allow in-depth studies of its structure and composition that could prove invaluable to a future deflection effort. Finally, such an achievement would garner national prestige and inspire the nation. "It would be mankind's first attempt at modifying the heavens to enable the permanent settlement of humans in space."[44]

This emphasis on multi-use technologies was carried through into what became the Asteroid Redirect Mission (ARM). Based on the KISS report, NASA commissioned a three-month study in 2013 to determine whether the feasibility of a near-Earth asteroid retrieval mission concept could withstand more detailed scrutiny. Conducted from January through March by JPL in

---

41. John R. Brophy, interview by Conway, 31 October 2018, transcript in NASA History Division HRC; John Brophy, "Asteroid Retrieval Concept History," 9 April 2013, unpublished document. Lo had to step out of the study due to a family emergency.

42. J. R. Brophy et al., *Asteroid Retrieval Feasibility Study* (Pasadena, CA: Keck Institute for Space Studies, 2 April 2012), *http://kiss.caltech.edu/final_reports/Asteroid_final_report.pdf* (accessed 12 June 2019).

43. For the KISS study, the concept design for the propulsion system was done by Glenn Research Center's COMPASS group and led by Steve Olson.

44. Brophy et al., "Asteroid Retrieval Feasibility Study."

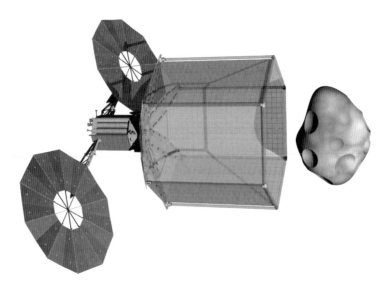

Figure 9-2. Asteroid Redirect Mission Option A concept: Capture an asteroid less than 8 meters in diameter. (Brian Muirhead, "Asteroid Redirect Robotic Mission [ARRM] Concept Overview: Briefing to SBAG," 30 July 2014, p. 21)

collaboration with Glenn Research Center, Johnson Space Center, Langley Research Center, and Marshall Space Flight Center, the study concluded that "the key aspects of finding, capturing and redirecting an entire small, near-Earth asteroid to the Earth-Moon system by the first half of the next decade are technically feasible."[45]

The ARM mission was almost immediately unpopular in both Congress and the scientific community. The Small Bodies Assessment Group argued in July 2013 that it had not been defined as a science mission and should not be funded by the Science Mission Directorate for that reason; dedicated robotic sample-return missions like Japan's Hayabusa mission or NASA's recently approved OSIRIS-REx mission would be more cost-effective.[46] In 2014, MIT's Richard Binzel presented a harsh assessment of the mission to SBAG, comparing it to a fantasy mission he called FARCE—the Far Away Robotic sandCastle Experiment. He contended that the "scientific" goals were being invented as post hoc justification and that NASA's resources would be

45. John R. Brophy and Brian Muirhead, "Near-Earth Asteroid Retrieval Mission (ARM) Study" (paper presented at the 33rd International Electric Propulsion Conference, Washington, DC, 6–10 October 2013).

46. Findings from SBAG 9 (Small Bodies Assessment Group), *https://www.lpi.usra.edu/ sbag/findings/index.shtml#sbag9* (accessed 12 June 2019).

better spent using the solar-electric propulsion module to move NASA hardware into lunar orbit instead—a habitat, or outpost, of some kind.[47] Later in the year, he wrote an editorial for *Nature* that argued for a spaceborne survey to find the huge number of small, near-Earth objects that would come close enough to permit human exploration without having to capture and move one. (About six per month pass within the Moon's orbit.)[48] That idea got him an invitation to present to NASA Administrator Charles Bolden, which he used to try to "change the conversation" toward the space-based survey that had dominated all the prior planetary defense reports but still had not been funded. About two months after Binzel's presentation to Bolden, the NASA Advisory Committee, chaired by Cornell planetary scientist Steve Squyres, voted in favor of a "finding" against the redirect mission, too. In addition to Binzel's arguments against ARM, Squyres's group did not believe that the cost cap imposed on the mission, $1.25 billion, was realistic.[49] The KISS study's estimate had been $2.6 billion.[50]

The scientific community's lack of enthusiasm for ARM in its first few years was mirrored in Congress. At hearings before the Space Subcommittee of the House Science, Space, and Technology Committee held in May 2013, only one of four witnesses (Louis Friedman of the Planetary Society, cochair of the 2012 KISS study) actually supported it. The subcommittee chair, Congressman Steven Palazzo of Mississippi, commented that he thought "it may prove a detour for a Mars mission," not a steppingstone.[51] Representative Frank Wolf, chair of the NASA appropriations subcommittee in the House,

---

47. Richard P. Binzel, "Asteroids in the Context of Human Exploration: A Sustainable Path," 30 July 2014, *https://www.lpi.usra.edu/sbag/meetings/jul2014/presentations/0200_Wed_Binzel_Asteroids.pdf* (accessed 12 June 2019). Also see Len Ly, "Asteroid Expert Richard Binzel: ARM Is 'Emperor With No Clothes,'" Spacepolicyonline.com (blog), 1 August 2014, *https://spacepolicyonline.com/news/asteroid-expert-richard-binzel-arm-is-emperor-with-no-clothes/* (accessed 12 June 2019).

48. Richard P. Binzel, "Find Asteroids To Get to Mars," *Nature* (30 October 2014): 559–560.

49. NASA, "NASA Advisory Council Meeting Minutes January 2015," 14 January 2015, *https://www.nasa.gov/sites/default/files/files/SSC_MeetingMinutes2_Jan2015-Tagged.pdf* (accessed 12 June 2019).

50. Brophy et al., *Asteroid Retrieval Feasibility Study*, p. 6.

51. Hearing before the Subcommittee on Space, Committee on Science, Space, and Technology, House of Representatives, "Next Steps in Human Exploration to Mars and Beyond," 113th Cong., 1st sess. (21 May 2013), p. 11.

wrote to President Obama in December 2013 to advocate sending astronauts to the Moon instead. ARM, he said, was "misguided." "While there may be some merit to developing technologies involved in capturing an asteroid, this is hardly compelling as a human mission and is a multi-billion dollar distraction."[52] His successor as chair of that committee, John Culberson of Texas, told reporter Eric Berger in December 2013 that ARM was not "gonna happen…I don't think pushing a rock around space is a productive use of their time and scarce resources."[53] The 2014 NASA Authorization Act required the NASA Administrator to "develop a Human Exploration Roadmap to define the specific capabilities and technologies necessary to extend human presence to the surface of Mars and the sets and sequences of missions required to demonstrate such capabilities and technologies," emphasizing a "stepping stone approach to exploration" while omitting mention of asteroid retrieval.[54]

In 2015, the ARM program changed the mission design and planned capture mechanism. Instead of deploying a "capture bag" that would completely encase a small (up to 8-meter) asteroid, the new mission architecture would instead involve landing on a larger asteroid and using robotic arms to carry a 4-meter boulder back to lunar orbit.[55] This transformation was motivated by the reality that the existing ground-based surveys could not reliably detect asteroids in the 8-meter size class that ARM was aimed at. Michele Gates, NASA's program executive for ARM, recalled that this fact was driven home by their inability to recover 2009 BD, one of the potential ARM targets, after

---

52. Frank R. Wolf to Hon. Barack H. Obama, 13 December 2013, attachment to *https://spacepolicyonline.com/news/wolf-asks-obama-to-hold-white-house-conference-in-2014-on-return-to-moon/* (accessed 4 November 2019), copy in NEO History Project collection.

53. Eric Berger, "Love Planetary Science? Dying To Explore Europa's Oceans? Meet the Man Who Can Make It Happen," 13 December 2013, *https://blog.chron.com/sciguy/2013/12/love-planetary-science-dying-to-explore-europas-oceans-meet-the-man-who-can-make-it-happen/* (accessed 23 October 2018).

54. NASA Aeronautics and Space Administration Authorization Act of 2014, H. Rept. 113-470, 113th Cong., 2nd sess., 5 June 2014, quoted from sec. 202, p. 5.

55. "NASA Announces Next Steps on Journey to Mars: Progress on Asteroid Initiative," NASA Release 15-050, 25 March 2015, *https://www.nasa.gov/press/2015/march/nasa-announces-next-steps-on-journey-to-mars-progress-on-asteroid-initiative* (accessed 12 June 2019).

its 2011 apparition.[56] Targeting a larger asteroid that could be tracked reliably was a pragmatic adjustment of the mission, while also reinforcing Binzel's argument for the need to carry out a thorough survey before undertaking ARM or any mission like it.

The shift in mission goal toward "Option B's" pick-up-a-rock profile offered the potential for a planetary-defense-oriented demonstration, which came to be called the "enhanced gravity tractor."[57] Since the ARM spacecraft would be grabbing a multi-ton boulder off the surface of the target asteroid, the combined spacecraft/boulder mass would be much larger than that of the spacecraft alone. And gravity is a function of mass. So the combined spacecraft/boulder would impose a larger gravitational attraction on the asteroid than the spacecraft alone would have. The combination should be able to alter the asteroid's trajectory more rapidly, too.

The enhanced gravity tractor demonstration was to occupy the mission for most of a year once the spacecraft reached its target asteroid. After picking up a convenient rock, the spacecraft would be directed to a "halo" orbit one asteroid radius from the surface. It would remain in that state for 30–90 days, then move farther away while scientists on the ground verified that it had, in fact, slightly altered the asteroid's orbit. This phase would last for four to five months. Then the spacecraft would be ordered to its final destination, a lunar orbit.[58]

The ARM concept was an opportunity to develop new technologies and capabilities that would be useful across a variety of areas, including everything from rendezvous and landing to planetary defense, from autonomous operations to resource mining. For instance, during a round of system concept studies, Associate Administrator for NASA's Human Exploration and Operations Mission Directorate William Gerstenmaier remarked, "[W]e are taking the next steps to develop capabilities needed to send humans deeper into space than ever before, and ultimately to Mars, while testing new techniques to

---

56. Michele Gates, interview by Yeomans and Conway, 22 October 2018, copy in NEO History Project collection.

57. Daniel D. Mazanek, David M. Reeves, Joshua B. Hopkins, Darren W. Wade, Marco Tantardini, and Haijun Shen, "Enhanced Gravity Tractor Technique for Planetary Defense" (paper presented at the 4th International Academy of Astronautics (IAA) Planetary Defense Conference, Frascati, Rome, Italy, 13–17 April 2015), *https://ntrs. nasa.gov/archive/nasa/casi.ntrs.nasa.gov/20150010968.pdf* (accessed 12 June 2019).

58. Mazanek et al., "Enhanced Gravity Tractor Technique for Planetary Defense," p. 9.

protect Earth from asteroids."[59] These capabilities would form a broad base of expertise available for missions to diverse locations with multiple objectives, preparing for a future of spaceflight that could be flexible and ready for anything. The ARM would fulfill the legacy of Apollo not by directly aiming for the Moon or Mars, but by providing a crucial steppingstone to those objectives and any others that might follow.[60]

As NASA was making this shift toward the ARM and the development of multi-use technologies, the Agency was also finding that it was no longer alone in its interest in—and perhaps even its ability to mount—a near-Earth asteroid mission. The second decade of the 21st century saw several private companies enter the arena to explore one branch of what then-Director of NASA Ames Research Center Pete Worden called "the key motivations of humanity...fear, greed, and curiosity."[61] All three had long been drivers of NEO research within government space agencies and international organizations, although the profit motive for mineral or water extraction was generally couched in terms of support for larger space operations rather than individual returns. Now, the potential for lucrative resource exploitation in the private sector was seemingly within reach, heralding the realization of ambitious dreams spanning more than a century.

Private Interest in Asteroid Mining

In 1979, an Associated Press article on the future of private asteroid mining was published in newspapers across the United States.[62] Featuring a profile of John Kraus, director of the Ohio State–Ohio Wesleyan radio observatory and a visiting professor of electrical engineering at Ohio University, the article presciently predicted the inevitable entrance of private companies into an

---

59. "NASA Selects Studies for the Asteroid Redirect Mission," 19 June 2014, *https://www.nasa.gov/content/nasa-selects-studies-for-the-asteroid-redirect-mission* (accessed 14 May 2021).

60. "President Barack Obama on Space Exploration in the 21st Century."

61. Pete Worden, interview by Yeomans and Conway, 7 October 2016, transcript in NEO History Project collection.

62. E.g., "Mankind's Future Lies in Mineral-Rich Asteroid," *Newark Advocate* (17 November 1979); also quoted in Matt Novak, "Asteroid Mining's Peculiar Past," BBC, 18 November 2014, *http://www.bbc.com/future/story/20130129-asteroid-minings-peculiar-past* (accessed 12 June 2019).

industry long dominated by governments. Kraus is quoted as saying, "You won't be able to keep them out once you turn a profit in space," and he draws upon the analogy of the American West to express the allure of the final frontier: "[it's] where mankind's future lies—it offers a tremendous potential for energy, materials and room for expansion."[63] While it would take decades for private companies to venture into space, the promise of abundant resources and near-term innovative technologies allowed this objective to be framed as a realizable ambition.

Throughout the 1970s and 1980s, closely tied to the broader goal of space colonization, prospecting on the Moon and small bodies of the solar system was a focal point for Gerard K. O'Neill's Space Settlement workshops and other space advocacy efforts. Convinced that the U.S. government would not pursue space resource utilization to a sufficient extent—or think it appropriate that it should—O'Neill founded the Space Studies Institute (SSI) in 1977.[64] As a nonprofit organization, SSI funded related research and organized conferences to facilitate the generation of knowledge and make it available to private enterprise. The initial research focus was O'Neill's mass driver concept, but many other small and medium grants were distributed as well, including one that funded Helin's first ground-based search for Earth-Sun Trojans.[65]

The L5 Society, an organization founded to promote O'Neill's concept for a space colony located at the L_4 or L_5 Lagrangian point in the Earth-Moon system, was incorporated in 1975. Although its members never achieved their ultimate goal "to disband the society in a mass meeting at L-5,"[66] the group wielded considerable influence during the 1980 Moon Treaty Hearings, ultimately succeeding in preventing the treaty's ratification by the United States.[67] Formally named the Agreement Governing the Activities of States on the Moon and Other Celestial Bodies, the Moon Treaty had been developed by the Legal Subcommittee of the United Nations' Committee on the Peaceful

63. Novak, "Asteroid Mining's Peculiar Past," n. 62.

64. "History," *Space Studies Institute, http://ssi.org/about/history/* (accessed 12 June 2019).

65. "SSI Newsletters: 1984 November December," *http://ssi.org/reading/ssi-newsletter-archive/ssi-newsletters-1984-1112-december/* (accessed 12 June 2019).

66. *L-5 News*, no. 1 (September 1975), *https://space.nss.org/l5-news-1975/* (accessed 15 September 2021).

67. Michael Listner, "The Moon Treaty: Failed International Law or Waiting in the Shadows?" *Space Review*, 24 October 2011, *http://www.thespacereview.com/article/1954/1* (accessed 14 May 2021).

Uses of Outer Space (COPUOS) starting in 1972. It articulated that the Moon and other celestial bodies of the solar system aside from Earth should be used for peaceful purposes only, that the United Nations should be made aware of any bases established on them, and that any activities on or near these bodies should not disrupt their environments.[68] The key section that raised red flags with the L5 Society was Article 11, which declared the Moon and its natural resources (and by extension asteroids and other small bodies) to be the "common heritage of mankind," not subject to ownership by any state, organization, entity, or natural person. Worried that this language would prevent private enterprise from pursuing extraterrestrial mining endeavors, and therefore hinder long-term goals of living in space, the society hired Washington lawyer-lobbyist Leigh Ratiner to help them generate concern among members of the Senate Foreign Relations Committee.[69] Ultimately, action on signing the treaty was suspended and the incoming Reagan administration shelved it.[70] As of now, none of the "Big Three" spacefaring nations—the United States, Russia, and China—have signed, acceded to, or ratified the Moon Treaty, which is currently considered a failed international law.[71]

With the failure of the Moon Treaty to be ratified by the major spacefaring nations, the legal status of resource extraction ventures on the Moon other bodies fell into limbo. Though not as restrictive as the Moon Treaty, the Outer Space Treaty of 1967 contained the basic provision that "Outer space, including the moon and other celestial bodies, is not subject to national appropriation by claim of sovereignty, by means of use or occupation, or by any other means."[72] That language left the treaty open to debate over what defines a

---

68. UN General Assembly Resolution 34/68, "Agreement Governing the Activities of States on the Moon and Other Celestial Bodies," 5 December 1979, *http://www.unoosa.org/pdf/gares/ARES_34_68E.pdf* (accessed 4 November 2019).

69. Michael A. G. Michaud, *Reaching for the High Frontier*, chap. 5 (Westport, CT: Praeger, 1986), *https://space.nss.org/reaching-for-the-high-frontier-chapter-5/* (accessed 12 June 2019).

70. Listner, "The Moon Treaty."

71. Ibid. See also *http://disarmament.un.org/treaties/t/moon* (accessed 12 June 2019). The L5 Society merged with Wernher von Braun's National Space Institute to found the National Space Society in 1987. The nonprofit hosts an annual International Space Development Conference and supports robotic and crewed missions in both the public and private sectors (see *https://space.nss.org/*).

72. The formal name of this treaty is "Treaty on Principles Governing the Activities of States in the Exploration and Use of Outer Space, including the Moon and Other

"celestial body," with opposing arguments hinging on whether a body is a *place* that cannot be physically transferred (as in the case of terrestrial land) or a *thing* that can be moved.[73] In 1997, entrepreneur and founder of the Space Development Corporation (SpaceDev) Jim Benson announced his plan to build the Near-Earth Asteroid Prospector (NEAP), a low-cost robotic mission that would travel to and claim a near-Earth asteroid in order to demonstrate that "profitable, commercial space exploration missions can be flown" to further the development and utilization of space resources.[74] This intention presented a challenge to space law because Benson would not be traveling to the asteroid himself to take ownership of it, but sending a robotic probe instead, constituting a claim of "telepossession."[75] Some terrestrial precedent for telepossession had been established, as in the 1989 case of a shipwreck claimed by a salvage company, so Benson's robotic mission remained in a legal gray area and went unchallenged.[76] However, funding challenges and business-related legal issues ultimately kept NEAP from getting off the ground, and SpaceDev switched its focus to other contracts in the early 2000s to stay afloat.[77]

As the first decade of the 21st century progressed, commercial spaceflight began to establish itself as a viable partner—or even alternative—to government-sponsored space organizations. Blue Origin was established by Jeff Bezos in 2000, and SpaceX followed in 2002, founded by Elon Musk and Tom Mueller. The Google Lunar XPRIZE was announced in 2007, offering $20 million dollars to the first group to send an autonomous rover to

Celestial Bodies." Outer Space Treaty, Article II, *http://www.unoosa.org/pdf/gares/ARES_21_2222E.pdf* (accessed 4 November 2019).

73. Virgiliu Pop, "Legal Considerations on Asteroid Exploitation and Deflection," in *Asteroids: Prospective Energy and Material Resources* (Berlin and Heidelberg: Springer, 2013), pp. 659–680.

74. James Benson, "Near Earth Asteroid Prospector" (presented at the 36th AIAA Aerospace Sciences Meeting and Exhibit, Reno, NV, 12–15 January 1998), doi:10.2514/6.1998-646.

75. Pop, "Legal Considerations on Asteroid Exploitation and Deflection," p. 670.

76. *The Boston Globe* asked Helin for comment on this in 1998; here is her answer: "'My visceral reaction was "heavens forbid, not on your life,"' said Eleanor Helin, a NASA astronomer who has discovered literally hundreds of asteroids—including one that Benson is considering as a target." David Chandler, "Staking a Claim," *Boston Globe* (8 June 1998): C1.

77. Rex Ridenoure, "NEAP: 15 Years Later," *Space Review*, 17 June 2013, *http://www.thespacereview.com/article/2315/2* (accessed 12 June 2019).

the Moon, land on the surface and travel about 1,000 feet, and send high-definition images and video back to Earth.[78] The Obama administration's focus on sending astronauts to an asteroid was also accompanied by a strategic shift toward public-private partnerships that would allow NASA to collaborate with private companies to develop and operate critical technologies.[79] In 2008, NASA awarded hefty contracts—$1.6 and $1.9 billion, respectively—to both SpaceX and Orbital Sciences (which had been working with NASA in a public-private partnership since the 1990s) to send cargo to the International Space Station.[80]

For many other private spaceflight companies, resource extraction was the primary motivator, and objectives for space prospecting fell into two broad categories: the mining of valuable minerals to be returned to and sold on Earth versus the extraction of materials—such as water, structural metals, or helium-3—for use in space to support ongoing operations. Key to the latter emphasis on space manufacturing was the idea of mining water to produce hydrogen and oxygen for spacecraft propellant. Moon Express was co-founded in 2010 by Robert Richards, Naveen Jain, and Barney Pell, with the long-term goal of mining valuable minerals on the Moon, particularly platinum-group metals believed to fetch a tidy profit in terrestrial markets.[81] This was followed by the announcement of several more commercial developments in 2012 and 2013. Arkyd Astronautics, which had been founded in early 2009 by Peter Diamandis and Eric Anderson, changed its name to Planetary Resources on 1 January 2012 and held a press conference in April outlining plans for a near-Earth asteroid mining venture.[82] Starting with a series of small, inexpensive space telescopes to characterize potential target asteroids, they would later

---

78. Adam Mann, "The Year's Most Audacious Private Space Exploration Plans," Wired.com, 27 December 2012, *https://www.wired.com/2012/12/audacious-space-companies-2012/* (accessed 12 June 2019).

79. "President Barack Obama on Space Exploration in the 21st Century."

80. See, e.g., NASA Office of Inspector General, "Commercial Cargo: NASA's Management of Commercial Orbital Transportation Services and ISS Commercial Resupply Contracts," IG-13-016, 13 June 2013, *https://oig.nasa.gov/docs/IG-13-016.pdf* (accessed 12 June 2019).

81. E.g., Saki Knafo and A. J. Barbosa, "The New Space Biz: Companies Seek Cash in the Cosmos," *Huffington Post*, 22 July 2011, *https://www.huffingtonpost.com/2011/07/22/new-space-business_n_907358.html* (accessed 12 June 2019).

82. "Timeline," PlanetaryResources.com, *https://www.planetaryresources.com/company/timeline/* (accessed 15 December 2017).

adapt these spacecraft for prospecting missions looking for valuable minerals like platinum-group metals.[83]

More support for asteroid mining operations came from the B612 Foundation's 2012 announcement of their proposed Sentinel mission. As we saw in the previous chapter, the B612 Foundation emerged from a 2001 planetary defense meeting at NASA's Johnson Space Center. Founded by astronauts Edward T. Lu and Rusty Schweickart, together with planetary scientists Clark Chapman and Piet Hut, the organization sought to promote low-thrust (and non-nuclear) techniques for NEO deflection. In the Sentinel announcement and subsequent communications, B612 played up the telescope's usefulness to asteroid mining endeavors, pointing out that improved orbit determination resulting from their mapping campaign would give private companies a leg up on where to look for potential targets.[84] However, as chapter 8 describes, private funding efforts for the telescope fell short, and NASA terminated its agreement with B612 in 2015.

A year after the Planetary Resources announcement and six months after Sentinel was announced, Deep Space Industries (DSI) entered the private spaceflight market. On 22 January 2013, the company announced plans to launch a fleet of mining spacecraft based on low-cost CubeSat components in 2015, with the goal of beginning resource extraction operations targeting metals and water by the mid-2020s.[85] Predicting an abundance of wealth to go around, Planetary Resources president Chris Lewicki welcomed the new competitor, stating, "Deep Space Industries also sees the importance of accessing and utilizing the resources of space. Asteroid mining will open a trillion-dollar industry and provide a near-infinite supply of space-based resources to support our growth both on this planet and off."[86]

---

83. Dan Leone, "Asteroid Mining Venture To Start with Small, Cheap Space Telescopes," SpaceNews.com, 24 April 2012, *http://spacenews.com/asteroid-mining-venture-start-small-cheap-space-telescopes/* (accessed 12 June 2019).

84. Mike Wall, "Private Space Telescope Project Could Boost Asteroid Mining," Space.com, 10 July 2012, *https://www.space.com/16501-private-space-telescope-asteroid-mining.html* (accessed 12 June 2019).

85. Mike Wall, "Asteroid-Mining Project Aims for Deep-Space Colonies," Space.com, 22 January 2013, *https://www.space.com/19368-asteroid-mining-deep-space-industries.html* (accessed 12 June 2019).

86. "Deep Space Industries' Lofty Asteroid Ambitions Face High Financial Hurdles," nbcnews.com, 22 January 2013, *https://www.nbcnews.com/science/cosmic-log/deep-*

Despite this rosy outlook, both companies faced skepticism on multiple fronts, including doubts about the true availability of the much-anticipated resources, as well as the cost-effectiveness of returning them to Earth. To the latter criticism, DSI responded that their main goal would be the production of extraterrestrial materials for use in space to support large-scale space operations, not to bring extracted minerals back for sale in Earth markets. Primarily, this meant finding and mining water ice, which could be chemically processed into fuel and oxygen. But achieving this objective would depend on a much-expanded private space industry, since the fantasy of convenient water-based space "fuel depots" would require significant demand to become a reality. These and other critical factors beyond DSI's direct control projected a daunting future for the nascent company.

The actual value of asteroid resources, as well as the economic feasibility of reaching them and extracting them, has been a subject of dispute since the settlement studies of the 1970s and earlier. Planetary scientist John S. Lewis's 1998 book *Mining the Sky* painted an optimistic picture of abundant and practically minable minerals, as well as water resources that could be reduced to hydrogen and oxygen and offered to spacecraft at refueling stations. A nine-month study for a NASA Innovative and Advanced Concepts (NIAC) investigation of the Robotics Asteroid Prospector (RAP), however, contradicted this outlook, with one paper concluding: "Despite all these inspiring and idealistic purposes for the retrieved asteroid, the RAP team could not find any scenario for a realistic commercial economic return from such a mission. The only scenario for making a profit appears to be all in situ mining, extraction, and processing to enable the delivery of finished products or commodities to the customers who will want them and pay for them."[87] The study reported a range of "fairly dispositive findings," including the prohibitively high cost of returning asteroid resources to low-Earth orbit or to Earth's surface, the lack of a market for rare-earth elements (long considered a major potential asset of

space-industries-lofty-asteroid-ambitions-face-high-financial-hurdles-flna1b8077194 (accessed 12 June 2019).

87. K. Zacny et al., "Asteroid Mining," *Proceedings of the AIAA SPACE 2013 Conference and Exposition*, AIAA paper 2013-5304 (San Diego, CA, 12–13 September 2013), doi:10.2514/6.2013-5304.

mining small solar-system bodies), and the infeasibility of solar-electric propulsion as a long-term technology sustaining asteroid exploitation.[88]

On the legal front, asteroid mining prospects fared better. In November 2015—thanks in part to lobbying efforts by Planetary Resources, DSI, and other private space industry companies—Congress passed H.R. 2262, the U.S. Commercial Space Launch Competitiveness Act, making it explicitly legal for U.S. companies to "engage in the commercial exploration and exploitation of 'space resources' [including...water and minerals]."[89] Also known as the SPACE Act, the provision asserts that such activities do not constitute "appropriation" by the United States, which is banned under the Outer Space Treaty, but it remains undetermined whether it violates or is consistent with international law.[90]

Even with this legal boost, private asteroid mining did not become a successful business venture in the 2000s. The best known of these private companies, Planetary Resources, shifted its focus to Earth observation technology in 2016, an area with immediate applications to terrestrial resource extraction, though it also claimed a successful Earth-orbit test of some of its asteroid-hunting technology in April 2018.[91] But the company failed to secure enough venture capital to sustain itself and was acquired by a cryptocurrency firm,

---

88. Marc Cohen et al., "Robotic Asteroid Prospector (RAP) Staged from L-1: Start of the Deep Space Economy, Final Report 9 July 2013, Corrected 17 July 2013," *https://www.nasa.gov/directorates/spacetech/niac/2012_phase_I_fellows_cohen.html* (accessed 12 June 2019).

89. H.R. 2262, U.S. Commercial Space Launch Competitiveness Act, H. Rept. 114-119, 114th Cong., *https://www.congress.gov/bill/114th-congress/house-bill/2262* (accessed 12 June 2019).

90. James Rathz, "Law Provides New Regulatory Framework for Space Commerce," *Regulatory Review*, 31 December 2015, *https://www.theregreview.org/2015/12/31/rathz-space-commerce-regulation/* (accessed 12 June 2019).

91. "Planetary Resources Raises $21.1 Million in Series A Funding; Unveils Advanced Earth Observation Capability," PlanetaryResources.com, 26 May 2016, *https://www.planetaryresources.com/2016/05/planetary-resources-raises-21-1-million-in-series-a-funding-unveils-advanced-earth-observation-capability/* (accessed 17 May 2021); Mike Wall, "Asteroid Miners' Arkyd-6 Satellite Aces Big Test in Space," Space.com, 25 April 2018, *https://www.space.com/40400-planetary-resources-asteroid-mining-satellite-mission-accomplished.html* (accessed 25 April 2018).

ConsenSys, Inc., in October 2018.[92] Cryptocurrency "mining" converts electricity into data (and a lot of waste heat), which is not at all what asteroid miners would have to do. Deep Space Industries also changed its strategic focus from asteroid mining toward spacecraft technology more generally, and it was acquired by Bradford Space in 2019.

The demise of the two significant U.S. private asteroid mining firms paralleled the end of NASA's Asteroid Redirect Mission. A change of administration in January 2017 put lunar advocates in charge of NASA's future human spaceflight planning, and ARM was canceled in April.[93] Cancellation was followed by a period of uncertainty over the program's new direction, but in 2019, the administration announced a new lunar landing goal: to return to the Moon with astronauts by 2024.[94] Beyond that, details of the new program were unclear. Louis Friedman, who had pitched the robotic asteroid retrieval mission to NASA Deputy Administrator Lori Garver and on Capitol Hill, reflected that congressional dislike reflected promoters' inability to convince critics that the retrieval mission was not the end goal of the human spaceflight program, but merely a steppingstone. He thought it should have been seen as the Gemini missions had been during the Apollo Moon campaign—as tests of equipment and methods that were meant for a larger goal.[95] NASA had developed a "Journey to Mars" promotion campaign to try to make that point publicly during the last years of the Obama administration

---

92. "Asteroid Mining Company Planetary Resources Acquired by Blockchain Firm," SpaceNews.com, 31 October 2018, *https://spacenews.com/asteroid-mining-company-planetary-resources-acquired-by-blockchain-firm/* (accessed 12 June 2019).

93. "NASA Closing Out Asteroid Redirect Mission," SpaceNews.com, 14 June 2017, *http://spacenews.com/nasa-closing-out-asteroid-redirect-mission/* (accessed 12 June 2019).

94. The program was named Artemis, after the Greek goddess of the hunt and sister to Apollo. See, e.g., Eric Berger, "NASA's Full Artemis Plan Revealed: 37 Launches and a Lunar Outpost," *Ars Technica*, 20 May 2019, *https://arstechnica.com/science/2019/05/nasas-full-artemis-plan-revealed-37-launches-and-a-lunar-outpost/* (accessed 17 May 2021). The program was announced after submission of the President's FY 2020 budget proposal and thus was not part of the Agency's budget request. See Jeff Foust, "NASA Seeks Additional $1.6 Billion for 2024 Moon Plan," SpaceNews.com, 13 May 2019, *https://spacenews.com/nasa-seeks-additional-1-6-billion-for-2024-moon-plan/* (accessed 17 May 2021); Jeff Foust, "House Bill Restores Funding to Earth Science and Astrophysics Missions," SpaceNews.com, 22 May 2019, *https://spacenews.com/house-bill-restores-funding-to-earth-science-and-astrophysics-missions/* (accessed 12 June 2019).

95. Lou Friedman, interview by Conway, 6 November 2018, transcript in NEO History Project collection.

and had at least begun to make congressional inroads. In the fiscal year 2016 Commerce, Justice, Science and Related Agencies Appropriations bill, House legislators wrote:

> While questions remain about the overarching mission of the asteroid redirect mission, the Committee understands that it has been useful to the extent that it has motivated NASA to develop new rocket propulsion technology to be used in interstellar travel and methods to deflect near earth objects that threaten the Earth. The Committee is particularly supportive of these portions of the mission….[96]

NASA's Lindley Johnson reflected later that the ARM saga had two salutary effects. As we saw in chapter 8, it resulted in greater resources being provided to support survey efforts, though not enough to launch a new space-based survey effort beyond NEOWISE. And it broadened knowledge about NEOs at NASA Headquarters. More people understood more about them and what it would take to exploit them.

96. Commerce, Justice, Science and Related Agencies Appropriations Bill, 2016, H. Rept. 114-130, 114th Cong., 1st sess., 27 May 2015, p. 58, *https://www.congress.gov/ congressional-report/114th-congress/house-report/130/1?q=%7B%22search%22%3A%5B %22nasa+appropriations+2016%22%5D%7D* (accessed 12 June 2019).

CHAPTER 10
ORGANIZING PLANETARY DEFENSE

As we have seen in prior chapters, there was considerable resistance within the U.S. federal government to having any formal role in addressing the NEO hazard. Neither the U.S. Air Force nor NASA wanted the role. Natural hazards did not fall within the Defense Department's legal authority in any case, and NASA did not accept the Spaceguard Survey's tasking of finding 90 percent of the largest NEOs until Congress had asked for it several years running. These requests had followed from the 1989 FC close call and the more dramatic 1994 Shoemaker-Levy 9 comet impact with Jupiter, products of the issue entrepreneurship of Congressmen George E. Brown and Dana Rohrabacher.

The formation of the NASA Near-Earth Objects Observations Program in 1998 provided an institutional home for this new hybrid science-policy effort. NEOOP supported the NEO surveys, follow-up observations, and characterization efforts discussed in earlier chapters. Its support contributed to the understanding of near-Earth asteroids and comets as well as to the policy goal of finding 90 percent of the 1-kilometer-diameter and larger NEAs expected to exist. As we saw in chapter 8, by 2010, the program's officials believed that they had probably achieved that goal. In anticipation of that achievement, in 2003 NEOOP had convened a science definition team to evaluate a new goal. That became the basis of the 2005 Brown Act requirement, to find 90 percent of the NEAs larger than 140 meters in diameter by the end of 2020.[1]

---

1. NASA Authorization Act, Pub. L. No. PL 109-155 (2005), *https://www.gpo.gov/fdsys/pkg/PLAW-109publ155/pdf/PLAW-109publ155.pdf* (accessed 22 July 2019). This law also provided NASA with the legal authority (and responsibility) to conduct NEO surveillance.

The 2005 NASA Authorization Act also required NASA to carry out an analysis of survey options to accomplish the Brown Act requirement as well as an "analysis of possible alternatives that NASA could employ to divert an object on a likely collision course with Earth."[2] This congressional demand launched the Analysis of Alternatives study discussed in chapter 8 from the perspective of next-generation surveys; in this chapter, we will look at the mitigation analysis and a National Research Council (NRC) review that Congress requested after this study's publication.[3]

While the NRC review was in progress, Congress added still another requirement that wound up expanding NASA's role substantially. In the NASA Authorization Act of 2008, Congress requested that the White House Office of Science and Technology Policy develop a policy for "notifying Federal agencies and relevant emergency response institutions of an impending near-Earth object threat" and recommend a lead Federal agency for planetary defense.[4] The mandate triggered the first explicit civil defense exercises for impacts and, combined with the Chelyabinsk bolide event in 2013, eventually resulted in the formation of the NASA Planetary Defense Coordination Office.

Rocks from the Sky

The 2000s witnessed a handful of celestial impacts that received public attention, starting with a meteor that disintegrated over Canada's Yukon Territory near Tagish Lake on 18 January 2000. The object was not observed

---

2. NASA Authorization Act, Pub. L. No. PL 109-155 (2005), section 321, subsection (d), paragraph 4C, *https://www.gpo.gov/fdsys/pkg/PLAW-109publ155/pdf/PLAW-109publ155.pdf* (accessed 22 July 2019).

3. National Research Council, *Defending Planet Earth: Near-Earth Object Surveys and Hazard Mitigation Strategies* (Washington, DC: National Academies Press, 2010). The NRC study references the Consolidated Appropriations Act of 2008 (PL 110-161) as its basis, but the original text of the law, dated 17 December 2007, does not contain the request. The law was amended in January 2008, and that amendment contains the request. House Appropriations Committee, Consolidated Appropriations Act, 2008, Pub. L. No. PL 110-161 (2008), *https://www.govinfo.gov/content/pkg/CPRT-110HPRT39564/pdf/CPRT-110HPRT39564-DivisionB.pdf* (accessed 22 July 2019).

4. House Committee on Science and Technology, National Aeronautics and Space Administration Authorization Act of 2008, H. Rept. 110-702, sec. 804, 110th Cong., 2nd sess., 9 June 2008, p. 14, *https://www.congress.gov/congressional-report/110th-congress/house-report/702/1* (accessed 8 July 2019).

by astronomers, but its explosion was witnessed by dozens of people, surveillance cameras, cell phone cameras, U.S. intelligence satellites, and seismic and infrasound monitors operated for nuclear nonproliferation purposes. It deposited around 10 kilograms of fragments on the frozen surface of Tagish Lake. A local citizen found the first fragments a week after the impact; Alan Hildebrand and Peter G. Brown assembled an expedition to collect more and interview as many witnesses as could be found to help reconstruct its trajectory.[5] This former 4-meter-diameter near-Earth asteroid proved to be a carbonaceous chondrite.

Late in the busy year of 2008, an unexpected bolide, 2008 TC3, was identified shortly before its entry and explosion above Sudan, permitting a test of new communications protocols. On 6 October 2008, Richard Kowalski, using the Catalina Sky Survey's 1.5-meter telescope on Mount Lemmon in Arizona, observed a small asteroid, only 3 or 4 meters in diameter, very close to Earth. It was also soon observed by Gordon Garradd at the Siding Spring Observatory in Australia, then still active in the Catalina Sky Survey, and two amateur observers in Tucson and South Australia. The Minor Planet Center's orbit solution placed it on a collision course with Earth, and JPL's Steven Chesley projected it entering the atmosphere over northern Sudan on 7 October.

The small size of the object, which the MPC designated 2008 TC3, meant that no

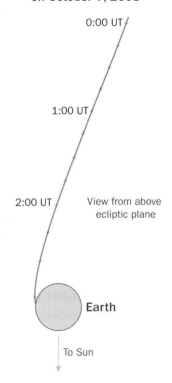

Impact Trajectory of 2008 TC3 on October 7, 2008

0:00 UT

1:00 UT

2:00 UT

View from above ecliptic plane

Earth

To Sun

Figure 10-1. Impact trajectory of 2008 TC3, as it would be seen from above the ecliptic plane (or, from above Earth's north geographic pole). (Image courtesy of Donald K. Yeomans)

5. James Berdahl, "Morning Light: The Secret History of the Tagish Lake Fireball," MIT Comparative Media Studies/Writing, 4 June 2010, *http://cmsw.mit.edu/tagish-lake-fireball-secret-history/* (accessed 18 May 2021); Alan R. Hildebrand et al., "The Fall and Recovery of the Tagish Lake Meteorite," *Meteoritics & Planetary Science* 41, no. 3 (2006): 407–431.

one expected it to reach the ground. But NASA's Lindley Johnson exercised notification protocols that were in draft at the time.[6] These required that the NASA Office of External Affairs handle the notification of other government agencies. Since the expected entry was over Sudan, not the United States, notification went to the U.S. State Department with a request to inform Sudan's government, as well as to the National Security Council, the White House Office of Science and Technology Policy (OSTP), the Joint Space Operations Center, and the National Military Command Center.[7] Late on 6 October, NASA also released a media advisory quoting Yeomans: "The unique aspect of this event is that it is the first time we have observed an impacting object during its final approach."[8] It also cautioned that there was little chance that sizable fragments would reach the ground.

The European weather satellite Meteosat 8 witnessed 2008 TC3's explosion, as did U.S. intelligence satellites. Edward Tagliaferri, who had prepared the AIAA's white paper on the NEO threat back in 1990, was compiling and analyzing data from classified sensors on bolides entering Earth's atmosphere and heard about the upcoming impact.[9] He managed to get details of the explosion out to Peter G. Brown of the University of Western Ontario.[10] Brown operated a website dedicated to near-real-time meteor tracking, and he put the information out on his meteor website.[11] The explosion of 2008 TC3 had happened at an altitude of around 37 kilometers. This was much higher

---

6. These became NASA Policy Directive 8740.1, "Notification and Communications Regarding Potential Near-Earth Object Threats," 27 January 2017, *https://nodis3.gsfc.nasa.gov/displayDir.cfm?t=NPD&c=8740&s=1* (accessed 12 November 2019).

7. A complete list is given in AF/A8XC, "Natural Impact Hazard (Asteroid Strike) Interagency Deliberate Planning Exercise After Action Report," December 2008, p. 5.2-27, *https://cneos.jpl.nasa.gov/doc/Natural_Impact_After_Action_Report.pdf* (accessed 18 May 2021).

8. "NASA—Small Asteroid To Light Up Sky over Africa," NASA Press Release 08-254, 6 October 2008, *https://www.nasa.gov/home/hqnews/2008/oct/HQ_08254_Small_asteroid.html* (accessed 10 July 2019).

9. See AIAA Space Systems Technical Committee, "Dealing with the Threat of an Asteroid Striking the Earth: An AIAA Position Paper," April 1990, *https://space.nss.org/wp-content/uploads/1990-AIAA-Position-Paper-Threat-Of-Asteroid-Strike.pdf* (accessed 15 September 2021).

10. Edward Tagliaferri, interview by Yeomans, 14 December 2016, transcript in NASA History Division HRC.

11. Brown's website posting is at *http://aquarid.physics.uwo.ca/~pbrown/usaf/usg282.txt* (accessed 3 October 2017).

than expected, and nothing was likely to reach the ground from that height. Nevertheless, staff members of the University of Khartoum made an initial, unsuccessful attempt to find fragments the next day.

In addition to the satellite observations, there were many eyewitnesses to 2008 TC3's demise. The asteroid had been large enough to create a dense dust cloud, captured on numerous cell phone images. A cell phone image of the asteroid's fiery trail soon emerged, too.[12] Muawia Shaddad at the University of Khartoum invited meteor specialist Petrus "Peter" Jenniskens of the SETI Institute to give a talk in early December. They examined weather data to understand where the asteroid's remnants might have fallen. They also visited several eyewitnesses, and with a group of students and staff searched an area near a train station (known as Station 6, or Almahata Sitta in Arabic) that was along the breakup path and from which the object's disintegration had been witnessed. On 6 December, a student found the first pieces of debris; ultimately, searchers found hundreds of fragments during several different search expeditions. The meteorites were collectively named for the train station.[13]

The 2008 TC3 event provided a coda, of a sort, to the Spaceguard Survey goal of discovering 90 percent of all the 1-kilometer or larger near-Earth asteroids expected to exist. This was the year the goal should have been met, but it was not. As of May 2008, about 80 percent of the expected population of these near-Earth asteroids had been discovered; it would take until the end of 2010 before Yeomans was willing to report that the goal had been met.[14]

A NEO discovered in early 2011 triggered another campaign by the B612 Foundation's Russell "Rusty" Schweickart to get NASA to launch a transponder mission to investigate, and it produced a flurry of news reports as well. This was 2011 AG5, discovered by the Catalina Sky Survey on 8 January 2011. JPL's Sentry program initially calculated a 0.2 percent chance of impact in 2040, based on 213 observations. Lost in September 2011, 2011 AG5 was recovered by David Tholen, Richard Wainscoat, and Marco Micheli of the Institute for Astronomy using the Gemini 8-meter telescope on Maunakea

12. See *https://apod.nasa.gov/apod/ap081108.html* (accessed 22 July 2019).

13. P. Jenniskens et al., "The Impact and Recovery of Asteroid 2008 TC3," *Nature* 458, no. 7237 (26 March 2009): 485–488, doi:10.1038/nature07920; Muawia Shaddad et al., "The Recovery of Asteroid 2008 TC3," *Meteoritics & Planetary Science* 45, nos. 10–11 (October 2010): 1557–1589, doi:10.1111/j.1945-5100.2010.01116.x.

14. D. Yeomans et al., "Status Report of the NEO Program Office," May 2008, courtesy of D. K. Yeomans, copy in NEO History Project collection.

in October 2012, and the new observations eliminated the risk.[15] But during the year 2011, when AG5 was unobserved, Schweickart conducted a deflection analysis that he presented to COPUOS's 49th meeting in February 2012. In order to strike Earth, 2011 AG5 would have to pass through a "keyhole" in 2023, and it was at that point that the asteroid could most easily have its course changed by a kinetic impactor. So Schweickart argued that a decision to send a tracker mission needed to be made immediately. He wrote to NASA Administrator Bolden in January 2012 and again in March 2012 to argue his point.[16]

As had been done for Apophis, Yeomans's group at JPL evaluated the evolution of position uncertainty as ground-based observations from apparitions in 2013, 2015, and future years were added to refine the asteroid's position. If 2011 AG5 was not on an impact trajectory, observations in 2012–13 had a 95 percent chance of eliminating the risk, and observations in 2015–16 could raise that chance to 99 percent. Conversely, if the asteroid *was* on a collision trajectory, then the 2012–13 observation opportunities would raise the impact probability to 10–15 percent, and the 2015–16 observations to "around 70%."[17] Since observations in the next few years were likely to be definitive, no deflection decision needed to be made yet. And there were potential deflection mission opportunities with launches between 2018 and 2020.

The NASA NEO Observations Program hosted a workshop at Goddard Space Flight Center in late May 2012 to review and explain these findings. The consensus message they released was, in effect, to wait for the 2013 observations before taking further action. "In the unlikely event that observations made in Fall 2013 show a significant increase in the Earth impact

---

15.   Center for NEO Studies, "'All Clear' Given on Potential 2040 Impact of Asteroid 2011 AG5," *https://cneos.jpl.nasa.gov/news/news176.html* (accessed 16 July 2019).

16.   Russell L. Schweickart to Charles Bolden, 3 March 2012, copy in NEO History Project collection; Phil Plait, "Asteroid 2011 AG5: A Football-Stadium-Sized Rock To Watch Carefully," 6 March 2012, *https://slate.com/technology/2012/03/asteroid-2011-ag5-a-football-stadium-sized-rock-to-watch-carefully.html* (accessed 15 September 2021); and see Linda Billings, "Impact Hazard Communication Case Study: 2011 AG5," August 2014, presented to the First IAWN [International Asteroid Warning Network] Communications Workshop, Colorado, September 2014, *http://iawn.net/meetings/communications.shtml* (accessed 22 July 2019).

17.   Yeomans et al., "Report on Asteroid 2011 AG5 Hazard Assessment and Contingency Planning," 1 June 2012, *https://cneos.jpl.nasa.gov/doc/2011_AG5_Deflection_Study_report_13.pdf* (accessed 17 July 2019).

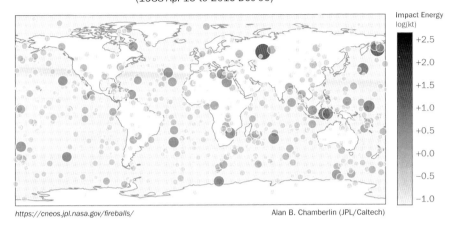

Figure 10-2. Fireballs (bolides) detected by U.S. government sensors between mid-1988 and December 2019. (Generated from *https://cneos.jpl.nasa.gov/fireballs/*)

probability, there is still sufficient time to plan and carry out a successful deflection campaign."[18] As it turned out, the observations made later that year by Tholen's team with the Gemini 8-meter telescope on Maunakea eliminated the remaining impact risk.[19]

The discovery of 2008 TC3 about 20 hours prior to impact, the 2011 AG5 controversy, and still another short-warning impact event—the discovery of the small asteroid 2014 AA only 21 hours prior to Earth impact on 1 January 2014—underscored the need for a very rapid orbit determination and impact prediction process.[20] The JPL Sentry system and its European

18. "Summary of Potentially Hazardous Asteroid Workshop Findings," 29 May 2012, *https://cneos.jpl.nasa.gov/doc/2011_AG5_workshop_sum.pdf* (accessed 17 July 2019).

19. The observers were David Tholen, Richard Wainscoat, and Marco Micheli. See Center for NEO Studies, "'All Clear' Given on Potential 2040 Impact."

20. The asteroids 2008 TC3 and 2014 AA were discovered by Richard Kowalski at the Catalina Sky Survey, and since they were only a few meters in extent, neither posed a substantial threat. Using impact constraints based upon the optical positional observations, Peter Brown utilized the low-frequency infrasound component of the International Monitoring System operated by the Comprehensive Nuclear Test Ban Treaty Organization to place the impact location of 2014 AA at about 15 degrees north latitude and a longitude of about 43 degrees west. D. Farnocchia, S. R. Chesley, P. G. Brown, and P. W. Chodas, "The Trajectory and Atmospheric Impact of Asteroid 2014 AA," *Icarus* 274 (2016): 327–333.

cousin NEODyS could provide Earth approach and impact predictions over the next century for objects whose orbits were well established, but not for newly discovered objects that might be very close to impact. In 2016, Davide Farnocchia and his JPL colleagues developed the Scout system, designed to provide the earliest possible warning for short-term impacts and close approaches of newly discovered objects.[21] Scout takes advantage of the Minor Planet Center's NEO Confirmation web page (NEOCP), which posts the position observations of newly discovered, potential near-Earth objects. Every 5 minutes, Scout fetches information from this page, and though the observational data may be sparse, Scout can produce a suite of possible orbits and assess whether the object might approach, or even reach, Earth. The Minor Planet Center also provides prompt alerts of possible close approaches and impacts so that the Scout and NEOCP systems can provide mutual verification—much as the Sentry and NEODyS system do. By providing an early warning system for possible Earth impactors, Scout and the NEOCP allow prioritization for those objects that require immediate follow-up observations to improve knowledge of their orbits. They therefore also serve as informal coordination and communications tools, enabling astronomers, including amateurs, to contribute observations.

Organizing for Planetary Defense

As we saw in chapter 8, the discovery in late 2004 of near-Earth asteroid 99942 Apophis and the short-lived possibility that it might collide with Earth in 2036 caused a media firestorm and provided an extra boost to secure the inclusion of the Brown Act in the 2005 NASA Reauthorization Act. Congress had also asked for the Analysis of Alternatives study prepared by NASA's Office of Program Analysis and Evaluation (OPAE), which had resulted in criticism from Schweickart and Clark Chapman due to its endorsement of nuclear-based mitigation strategies. Both Schweickart and Chapman had felt that the more refined non-nuclear options had not been properly considered, which exacerbated the frustration caused by the summary report's dismissal

---

21. D. Farnocchia, S. R. Chesley, and A. B. Chamberlin, "Scout: Orbit Analysis and Hazard Assessment for NEOCP Objects" (presented at the American Astronomical Society, DPS meeting no. 48, id.305.03, October 2016). Alan Chamberlin is largely responsible for developing and maintaining the JPL Center for Near-Earth Object Studies web processes, where Sentry and Scout reside; see *https://cneos.jpl.nasa.gov*.

of added funding to support a new program.[22] But still another congressional request in 2008 prodded NASA further along a path toward a formal planetary defense organization.

In the language of the 2008 omnibus appropriations bill, Congress included a directive that the National Research Council (NRC) review the Analysis of Alternatives report and other recent analyses of NEO detection and deflection efforts and provide recommendations for the best way to complete the survey goal set in the 2005 Brown Act and the optimal strategy to devise a mitigation plan. But the specific context for the request was not the criticism of the study by Schweickart and Chapman. Instead, it was the product of a National Science Foundation review panel's decision to reduce the Arecibo Observatory's funding by half—and even close it in 2011—if a new source of funds to operate it could not be found.[23] As we have seen in prior chapters, though, Arecibo was important for characterizing and refining the orbits of nearby asteroids, so Congress directed NASA to provide enough funding to keep Arecibo operating while the National Research Council produced its review.[24]

The resulting study, *Defending Planet Earth: Near-Earth-Object Surveys and Hazard Mitigation Strategies*, was released in 2010.[25] From the mitigation perspective, this document included little new information, and concluded that

[a]lthough the committee was charged in its statement of task with determining the "optimal approach to developing a deflection capability," it concluded

---

22. Russell L. Schweickart, "Technical Critique of NASA's Report to Congress and associated of [*sic*] '2006 Near-Earth Object Survey and Deflection Study: Final Report,' Published 28 December 2006," 1 May 2007, copy in NEO History Project collection; Clark R. Chapman, "Critique of '2006 Near-Earth Object Survey and Deflection Study: Final Report,' Published 28 December 2006," by NASA Headquarters Program Analysis and Evaluation Office, 2 May 2007, copy in NEO History Project collection.

23. Yudhijit Bhattacharjee, "Arecibo To Stay Open Under New NSF Funding Plan," *Science* 328 (18 June 2010): 1462–1463. However, in 2020, the Arecibo telescope's instrument platform collapsed into its bowl, destroying the antenna. As of this writing, no decisions about a replacement had been made. See Eric Hand, "Arecibo Telescope Collapses, Ending 57-Year Run," *Science*, doi:10.1126/science.abf9573.

24. H.R. 110-702, National Aeronautics and Space Administration Authorization Act of 2008, sec. 806, 11th Cong., 2nd sess., 9 June 2008.

25. National Research Council, *Defending Planet Earth: Near-Earth-Object Surveys and Hazard Mitigation Strategies* (Washington, DC: National Academies Press, 2010).

that work in this area is relatively new and immature. The committee therefore concluded that the "optimal approach" starts with a research program.[26]

The committee also argued that this research program should not be funded from existing science budgets. "Because this is a policy-driven, applied program, it should not be in competition with basic scientific research programs or funded from them."[27] In other words, this group was arguing that cosmic impact mitigation research was not a scientific priority. It was "applied research." To this point, all NEO research funding had come from NASA's planetary science funds, and researchers wanted this to stop. Their view was that planetary defense should not come at the expense of planetary science; it should have its own funding.

The NRC contended that there were four key areas upon which NASA should focus its mitigation research program. One of these had not been addressed in the earlier studies: civil defense. The most likely form of Earth impact would come from smaller asteroids, which would explode while still airborne, as the 1908 Tunguska bolide and 2008 TC3 had done. For those, shelter-in-place strategies, evacuation plans, and survivable emergency infrastructure would all provide life-saving support in a cost-efficient manner. These smaller impacts would occur every couple of centuries, on average. The necessary procedures and infrastructure would also be valuable in the event of even larger impacts.[28]

The other three areas involved in-space capabilities. Slow-push methods were recognized to be the most controllable techniques for diverting small asteroids with a sufficiently advanced warning time of decades. Of these, the gravity tractor was identified as the method least sensitive to the variations in physical characteristics of the asteroid and by far the most technologically feasible. Kinetic impactors were also determined to be adequate deflection techniques for NEOs up to 1 kilometer in diameter, again with years to decades of lead time. This method had the added advantage that the ability to hit an asteroid with a spacecraft had recently been demonstrated in space during

---

26. NRC, *Defending Planet Earth*, p. 4.
27. Ibid., p. 5.
28. Ibid., pp. 69–70.

the Deep Impact mission's collision with Comet Tempel-1 in July 2006.[29] However, the kinetic impactor strategy is also very sensitive to the properties of the asteroid, an aspect that the NRC report deemed worthy of further study. The authors recommended that if Congress wished to fund "mitigation research at an appropriately high level," first priority should be testing a kinetic impactor with an accompanying characterization effort.[30]

The final area of interest outlined in the study focused on nuclear deflection techniques. Echoing the language of prior deflection studies, the report found that "[o]ther than a large flotilla (100 or more) of massive spacecraft being sent as impactors, nuclear explosions are the only current, practical means for changing the orbit of large NEOs (diameter greater than about one kilometer). Nuclear explosions also remain as a backup strategy for somewhat smaller objects if other methods have failed. They may be the only method for dealing with smaller objects when warning time is short, but additional research is necessary for such cases."[31] This group contended that the most effective approach would be the use of nuclear stand-off detonations, relying primarily on the neutron radiation produced by the device to heat and vaporize the asteroid's surface material. The resulting ejecta jet would produce the desired velocity change. If necessary, a succession of stand-off devices could be used over several years to shift a particularly large impactor's orbit. This might also be necessary if an impactor appeared to be particularly fragile.

Following the release of *Defending Planet Earth*, the White House Office of Science and Technology Policy (OSTP) responded to the congressional mandate that had been included in the 2008 NASA Authorization Act.[32] Section 804 of the Authorization Act had given the Director of OSTP, John P. Holdren, two years to complete two tasks: first, to lay out a plan of action, in the event of an impending near-Earth object threat, to notify the relevant federal agencies and emergency response institutions; and second, to recommend the appropriate agencies that should be responsible for "(A) protecting the

29. Although the impact was much smaller than would be required to divert a large body, the demonstration of a kinetic impactor (e.g., one moving body successfully striking another on target) was significant.

30. NRC, *Defending Planet Earth*, p. 87.

31. Ibid., p. 79.

32. John P. Holdren, OSTP, to John D. Rockefeller IV et al., Senate Committee on Commerce, Science, and Transportation, 15 October 2010, *https://www.nasa.gov/sites/default/files/atoms/files/ostp-letter-neo-senate.pdf* (accessed 22 July 2019).

United States from a near-Earth object that is expected to collide with Earth; and (B) implementing a deflection campaign, in consultation with international bodies, should one be necessary."[33] To address the first task, OSTP outlined a plan to leverage existing communications protocols and resources employed by the Federal Emergency Management Agency (FEMA) in the Department of Homeland Security (DHS), with additional support from the Department of State (DOS) for international communications. These procedures had been thoroughly tested in recent years, notably in cases of reentering space objects, such as the U.S. reconnaissance satellite USA-193, which was destroyed by the Burnt Frost Joint Task Force in February 2008.[34] In sum, "FEMA considers these procedures to be well-understood and applicable to the emergency notifications needed for a potential NEO threat."[35]

Holdren's office also reaffirmed NASA's crucial role in the early detection of such threats to provide warnings as much in advance as possible. NASA would be responsible for notifying FEMA, the Executive Office of the President, the Defense Department's Joint Space Operations Center, the Department of State, and other federal agencies, which would then utilize existing communications protocols and resources for further notifications. The mitigation options, it acknowledged, were still at too early a stage of development, and therefore NASA would initiate a research program to assess mitigation and deflection technologies. In its budget proposal for fiscal year 2011, the administration had requested an increase in the NEO Observations

---

33. H.R. 6063, National Aeronautics and Space Administration Authorization Act of 2008, section 804, 110th Cong., 15 May 2008, *https://www.congress.gov/bill/110th-congress/house-bill/6063/text* (accessed 22 July 2019).

34. Debris from this event reportedly reentered and burned up within a few months. See "U.S. Satellite Shootdown Debris Said Gone from Space," Reuters, 27 February 2009, *https://www.reuters.com/article/us-space-usa-china-idUSTRE51Q2Q220090227* (accessed 19 August 2019).

35. John P. Holdren, OSTP, to John D. Rockefeller IV et al., Senate Committee on Commerce, Science, and Transportation, 15 October 2010, p. 6, *https://www.nasa.gov/sites/default/files/atoms/files/ostp-letter-neo-senate.pdf* (accessed 22 July 2019); "Navy Missile Hits Dying Spy Satellite, Says Pentagon," CNN, 21 February 2008, *http://www.cnn.com/2008/TECH/space/02/20/satellite.shootdown/index.html* (accessed 10 December 2017). The unusual action was taken due to fear that the spacecraft's fuel tank of toxic hydrazine might survive reentry, though it also served as a highly visible demonstration of the Navy's antiballistic missile capabilities.

Program's budget from $5.8 million to $20.3 million to pay for this expanded research effort.

The 2010 OSTP memo represented something of a milestone for NASA's NEO efforts, in that it marked the early stages of a true plan of action—if not on the mitigation and deflection side, at least on the civil defense front. As Lindley Johnson put it, "[T]hat got us started looking at it more as an overall program for planetary defense versus just detection and tracking, and got us working with FEMA on these emergency response exercises."[36] The first of these exercises had already taken place in 2008, conducted by the Future Concepts and Transformation Division of the U.S. Air Force under the leadership of Lieutenant Colonel Peter Garretson. The Air Force's Future Concepts and Transformation Division existed to do war-gaming simulations of potential future threats, and Garretson, Lindley Johnson, and others had unsuccessfully attempted to get an asteroid impact scenario adopted as one of its exercises three years earlier. Garretson wrote in the 2008 exercise's after-action report that Congress's tasking to the NRC and to OSTP had influenced the decision to adopt the impact scenario for 2008.[37]

This simulation had postulated the discovery of a binary asteroid, with a 370-meter primary body orbited by a 50-meter "moon," on a trajectory that mirrored Apophis's.[38] Two teams of U.S. government experts were given different scenarios: one team had to plan for an impact expected 72 hours after discovery, while the other team had 7 years before impact in which to mount a deflection campaign. The tabletop exercise lasted only a day, so the group was expected to produce "straw man" plans, not highly detailed ones. They concluded that while it was clear that the NEO hazard required "advance delineation of responsibilities, formalization of the notification process, and clarification of authorities and chains of command," they were not able to figure out which agency should lead a deflection campaign.[39] That decision had to be made by higher authorities. Garretson also explained in his report that the group had noted that there were no decision support tools to facilitate

---

36. Lindley Johnson, interview by Yeomans and Conway, 29 January 2016, transcript in NEO History Project collection.

37. AF/A8XC, "Natural Impact Hazard," p. 2.8-5.

38. Yeomans drafted the impact scenario with help from Mark Boslough of Sandia National Laboratory, Jay Melosh at the University of Arizona, and Steve Ward from UC Santa Cruz. The fictional asteroid was named 2008 Innoculatus.

39. AF/A8XC, "Natural Impact Hazard, p. iii.

rapid assessments, which they would need for short-notice impacts. Initial disaster response actions in the United States were local and state responsibilities; the federal government's job was the provision of information and support to the states, and it was not equipped or organized in 2008 to provide those things effectively.

The Office of Science and Technology Policy's 2010 memorandum clarified some of the roles that had been unclear to Garretson's panel. It assigned to NASA the detection, tracking, and notification activities that the Agency was already undertaking and directed NASA to utilize already-existing emergency communications protocols that belonged to FEMA and the State Department. It also recommended that NASA retain responsibility for mitigation research, and in particular that it undertake assessment of technological capabilities that might be needed for future deflection campaigns. This work would have to be coordinated with other agencies, too, as inevitably any actual deflection effort would draw upon Defense Department resources.[40]

The NASA Administrator, former astronaut and retired U.S. Marine Corps General Charles Bolden, also tasked the NASA Advisory Committee (NAC) with advising him on the contents of the NEO program and how it should be organized within the Agency. This was done via a subgroup of the NAC, the Ad-Hoc Task Force on Planetary Defense, which met in April 2010.[41] This group recommended that NASA establish a Planetary Defense Coordination Office "to coordinate the necessary expertise and internal resources to establish a credible capability to detect any NEO impact threat, as well as plan and test measures adequate to mitigate such a threat."[42] This coordination role would encompass the intragovernmental communications and planning role already discussed above. The Ad-Hoc Task Force contended that this organization would need a budget of $250 to $300 million per year for the next decade to meet the Brown Act search goal, develop and demonstrate

---

40. Holdren to Rockefeller.

41. Tom Jones and Rusty Schweickart, cochairs, "NAC Ad-Hoc Task Force on Planetary Defense Corrected Minutes" NASA, 15–16 April 2010, *https://www.nasa.gov/sites/default/files/466630main_20100415_AdHocTaskForce_PlanetaryDefense_CorrectedMinutes.pdf* (accessed 22 July 2019).

42. Thomas D. Jones and Russell L. Schweickart, cochairs, *Report of the NASA Advisory Council Ad Hoc Task Force on Planetary Defense* (Washington, DC: NASA Advisory Committee, 6 October 2010), p. 9, *http://www.boulder.swri.edu/clark/tf10pub.pdf* (accessed 22 July 2019).

deflection capabilities, and improve NASA's analytic and simulation capabilities. Two key recommendations called for a space telescope, in a Venus-like orbit, to complete the Brown Act survey quickly, as well as a spaceborne test of a kinetic impact deflection.

The Expanding Role of International Partners

The Ad-Hoc Task Force also recommended that NASA's planetary defense officer should "proactively challenge the international community to join in the analytical, operational, and decision-making aspects of Planetary Defense."[43] There was already movement in the direction of planetary defense within the European Space Agency. The most significant was a decision to initiate a space situational awareness program in 2009. As ESA chose to define it, "space situational awareness" included space weather, surveillance and tracking of artificial satellites, and detection of near-Earth objects.[44] In the United States, different agencies were responsible for these functions: the National Oceanic and Atmospheric Administration shared responsibility for space weather forecasting with the Air Force; the Air Force handled space surveillance; and NASA was responsible for near-Earth objects, though with some dependence on Air Force assets and technology development, as we have seen. Without the motivation provided by the United States' world-spanning military obligations, European nations had not invested in the capability to monitor space traffic, perhaps because it seemed largely a military activity. But an anti-satellite weapon test by China had produced an expanding cloud of debris in 2007; in 2009, an Iridium communications satellite was destroyed in a collision with a derelict Russian satellite, Kosmos 2251, creating still more orbiting debris.[45] The International Space Station was also frequently

---

43. Ibid.

44. European Space Agency, "SSA Preparatory Programme Highlighted in ESA Bulletin," European Space Agency Bulletin no. 147 (August 2011), *https://esamultimedia.esa.int/ multimedia/publications/ESA-Bulletin-147/pageflip.html* (accessed 15 September 2021).

45. Becky Iannotta and Tariq Malik, "U.S. Satellite Destroyed in Space Collision," Space.com, 11 February 2009, *https://www.space.com/5542-satellite-destroyed-space-collision.html* (accessed 27 September 2017); Leonard David, "China's Anti-Satellite Test: Worrisome Debris Cloud Circles Earth," Space.com, 2 February 2007, *https:// www.space.com/3415-china-anti-satellite-test-worrisome-debris-cloud-circles-earth.html* (accessed 12 November 2019).

in the news for having to be maneuvered to avoid what is often called "space junk." These events underscored the vulnerability of spaceborne infrastructure regardless of military intent. Constructing a European capability to avoid collisions became a prudent idea.

As part of this Space Situational Awareness (SSA) program, the European Space Agency put in place a near-Earth object program. The NEO portion of this program, led by Detlef Koschny, included a European Space Research Institute (ESRIN) NEO coordination center, which was established in 2013, outside Rome in Frascati, Italy. This center provides priorities for NEO follow-up observations, NEO Earth close-approach tables and risk analysis, and an orbit visualization tool.[46] Space Situational Awareness personnel also distribute NEO information and coordinated international discussions of NEO topics, including efforts within the United Nations' Committee on the Peaceful Uses of Outer Space. The SSA program also provided support to a number of European observatories, primarily for follow-up observations, but also for some discovery and light-curve observations. These supported observatories include the ESA 1-meter telescope at Tenerife in the Canary Islands, the 0.8-meter telescopes in Calar Alto, Spain, and the 1-meter telescope at Klet' in the Czech Republic. ESA's SSA program also supported the ongoing development of a 1-meter-effective-aperture, wide-field, multiple-fields-of-view (fly-eye) telescope designed for NEO observations and the development of systems to allow robotic observations of NEOs and space debris. Observing collaborations, without support, included the European Southern Observatory's Very Large Telescope (8.2-meter aperture) in northern Chile, the Large Binocular telescope on Mount Graham in Arizona (dual 8.4-meter apertures), and observatories in Brazil and South Korea.

The European Space Agency had also funded a study of a deflection demonstration mission called Don Quijote, with an architecture like NASA's Deep Impact, using an impactor and a stand-off spacecraft to provide

---

46. ESA's SSA/NEO website is *http://neo.ssa.esa.int*. NEO close-Earth passages and impact predictions are provided by the NEODyS system in Pisa, Italy, which is partially supported by the ESA-SSA program. The database of asteroid physical characteristics is maintained at the German Aerospace Center (DLR) by Gerhard Hahn and Stefano Mottola. A detailed chronology of NEO close-Earth approaches and milestone events, maintained by Karel van der Hucht, is also available at this site.

monitoring and data return.[47] While this mission did not progress beyond the study phase, it reflected growing interest in the NEO problem. In 2011, the European Council selected a proposal from Alan W. Harris of DLR called NEOShield to investigate deflection techniques as well as strategies for international implementation of deflection missions, should the need arise.[48] When the Ad-Hoc Task Force advocated greater attention to international coordination and possible partnerships by NASA, ESA was beginning to emerge as a viable option.

Reflecting growing international interest in the NEO hazard, Canada launched its first satellite devoted to searching for near-Earth objects and other satellites in 2013. The Near-Earth Object Surveillance Satellite (NEOSSat) mission, as it was called, was designed primarily to search for asteroids that spend most of their orbits sunward of Earth's. Its Principal Investigator was Alan Hildebrand of the University of Calgary. After a launch that was delayed by several years, NEOSSat proved to have inadequate image quality to achieve its scientific objectives.[49]

The Chelyabinsk Impactor

Another bolide event, the explosion of a meteor near the Russian city of Chelyabinsk on Friday, 15 February 2013, provided renewed impetus toward resolving the budgetary and policy questions that still had not been resolved. Unlike 2008 TC3, the Chelyabinsk bolide was not detected before impact.[50]

---

47. Ian Carnelli, Andres Galvez, and Dario Izzo, "Don Quijote: A NEO Deflection Precursor Mission," 2006, *https://www.researchgate.net/profile/Ian_Carnelli/publication/233721657_Don_Quijote_A_NEO_deflection_precursor_mission/links/53fddc430cf22f21c2f85679/Don-Quijote-A-NEO-deflection-precursor-mission.pdf* (accessed 10 July 2019).

48. A. W. Harris et al., "The European Union Funded NEOShield Project: A Global Approach to Near-Earth Object Impact Threat Mitigation," *Acta Astronautica* 90, no. 1 (September 2013): 80–84, doi:10.1016/j.actaastro.2012.08.026.

49. Elizabeth Howell, "Asteroid Hunter: An Interview with NEOSSat Scientist Alan Hildebrand," Space.com, *http://www.space.com/19926-asteroid-hunter-interview-alan-hildebrand.html* (accessed 13 January 2016); "The NEOSSat Satellite Is Meeting Some of Its Objectives But Still Has Issues To Resolve," *https://spaceq.ca/the_neossat_satellite_is_meeting_some_of_its_objectives_but_still_has_issues_to_resolve/* (accessed 13 February 2018).

50. Coincidentally, a known asteroid, 2012 DA14, had a close approach to Earth inside the ring of geosynchronous satellites on the same date. It was in a very different orbit and was unrelated to the Chelyabinsk bolide.

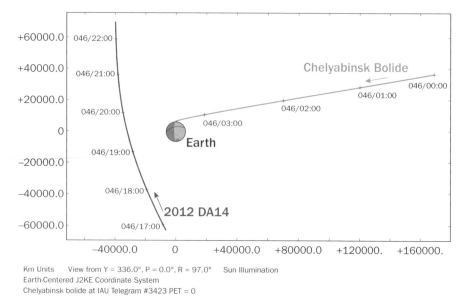

Figure 10-3. Orbit of 2012 DA14 compared to the impact trajectory of the Chelyabinsk bolide. (Image courtesy of Donald K. Yeomans)

Like 2008 TC3, though, it left many traces of its brief life in Earth's atmosphere: satellite detections; infrasound detector records; and many video recordings by surveillance cameras, cell phones, and automobile "dash cams." Its breakup and eventual detonation at around 23 kilometers' altitude broke windows in more than 3,000 buildings and injured more than a thousand people, though there were no fatalities. Fragments were quickly found by investigators, the largest of which was 1.5 meters in diameter. The bolide's estimated mass was around 10,000 tons, and the explosive energy was estimated to be a half megaton (or 0.05 Barringer crater units). Like 2008 TC3, the Chelyabinsk bolide had been relatively weak structurally, an ordinary chondrite that had been subject to repeated collisions and fracturing during its eons in space before encountering Earth.[51]

The Chelyabinsk impactor had two effects beyond the immediate damage it caused. Because it was so well documented, its breakup pattern could be reconstructed very accurately. In turn, that led to reconsideration of the

---

51. David A. Kring and Mark Boslough, "Chelyabinsk: Portrait of an Asteroid Airburst," *Physics Today* 67, no. 9 (1 September 2014): 35, *https://doi.org/10.1063/PT.3.2515* (accessed 20 May 2021).

potential damage from impacts more generally. Instead of remaining intact until it exploded, and therefore depositing its explosive energy in a single point in space (as bombs do) and generating a circular damage pattern on the surface, the Chelyabinsk bolide's breakup occurred gradually (relatively speaking), over about 4 seconds. So the actual damage pattern was more linear. In their review of the incident, David Kring and Mark Boslough wrote that the pattern "looked more like an inclined cylindrical bow shock than a spherical explosion."[52] This was positive, in a sense. If the bolide had acted more like a bomb, it would have done greater damage to a smaller area. This unexpected damage pattern was a result of the bolide having entered the atmosphere at a very low angle to the horizon. Kring and Boslough commented that most bolides did not enter at such a low angle (the Tunguska bolide apparently had not, for example), so this finding was not of great comfort. But the data would enable better modeling of impact damage in the future.

The second consequence of the Chelyabinsk event was two days of congressional hearings, titled "Threats from Space: A Review of U.S. Government Efforts To Track and Mitigate Asteroids and Meteors." The first session, held 19 March 2013, heard from John Holdren of OSTP, General William L. Shelton of the U.S. Air Force Space Command, and NASA Administrator Bolden. Most of the questioning went to Holdren and Bolden and focused on two subject areas: what the NASA program was accomplishing in terms of surveys and characterization, and who was ultimately in charge of the nation's NEO preparations. As Representative Bill Posey (R-FL) put it:

> Now, the Ranking Member [Rep. Eddie Bernice Johnson of Texas] asked about protocol, you know, who is in charge, and we got about three or four minutes of chatter but we never got an answer about who is in charge…. I would just like to recommend that the next time that you all come before us you give us a protocol and say this is who is in charge here…just a very clear matter of protocol who is in charge in various instances….[53]

---

52. Ibid.
53. "Threats from Space: A Review of U.S. Government Efforts To Track and Mitigate Asteroids and Meteors (Part I & Part II)," hearings before the Committee on Science, Space, and Technology, 113th Cong., 1st sess. (2013), p. 51.

As we have seen above, notification, reporting, and research responsibilities had been clarified by OSTP in 2010, and tabletop simulations to train personnel and exercise the various protocols were being held. (One was scheduled the month following this hearing.) But neither Holdren nor Bolden would commit to answering the who's-in-charge-of-deflection question that was being asked. Here the issue simply came down to this: who should be in charge depended on the nature of the threat. If a nuclear-based deflection campaign needed to be mounted, the U.S. Air Force might be a better choice; for a kinetic deflection campaign, NASA might be. Policy-makers did not want to make this decision in the absence of a specific threat.

This hearing highlighted an important non-asteroid context as well. In 2011, conflict between Congress and the administration over the mounting federal debt resulting from the Great Recession that had started in 2008 had led to the initiation of automatic, across-the-board spending cuts in all federal agencies known as "sequestration."[54] These cuts took effect in 2013. For NASA, sequestration caused the Agency budget to shrink from $21 billion in 2010 to $18.4 billion in FY 2014, and members of Congress wanted to hear how this was impairing the NEO program.[55] While Bolden (and General Shelton, who was having to furlough his civilian workforce) were frustrated by the mindless sequestration cuts and were clear in telling the committee members how they felt about those cuts, Bolden had to tell them it was not really affecting the NEO program. Despite sequestration, the administration had committed to expanding the NEO Observation Program's budget from $3.5 million in FY 2008 to $20.4 million in FY 2012 and to $40.5 million in FY 2014.[56] So the NEO Observations Program was getting an expanding piece of a shrinking pie—though still not a very large piece in absolute terms.

The rapidly increasing NEO program budget, finally, drew the attention of NASA's Inspector General (IG). At the time, Program Executive Lindley Johnson was the only Headquarters employee dedicated to the program. He had no staff to help manage his $20.5 million budget (in FY 2013) and 64 funding instruments, or to help establish appropriate performance metrics for

---

54. See, e.g., Congressional Budget Office, "Automatic Reductions in Government Spending—aka Sequestration," *https://www.cbo.gov/publication/43961* (accessed 22 July 2019).

55. In adjusted FY 2017 dollars. Source: *Aeronautics and Space Report of the President: Fiscal Year 2017 Activities* (Washington, DC: NASA, 2018).

56. Holdren et al., "Threats from Space," p. 64.

the programs he oversaw, or to deal with the interagency and international agreements and partnerships with which he was supposed to engage. NASA's own management rules required separation of financial duties to reduce the possibility of error and fraud. So Johnson should have had someone to help manage his grants and contracts. NASA rules also required that program executives develop and maintain performance metrics relevant to their programs and that programs develop schedules with milestones to be accomplished and estimates of the resources necessary to achieve them. The Inspector General concluded that "[w]ithout an appropriate management and staffing structure, the Program Executive is unable to evaluate needs compared to requirements and effectively communicate those needs to stakeholders."[57]

The IG report was not an indictment of Johnson. Instead, it was a criticism of NASA's support of the NEO program. NASA senior management had not committed the personnel necessary to appropriately managing the program as it had scaled up rapidly in response to Congress's demands and to the need to support the administration's Asteroid Redirect Mission. And NASA leaders agreed. The Associate Administrator (AA) for the Science Mission Directorate, former astronaut John Grunsfeld, largely concurred with these findings. While contending that the program's internal financial controls were already adequate, the AA agreed that the NEO Observations program office needed to be expanded. A new program plan would be in place by September 2015.[58] Johnson commented later, "[T]hat's [the IG report] what we used a lot for justification of why we needed to stand up the Planetary Defense Coordination Office and bring more people into working up here. So it was quite a bit of benefit in that respect."[59]

The expanded NEO program office was announced to the public in January 2016, with a new name: the Planetary Defense Coordination Office (PDCO).[60] The older NEO Observations Program continued as one of the units of the new office, while interagency and international partnerships and

---

57. Office of Inspector General, "NASA's Efforts To Identify Near-Earth Objects and Mitigate Hazards," report no. IG-14-030, NASA, 2014, p. 12.

58. Ibid., appendix D, p. 29.

59. Lindley Johnson, interview by Yeomans and Conway, 12 September 2017, transcript in NEO History Project collection.

60. NASA, "NASA Office To Coordinate Asteroid Detection, Hazard Mitigation," 7 January 2016, *http://www.nasa.gov/feature/nasa-office-to-coordinate-asteroid-detection-hazard-mitigation* (accessed 22 July 2019).

public communications all became offices under the new organization, as did flight project management. Lindley Johnson gained the title Planetary Defense Officer, and astronomer Kelly Fast took over the NEO Observations Program.

Finally, the National Science and Technology Council also established an interagency working group cochaired by the Office of Science and Technology Policy and the PDCO known as DAMIEN: the Interagency Working Group for Detecting and Mitigating the Impact of Earth-bound Near-Earth Objects. This group was tasked with preparing a national NEO preparedness strategy, followed by an action plan. The strategy document was published in December 2016, with the action plan following in June 2018.[61]

Pursuing International Collaborations

One of the criticisms of the Inspector General's report was that the older NEO Observations Program had not been able to take maximum advantage of potential international collaborations. There were a few opportunities, most importantly with the European Space Agency, as we saw earlier in the chapter. But there was also a United Nations–level effort to raise awareness of the NEO hazard and to develop warning and response strategies.

At the third United Nations Conference on the Exploration and Peaceful Uses of Outer Space (UNISPACE III), held in Vienna, Austria, during late July 1999, participating nations had resolved to "improve the international coordination of activities related to near-Earth objects, harmonizing the worldwide efforts directed at identification, follow-up observation and orbit prediction, while at the same time giving consideration to developing a common strategy that would include future activities related to near-Earth objects."[62] The task was assigned to the Scientific and Technical Subcommittee of the Committee

---

61. See Fred L. Kennedy and Lindley N. Johnson, cochairs, "National Near-Earth Object Preparedness Strategy," Washington, DC: National Science and Technology Council, December 2016; Aron R. Miles and Lindley N. Johnson, cochairs, "National Near-Earth Object Preparedness Strategy and Action Plan," Washington, DC: National Science and Technology Council, June 2018, *https://www.nasa.gov/sites/default/files/atoms/files/national_near-earth_object_preparedness_strategy_tagged.pdf* (accessed 15 July 2021).

62. "Report of the Third United Nations Conference on the Exploration and Peaceful Uses of Outer Space, 19–30 July 1999, Vienna, Austria," report A/CONF.184/6, p. 7. *http://www.unoosa.org/oosa/en/ourwork/psa/schedule/1999/unispace-iii.html* (accessed 16 July 2019).

on the Peaceful Uses of Outer Space (COPUOS). This initiative had been 14th on the prioritized list of activities for COPUOS to undertake, so the committee established Action Team 14 to make recommendations to it.[63]

Reflecting the relatively low priority of the issue, not much happened for years afterward. Somewhat in parallel, in 2005 Rusty Schweickart used the platform of the Association of Space Explorers (ASE), which had observer status to COPUOS, to organize a Panel on Asteroid Threat Mitigation to develop a "plan to draft a document on a NEO decision-making process." In 2008, they delivered this document to Action Team 14. It contained three principal recommendations: an Information, Analysis, and Warning Network should be established; a Mission Planning and Operations Group should be created to assess existing deflection capabilities and develop plans in the event of detected threats; and the United Nations should establish oversight of these other two organizations through an intergovernmental Mission Authorization and Oversight Group.[64] The chair of Action Team 14, Sergio Camacho of Mexico, was also a member of Schweickart's committee, so the report became a basis of discussion within the Action Team.

In February 2013, Action Team 14 finally made its own recommendations to COPUOS. Two of its recommendations were quite similar to those of Schweickart's panel: that an International Asteroid Warning Network should be assembled out of the various organizations already doing that work and that a "space mission planning advisory group" should be developed. This differed from the ASE proposal in being broader in scope, incorporating NEO-related science missions as well as potential future deflection missions. That way, the planning advisory group could serve a coordinating function for national science missions, too. But the third Action Team recommendation was quite different: that an Impact Disaster Planning Advisory Group should be formed. This one, though, was rejected by the Scientific and Technical Subcommittee of COPUOS.[65] Ultimately, the two surviving recommendations were implemented, with both new organizations being launched in 2014.

---

63. Action Teams were composed of subject matter experts, unlike COPUOS itself, which consisted of member state representatives.

64. Russell L. Schweickart, Thomas D. Jones, Frans von der Dunk, and Sergio Camacho-Lara, "Asteroid Threats: A Call for Global Response," Association of Space Explorers, 25 September 2008, p. 18.

65. Sergio Camacho, "Report of the Action Team on Near-Earth Objects: Recommendations for an International Response to a NEO Threat" (presented at the 50th Session

The International Asteroid Warning Network's (IAWN) steering committee first met at the Minor Planet Center in January 2014. IAWN was to be composed of individuals and organizations that contributed to NEO observation and communications activities, not of member states. The steering committee decided that their first task should be convening a communications workshop. The Secure World Foundation hosted the event that fall in Colorado. Participants in the invitation-only event reviewed the communications surrounding four recent asteroid events, including Apophis and 2011 AG5. Reporter Leonard David presented the Apophis case, commenting that many of his peers never quite understood that the projected risk had disappeared in a few days, and they also did not understand the keyhole issue that Apophis had introduced. "The media tend to feed off each other rather than to go to the few people who could legitimately inform them," he said.[66] Linda Billings, who served as a communications consultant for the Near-Earth Objects Observations Program, reviewed the 2011 AG5 communications effort. She contended that official statements from ESA and NASA were "clear, concise, and correct.... The media predictably sensationalized the risk of impact. In much of the media coverage of 2011 AG5 from January 2011 to December 2012, sensational headlines led into stories that accurately explained knowledge about 2011 AG5."[67]

One important recommendation from this meeting compelled NEO scientists to develop a non-probabilistic means of communicating risk to the public. The Torino and Palermo scales might make sense to the statistically savvy, but perhaps not to the general public. Workshop participants developed a "Broomfield Hazard Scale" that expressed likely levels of damage from

---

Scientific and Technical Subcommittee, Vienna, Austria, 11 February 2013), *http://www.unoosa.org/pdf/pres/stsc2013/2013neo-01E.pdf* (accessed 21 May 2021); Sergio Camacho, "Background in Action Team 14 and UN Recommendations on NEO Threat" (presentation to the First Meeting of the Space Mission Planning Advisory Group, European Space Operations Center, Darmstadt, Germany, 6–7 February 2014), copy in NEO History Project collection.

66. Leonard David, "What's Up or Down with Apophis?" (presented at the Workshop on Communicating About Asteroid Impact Warnings and Mitigation Plans," 9–10 September 2014), *http://iawn.net/documents/201409_Communications//apophis-case-study-iawn-david-9-141.pdf* (accessed 22 July 2019).

67. Linda Billings, "Impact Hazard Communication Case Study: 2011 AG5," *http://iawn.net/documents/201409_Communications//2011-ag5-case-study-iawn-billings-9-14.pdf* (accessed 22 July 2019).

an impactor in one of six size classes but did not address the likelihood of impact. This scale did not find much acceptance outside the members of this workshop, perhaps because the larger NEO community was more interested in assessing impact probabilities than in assessing potential damage. Without a near-term, specific threat, a damage assessment scale did not seem useful.

In addition to its focus on communications, IAWN developed observing campaigns as a means of exercising its coordination and data-sharing functions. The first of these campaigns took place in 2017 and was led by Vishnu Reddy of the University of Arizona's Lunar and Planetary Laboratory.[68] It was aimed at recovering a small, 10-meter-class Apollo asteroid that had been discovered in October 2012 by Pan-STARRS 1 and designated 2012 TC4 by the Minor Planet Center. Only a handful of observations of it had been made, and the resulting orbit solution suggested that it would fly past Earth well within the orbit of the Moon. It would not be an impact risk, but uncertainty in its orbit was still rather high. Observers using one of the 8.2-meter Very Large Telescopes at the European Southern Observatory first recovered 2012 TC4 on 27 July 2017.[69] A variety of other observers worldwide were able to identify the asteroid starting in September. Both Goldstone and Arecibo radars were able to observe its flyby on 12 October at a distance of about 50,000 kilometers from Earth.[70]

The second organization to come out of Action Team 14's recommendations was the Space Mission Planning Advisory Group (SMPAG). In addition to its focus on space missions, SMPAG differed from IAWN in its membership. IAWN was open to any organization or individual with a track record of effective NEO discovery, follow-up observations, or communications. But SMPAG was restricted to nation-states and national and international space organizations "that coordinate and fund space activities."[71] Since SMPAG's

---

68. Mikayla Mace Star, "University of Arizona To Lead Global Exercise on Response to Asteroid Threat," *Arizona Daily Star* (updated 29 June 2019), *https://tucson.com/news/science/university-of-arizona-to-lead-global-exercise-on-response-to/article_bc6fa864-366f-5ce1-9592-2ace77c430e1.html* (accessed 17 July 2019).

69. "M.P.E.C. 2017-P26," *http://2012tc4.astro.umd.edu/References/2012TC4_VLTrecovery.txt* (accessed 22 July 2019).

70. The observing campaign and its data are documented at "The 2012 TC4 Observing Campaign," *http://2012tc4.astro.umd.edu/* (accessed 17 July 2019).

71. "Terms of Reference for the Near-Earth Object Threat Mitigation Space Missions Planning Advisory Group, 1st SMPAG meeting 6/7 February 2014," European

focus was on NEO deflection and mitigation, membership was limited to those organizations that could contribute to that task.

An early activity of SMPAG was the discussion of decision criteria to be jointly developed with IAWN. These needed to be chosen in advance of an actual threat so that the various organizations and personnel involved would be working from the same playbook, minimizing public confusion. These decision criteria were agreed upon in 2017. IAWN agreed to make formal warnings of predicted impacts "exceeding a probability of 1% for all objects characterized to be more than 10 meters in size."[72] Preparedness planning should begin if an impact were predicted to occur within 20 years, by an object assessed to be greater than 20 meters in size, with a probability of impact greater than 10 percent. For its part, SMPAG should start mission option planning if an impact were to be predicted within 50 years, with an impactor size exceeding 50 meters and a probability greater than 1 percent.[73]

Another SMPAG activity was the discussion of "reference missions"—conceptual deflection campaigns against various kinds of potential asteroid threats—and deflection test missions that were being funded by NASA and ESA. Both space agencies had funded conceptual studies over the previous decades and were pursuing a joint mission that was only loosely coupled. That way, in the event that either ESA or NASA did not get funding for its part of the mission, the other could continue. NASA chose a kinetic impactor test mission known as DART (Double Asteroid Redirection Test) to be developed by the Johns Hopkins Applied Physics Laboratory in Maryland, while ESA would send a later mission, the Asteroid Impact Mission (AIM), to evaluate the impactor's effect on the target asteroid. The conjoined mission was named AIDA (Asteroid Impact and Deflection Assessment).[74] As it happened, NASA's Johnson was able to get DART funded in the FY 2017 budget, but

---

Space Operations Centre (ESOC), Germany, *https://www.cosmos.esa.int/web/smpag/documents_and_presentations* (accessed 15 September 2021).

72. "SMPAG Action Item 5.1: Recommended Criteria & Thresholds for Action for Potential NEO Impact Threat," *https://www.cosmos.esa.int/documents/336356/1503750/SMPAG_5.1_Report_NASA.pdf/f399e4eb-5947-867c-2422-b9dcb7e3649c* (accessed 22 July 2019).

73. Ibid.

74. See, e.g., Cheryl Reed, Andy Cheng, Andy Rivkin, and Brian Kantsiper, "AIDA-DART Mission Overview," presented at the 7th Meeting of SMPAG, Pasadena, CA, 14 October 2016, *https://www.cosmos.esa.int/web/smpag/documents_and_presentations* (accessed 15 September 2021).

AIM was rejected by ESA's funders in December 2016.[75] However, in late 2019, ESA approved funding for the development of the less-expensive Hera mission designed to investigate, in 2026, the DART impact crater and refine the orbital changes to the asteroid's satellite induced by the DART impact in late September or early October 2022.[76]

The chosen target for the AIDA mission was a binary asteroid, whose primary, 1996 GT, had been discovered in 1996 by Spacewatch, and whose secondary "moon" had been discovered in 2003. This small asteroid "system" had been named 65803 Didymos. The primary is about 800 meters in diameter, while the moon is about 170 meters diameter. During their close approaches to Earth, the Didymos system is visible to ground observatories and sometimes the Goldstone and Arecibo radars; both radars had measured the two bodies during their 2003 flyby. The DART mission's target is the moon (often informally referred to as "Didymoon"). The mission's purpose is to assess a key issue in understanding the ability to deflect asteroids generally, the momentum transfer efficiency of an impactor. DART's impactor would possess a known amount of momentum the instant it struck the moon, and the question was how much of that momentum would be converted into thrust and change Didymoon's orbit around its primary, and how much would simply be absorbed. An intensive ground observation campaign after the 2022 impact would evaluate the orbit change, and therefore the momentum transfer efficiency.[77] If successful, the follow-on Hera mission will improve knowledge

---

75. NASA FY 2017 budget estimates, p. PS-6, *https://www.nasa.gov/sites/default/files/atoms/files/fy_2017_budget_estimates.pdf* (accessed 17 July 2019). Also see Jeff Foust, "NASA Presses Ahead with Asteroid Mission Despite ESA Funding Decision," SpaceNews.com, 13 December 2016, *https://spacenews.com/nasa-presses-ahead-with-asteroid-mission-despite-esa-funding-decision/*; Anon., "NASA'S First Asteroid Deflection Mission Enters Next Design Phase," NASA, 30 June 2017, *http://www.nasa.gov/feature/nasa-s-first-asteroid-deflection-mission-enters-next-design-phase* (accessed 22 July 2019).

76. Jeff Foust, "ESA Plans Second Attempt at Planetary Defense Mission," SpaceNews.com, 29 June 2018, *https://spacenews.com/esa-plans-second-attempt-at-planetary-defense-mission/* (accessed 22 July 2019); Steve Dent, "Europe's Space Agency Approves the Hera Anti-Asteroid Mission," Engadget, 2 December 2019, *https://www.engadget.com/2019-12-02-esa-approves-hera-asteroid-deflection-ission.html* (accessed 17 June 2020).

77. A. F. Cheng, P. Michel, M. Jutzi, A. S. Rivkin, A. Stickle, O. Barnouin, C. Ernst, J. Atchison, P. Pravec, and D. C. Richardson, "Asteroid Impact and Deflection

of the new orbit and provide better characterization of the asteroid pair than possible from the ground.

Both IAWN and SMPAG personnel also began participating in civil defense simulations like that held by the U.S. Air Force's Future Concepts and Transformation Division back in 2008. These tabletop exercises had continued in the United States at the Federal Emergency Management Agency's headquarters in April 2013 and May 2014. The 3rd International Academy of Astronautics Planetary Defense Conference, held in Flagstaff, Arizona, in April 2013, hosted the first international impact simulation.[78] These simulations became a staple of the Planetary Defense Conferences, which took place every odd-numbered year. For the 2017 meeting in Tokyo, for example, JPL's Paul Chodas designed a scenario in which a 200- to 300-meter asteroid was expected to impact somewhere in a corridor stretching from southern China through the Korean Peninsula to Japan in 2027.[79] The simulations served the dual purpose of education and recruiting; the planetary defense–interested international community remained small, and expanding it was a priority.

Reassessing the NEO Hazard

In 2013, Ames Research Center organized an impact effects modeling program that drew on high-energy modeling and supercomputing expertise at the Sandia and Lawrence Livermore laboratories of the Department of Energy. The existing impact models had been based on nuclear weapons detonations as the energy source, and as we saw above in the discussion of the Chelyabinsk airburst, nuclear explosions are not a perfect analog for asteroid impacts. Adapting the models to reflect these differences was one part of the

---

Assessment Mission: Kinetic Impactor," *Planetary and Space Science* 121 (February 2016): 27–35, *https://doi.org/10.1016/j.pss.2015.12.004* (accessed 30 June 2021).

78. Debbie Lewis and Richard Tremayne-Smith, "Asteroid 2013 PDC-E Post Exercise Report Addendum Response Activities," n.d., p. 40, *https://iaaspace.org/wp-content/uploads/iaa/Scientific%20Activity/postexercisereport2013.pdf* (accessed 15 September 2021).

79. The scenario was developed by William Ailor and Nahum Melamed of the Aerospace Corporation, Brent Barbee of NASA Goddard Space Flight Center, Mark Boslough and Barbara Jennings of Sandia National Laboratories, and Paul Chodas of JPL. See "Conference Summary and Recommendations," 2017 IAA Planetary Defense Conference, Tokyo, Japan, 15–19 May 2017, *https://iaaspace.org/wp-content/uploads/iaa/Scientific%20Activity/report2017pdc.pdf* (accessed 15 September 2021).

effort. Another was incorporating probabilistic scenario generation, so that many potential, but realistic, impact scenarios could be generated. With far more asteroids and their orbits known than had been the case in 2003, a more realistic set of impact scenarios could be generated.

A few years later, the Ames effort was reorganized into a project called ATAP, for Asteroid Threat Assessment Project, under Donovan Mathias. ATAP joined with the National Oceanic and Atmospheric Administration's (NOAA's) Pacific Marine Environmental Laboratory to host a workshop on tsunami generation by asteroid impacts in August 2016. The assessment of potential damage from tsunamis was a known weakness, having been identified in the 2003 Science Definition Team assessment, as discussed in chapter 8. Participants in this workshop drew much different conclusions from the ensuing decade of work on wave propagation and inundation models. For impacts of asteroids less than 250 meters in diameter far from shore, they concluded there was no danger to coastal populations and infrastructure. This hazard had been "substantially overestimated."[80] Moreover, near-shore airbursts of smaller impactors carried less risk to shore populations from inundation (as opposed to the direct blast effects) than previously assessed.

The ATAP project's model, really an ensemble of models, went by the name Probabilistic Asteroid Impact Risk (PAIR) model.[81] It drew on Alan W. Harris and Germano D'Abramo's 2015 assessment of the population of near-Earth asteroids for its scenario generation.[82] Recall that when the 2003 SDT had been written, only around 650 NEAs were known, and the population was strongly biased toward the largest asteroids, whereas by 2015, more than 11,000 had been discovered. Inclusion of these data also led to changes in the assessed impact hazard. While there were slightly more NEAs than anticipated in the smallest size classes (which do no damage at the surface),

---

80. David Morrison and Ethiraj Venkatapathy, "Asteroid Generated Tsunami: Summary of NASA/NOAA Workshop," NASA/TM-219463, January 2017, p. 3, *https://ntrs.nasa.gov/archive/nasa/casi.ntrs.nasa.gov/20170005214.pdf* (accessed 20 August 2019).

81. Donovan L. Mathias, Lorien F. Wheeler, and Jessie L. Dotson, "A Probabilistic Asteroid Impact Risk Model: Assessment of Sub-300m Impacts," *Icarus* 289 (1 June 2017): 106–119, *https://doi.org/10.1016/j.icarus.2017.02.009* (accessed 30 June 2021).

82. This was JPL's Alan W. Harris, who by this time had retired and founded his own research company, MoreData Inc.! D'Abramo was at IASF-Roma, Rome, Italy. Alan W. Harris and Germano D'Abramo, "The Population of Near-Earth Asteroids," *Icarus* 257 (September 2015): 302–312, *https://doi.org/10.1016/j.icarus.2015.05.004* (accessed 30 June 2021).

there were fewer NEAs than expected in the 50- to 500-meter categories, which meant less-frequent impacts, fewer resulting casualties, and less accumulated damage over long periods of time from that class of impactor. But the improved modeling efforts also led to a reduction in the minimum diameter necessary to achieve global-scale effects, from 1 kilometer to 700 meters; since that increased the number and frequency of potential impactors within that class, the effect was to drive the accumulated damage and casualties higher.[83]

NASA's Lindley Johnson commissioned an update of the 2003 NEO Science Definition Team report to incorporate the new information generated over the preceding dozen years. In chapter 8, we discussed the increasing NEO discovery rates as well as the development of new survey systems, like ATLAS and the LSST, that were incorporated into this update. But understanding of the risk to people and infrastructure by impacts had evolved too. This was driven by improved modeling, incorporating what had been learned about NEOs and impact processes over the preceding dozen years.

The 2017 Science Definition Team, again chaired by Grant Stokes of MIT's Lincoln Laboratory, integrated these changed assessments into its analysis of the risk reduction that could be achieved by completing the Brown Act Survey goal. By this time, it was clear that no new survey systems other than ATLAS would be available prior to 2023, so the SDT took that year as the starting point for a new survey system. Their analysis concluded that completing the Brown Act Survey in the 10 years following 2023 could be done only by including a space-based search component. A 1-meter-aperture infrared telescope at the L1 point would accomplish the task the fastest; a 0.5-meter-aperture infrared telescope at L1 would take a few years longer in completing the goal but would cost about half as much. No ground-based observatory by itself would reach the survey goal in under 25 years of operation.[84] This SDT also preferred the L1 orbit to the Venus-like orbit discussed in previous studies due to both lower costs and the ability of an L1-based telescope to offer a warning function.

Yet another National Academy of Sciences study in draft as of July 2019 came to the conclusion that an infrared spaceborne capability combined with

---

83. Near-Earth Object Science Definition Team, "Update To Determine the Feasibility of Enhancing the Search and Characterization of NEOs" (Washington, DC: NASA, September 2017), p. 58, *https://www.nasa.gov/sites/default/files/atoms/files/2017_neo_sdt_final_e-version.pdf* (accessed 30 July 2019).

84. Ibid., pp. 188–189.

the existing ground observatories was needed.[85] But this study also made the point, as the Discovery Program's rejection of the Near-Earth Object Camera (NEOCam) already had, that accounting for all the NEOs that threatened Earth was not high-priority science. Or, to use their words, "missions meeting high-priority planetary defense objectives should not be required to compete against missions meeting high-priority science objectives."[86] This was another way of saying that planetary defense was neither science nor science-driven in the minds of many scientists. It was applied science, designed not to expand human knowledge of the cosmos, but rather to provide a public service.

Applied research had been part of NASA's congressional mandate when it was founded in 1958. Congress had intended the Agency to make space technology useful, and for NASA's first two decades, it had a space applications program that was coequal organizationally to its science program.[87] That applications program had developed weather, communications, and land-use satellites. It was merged into NASA's science directorate during the 1980s, and applications fell off the Agency's organization chart, but NASA Earth science missions often have applied science components. NASA's sea-level–measuring missions, for example (Topography Experiment [TOPEX]/Poseidon and the joint U.S.-France Jason series), contribute to both physical oceanography and to understanding the impact of sea-level rise on coastal communities and infrastructure. NASA's Science Mission Directorate has also flown numerous meteorological research instruments in the past two decades that contribute to both understanding of weather processes ("basic" science) and advancing weather prediction, which has enormous economic value (and is therefore "applied" science).[88] After 2000, Congress required NASA to explicitly fund

---

85. This NRC study had significantly different membership from that of either of the SDT study committees or the prior NRC studies. See Committee on Near-Earth Object Observations in the Infrared and Visible Wavelengths, "Finding Hazardous Asteroids Using Infrared and Visible Wavelength Telescopes," July 2019 (pre-publication draft).

86. Ibid., p. S-5.

87. On the NASA Applications Program, see Erik M. Conway, "Bringing NASA Back to Earth," *Science and Technology in the Global Cold War*, ed. Naomi Oreskes and John Krige (Cambridge, MA: MIT Press, 2014), pp. 251–272; Pamela E. Mack and Ray A. Williamson, "Observing the Earth from Space," in *Exploring the Unknown: Selected Documents in the History of the U.S. Civil Space Program*, vol. 3, *Using Space*, ed. John M. Logsdon et al. (Washington, DC: NASA, 1998), pp 155–176.

88. For a history of NASA's atmospheric science research, see Erik M. Conway, *Atmospheric Science at NASA: A History* (Baltimore: Johns Hopkins University Press, 2008).

applied science and natural hazards research, too, which tasks were incorporated into its Earth science directorate. Congress's expansion of the National Aeronautics and Space Act in 2005 to include "detecting, tracking, cataloguing, and characterizing near-Earth asteroids and comets" as an explicit tasking for the Agency represented a recognition at the political level that an aspect of planetary science had become utilitarian.[89]

In this view, saving *Homo sapiens sapiens* from *Tyrannosaurus rex*'s fate is less a scientific task than a public service. Planetary scientists wanted to separate that role from planetary science to prevent loss of what they perceive as science funds to a new public service called planetary defense. Redefining NEO surveys as "applied science" was part of that effort.

The NASA fiscal year 2019 budget proposal reflected this demand, with the previous NEO Observation Program budget line replaced with an explicit planetary defense budget line and a proposed increase to $150 million, from $60 million, in fiscal year 2020. In the near term, the increase would fund the DART mission discussed earlier.[90] And despite the loss of the NEO program's principal patron in the House of Representatives, Dana Rohrabacher, in the 2018 mid-term election, the House appropriations subcommittee voiced its support for DART (and NEOCam) in its own fiscal year 2020 budget report.[91]

The American public supported the planetary defense task overwhelmingly, too. An April 2019 poll found that 62 percent of Americans thought planetary defense should be a top NASA priority, while sending astronauts to Mars was a top priority for only 18 percent.[92] A May 2019 poll showed that

89. Quoted from National Aeronautics and Space Act of 1958, as amended, 25 August 2008, *https://history.nasa.gov/spaceact-legishistory.pdf* (accessed 12 November 2019).

90. See NASA FY 2019 and FY 2020 Budget Estimates, *https://www.nasa.gov/sites/default/files/atoms/files/fy19_nasa_budget_estimates.pdf* and *https://www.nasa.gov/sites/default/files/atoms/files/fy_2020_congressional_justification.pdf* (accessed 12 November 2019).

91. Committee on Appropriations, Commerce, Justice, Science, and Related Agencies Appropriations Bill, 2020, H. Rept. 116-101, 116th Cong., 1st sess., 3 June 2019, *https://www.congress.gov/116/crpt/hrpt101/CRPT-116hrpt101.pdf* (accessed 12 November 2019).

92. "Poll Shows More Public Support for NASA Science Programs Than Human Exploration," SpaceNews.com, 6 June 2018, *https://spacenews.com/poll-shows-more-public-support-for-nasa-science-programs-than-human-exploration/* (accessed 1 July 2021); "Majority of Americans Believe Space Exploration Remains Essential," *Pew Research Center Science & Society* (blog), 6 June 2018, *https://www.pewresearch.org/science/2018/06/06/majority-of-americans-believe-it-is-essential-that-the-u-s-remain-a-global-leader-in-space/* (accessed 30 July 2019).

68 percent of Americans supported NASA's planetary defense mission, placing that activity above its scientific mission (59 percent), its International Space Station (42 percent), and its plans to send astronauts to Mars (27 percent).[93] Planetary defense had been accepted by the general public as a legitimate government task.

---

93. "Poll: Americans Want NASA To Focus More on Asteroid Impacts, Less on Getting to Mars," NPR.org, *https://www.npr.org/2019/06/20/734311961/poll-americans-want-nasa-to-focus-more-on-asteroid-impacts-less-on-getting-to-ma* (accessed 20 June 2019).

CHAPTER 11

CONCLUSION

O n 19 October 2017, Pan-STARRS 1 discovered a strange object on an unusual hyperbolic trajectory, already moving away from the Sun. Images from 18 October also showed the object, and initially the Minor Planet Center classified it as a comet, though a very strange one. In a call for more observations, Gareth Williams of the MPC commented that "if further observations confirm the unusual nature of this orbit, this object may be the first clear case of an interstellar comet."[1] Karen Meech of the University of Hawai'i organized follow-up observations with the Canada-France-Hawai'i Telescope and the European Southern Observatory's Very Large and Gemini South telescopes. Despite its proximity to the Sun when it passed perihelion on 9 September (0.25 au, inside Mercury's orbit), it showed no coma from volatiles being vaporized away. But it was clearly on an interstellar trajectory. Due to its lack of obvious activity, it received a temporary redesignation as an asteroid (A 2017 U1) while the MPC corresponded with the leadership of the International Astronomical Union to decide on a permanent designation. The naming rules for solar system objects did not apply to this first interstellar object, so a new rule was necessary. It became the first "I" (for Interstellar) object, 1I/2017 U1, avoiding the issue of whether it was actually an asteroid or comet. In 2018, it was reinterpreted as a comet again when its motion could only be explained if it is subjected to a comet-like, heliocentric, radial acceleration. The discovery of 1I/2017 U1 was credited to Robert Weryk, a postdoctoral fellow at the University of Hawai'i, who had first observed it

---

1. "MPEC 2017-U181: COMET C/2017 U1 (PANSTARRS)," *https://www.minorplanetcenter.net/mpec/K17/K17UI1.html* (accessed 27 December 2017).

in the Pan-STARRS imagery, and the object was named 'Oumuamua, the Hawaiian word for "scout."[2]

The brief visit from 'Oumuamua confirmed a longstanding expectation of astronomers that asteroids and comets must exist beyond our solar system. Just as Jupiter's gravity occasionally bounces comets and asteroids transiting toward the Sun from the outer solar system completely out of the solar system instead, planets around other stars must do the same. So there should be asteroids and comets wandering the galaxy, untethered to any star, that will occasionally fly through our solar system—but none had ever been seen before. The nearly full-time surveillance of the skies enabled by NASA's Near-Earth Objects Observations program's union of electronic detectors and inexpensive, powerful computing made the discovery possible.

Near-Earth object astronomy has clearly benefited from the dramatic digitalization of American science since 1980. For a few tens of millions of dollars per year, it is now possible to do what was essentially impossible a few decades ago. Extrinsic technological change drove this transformation, by and large—the NEO astronomers developed their own software, but their hardware, CCDs, computers, and telescopes were developed outside their own community. They were efficient, and effective, adopters.

The availability of inexpensive and sophisticated computing facilities also enabled the development and use of numerical simulations. Simulations were used to understand the evolution of small-body orbits and populations, to design search strategies and assess the progress of the discovery surveys, and to study impacts and their consequences. These simulations served both scientific and policy purposes, as suggested by the use of population models to assess progress toward policy goals.

---

2. "MPEC 2017-U183: A/2017 U1," *https://www.minorplanetcenter.net/mpec/K17/K17U13.html* (accessed 27 December 2017); "MPEC 2017-V17: New Designation Scheme for Interstellar Objects," *https://www.minorplanetcenter.net/mpec/K17/K17V17.html* (accessed 27 December 2017); Karen J. Meech et al., "A Brief Visit from a Red and Extremely Elongated Interstellar Asteroid," *Nature* 552, no. 7685 (December 2017): 378, *https://doi.org/10.1038/nature25020* (accessed 1 July 2021); M. Micheli et al., "Non-gravitational Acceleration in the Trajectory of 1I/2017 U1 ('Oumuamua)," *Nature* (27 June 2018), *https://doi.org/10.1038/s41586-018-0254-4* (accessed 1 July 2021).

NEOs and Policy

The first two chapters outlined the gradual realization that Earth and the Moon are inundated with impacting bodies and there exists a population of these bodies that orbit the Sun near the orbit of Earth. With the development of more and more sophisticated detectors, the known population of these near-Earth objects grew dramatically—all but erasing the reluctance of many to acknowledge the impact risk that these objects present. This change helped bring the field of near-Earth object research into policy salience in the 1990s, and its leaders adopted the language of risk and hazard. Governments' roles in mitigating natural hazards extend into the distant past and have taken a wide variety of forms.[3] Governments have facilitated insurance markets to provide financial remedies for accidents and hazards; deployed various kinds of warning systems (meteorological stations, stream gauges, etc.); and, in the specific case of flooding hazards, reengineered the landscape itself through flood-control systems—levees, dikes, dams, locks, concretization of streams and rivers, etc. In the case of cosmic risk, though, mitigation has been largely restricted to discovery, a form of early warning and risk reduction. The NASA-FEMA tabletop exercises represent a form of planning for future mitigation attempts (relocation of at-risk populations), but fortunately, one not yet necessary.

The Science Definition Team, Analysis of Alternatives, and Defending Planet Earth studies we reviewed in earlier chapters all framed the asteroid problem in terms of actuarial risk.[4] What is the economic damage expected from asteroid impacts, and thus what value can be assigned to reducing that risk? This is a narrative derived from utilitarianism, the philosophical doctrine built around the question of how to do the most good for the most people,

---

3. E.g., Arwen Mohun, *Risk: Negotiating Safety in American Society* (Baltimore: Johns Hopkins University Press, 2013); Theodore Porter, *Trust in Numbers: The Pursuit of Objectivity in Science and Public Life* (Princeton, NJ: Princeton University Press, 1995).

4. Near-Earth Object Science Definition Team, "Study to Determine the Feasibility of Extending the Search for Near-Earth Objects to Smaller Limiting Diameters" (Washington, DC: NASA, 22 August 2003), *https://cneos.jpl.nasa.gov/doc/neoreport030825.pdf* (accessed 1 July 2021); Office of Program Analysis and Evaluation, "2006 Near-Earth Object Survey and Deflection Study (DRAFT)" (Washington, DC: NASA, 28 December 2006); National Research Council, *Defending Planet Earth: Near-Earth Object Surveys and Hazard Mitigation Strategies* (Washington, DC: National Academies Press, 2010), *http://www.nap.edu/catalog/12842* (accessed 1 July 2021).

and the foundation of 20th century welfare economics. Cost-benefit analysis derives from the same source. The American administrative state relies heavily on this quasi-quantitative process in its decision-making in order to appear to be operating objectively and apolitically; by adopting this narrative, asteroid scientists were, in effect, arguing that cosmic risk should be treated by the state the same way as earthquake or flooding or wildfire risks.[5] They are natural hazards to be guarded against, not foreign enemies. Historically, military agencies have been an essential part of the American state's response to natural hazards, too. The U.S. Army Corps of Engineers has played a fundamental role in flood control since the 19th century, just to pick a well-studied historical example.[6] The addition of NEO surveillance and mitigation to the NASA charter in 2005 was an extension of the state's role in managing natural hazards.

But it was not a particularly welcome addition, as we have seen. The two agencies of the U.S. government with the most relevant expertise, the U.S. Air Force and NASA, did not really want the planetary defense assignment. One reason was the "giggle factor," that impacts were highly unlikely and belonged to a genre of risks that popular culture held in some disrepute. Within NASA, though, there was a more important cultural reason: the dominance of the pure science ideal. For two decades after the Agency was created, it had a high-level directorate devoted to developing useful space technologies—weather, communications, and associated remote sensing technologies. This was embedded in an Office of Applications, distinct from the Agency's Office of Science. NASA shut down its applications program during the Reagan administration for two reasons.[7] Agency management believed that it had not been very successful at getting its technologies accepted by users outside

---

5. See Theodore Porter, *Trust in Numbers: The Pursuit of Objectivity in Science and Public Life* (Princeton, NJ: Princeton University Press, 1996).

6. Martin Reuss, "Coping with Uncertainty: Social Scientists, Engineers, and Federal Water Resources Planning," *Natural Resources Journal* 32 (1992): 101–135; Martin Reuss, *Designing the Bayous: The Control of Water in the Atchafalaya Basin, 1800–1995* (College Station, TX: Texas A&M University Press, 2014).

7. On the demise of the NASA applications program, see Erik M. Conway, "Bringing NASA Back to Earth," *Science and Technology in the Global Cold War*, ed. Naomi Oreskes and John Krige (Cambridge, MA: MIT Press, 2015), pp. 251–272; for contents of the applications program, see Pamela E. Mack and Ray A. Williamson, "Observing the Earth from Space," in *Exploring the Unknown* vol. 3, *Using Space*, ed. John M. Logsdon et al. (Washington, DC: NASA SP-4407, 1998), pp. 155–176.

NASA, and Reagan administration officials believed that government should not be in the applications business in any case. Applied science and technology development were the realm of private industry, not the government.

NASA's surviving Office of Science, now known as the Science Mission Directorate, adopted the ideal that their activities should be driven only by curiosity—not by an effort to produce useful knowledge, not by a desire to produce a public good, and certainly not to serve a policy or political goal. Scholars call this the "pure science ideal."[8] We can see echoes of this ideal in the Discovery Program's evaluation of NEOCam as not producing "compelling science" and in the Committee on Near-Earth Object Observations in the Infrared and Visible Wavelengths argument that NEO discovery should not compete with "high-priority science objectives."[9] NEOCam and the larger issue of NEO discovery were viewed as "policy driven," not "science driven." And we can find echoes in the texts of the astronomy and planetary science decadal surveys written to justify NASA's science plans.[10] While both of the most recent decadal surveys discussed NEO discovery, particularly in the context of new ground observatories like the Large Synoptic Survey Telescope (LSST), neither made NEO discovery a high scientific priority. The space and planetary scientists inside and outside NASA who steer its programs have adopted an ideal under which utility and public service are not valid selection criteria for missions. Within NASA's Earth science areas, avoidance of "applied science" ended in the early 2000s, as the Agency's Earth Observing

---

8. See, e.g., George H. Daniels, "The Pure-Science Ideal and Democratic Culture," *Science* 156, no. 3783 (30 June 1967): 1699–1705, *https://doi.org/10.1126/science.156.3783.1699* (accessed 2 July 2021); David Kaldewey and Désirée Schauz, "Transforming Pure Science into Basic Research: The Language of Science Policy in the United States," chapter 3 in *Basic and Applied Research: The Language of Science Policy in the Twentieth Century* (New York: Berghahn Books, 2018), pp. 104–140, *https://doi.org/10.2307/j.ctv8bt0z7.9* (accessed 2 July 2021); David Kaldewey and Désirée Schauz, "'The Politics of Pure Science' Revisited," *Science and Public Policy* 44, no. 6 (1 December 2017): 883–886, *https://doi.org/10.1093/scipol/scx060* (accessed 2 July 2021).

9. Committee on Near-Earth Object Observations in the Infrared and Visible Wavelengths, "Finding Hazardous Asteroids Using Infrared and Visible Wavelength Telescopes," July 2019 (pre-publication draft), p. S-5.

10. See National Research Council, "Vision and Voyages for Planetary Science in the Decade 2013–2022" (Washington, DC: National Academies Press, 2011); NRC, "New Worlds, New Horizons in Astronomy and Astrophysics" (Washington, DC: National Academies Press, 2010).

System began returning terabytes of useful data. NASA's astronomers seem not to have noticed.

During the period covered in this book, NEO astronomy transformed from a minor research field with a small handful of practitioners—Gehrels, Helin, the Shoemakers—to a form of public service (though still with a relatively small handful of practitioners). Their funding sources were both military and civil (Air Force and NASA), as well as private donations (most notably to Spacewatch and the Minor Planet Center). In coming to fulfill a service function, NEO astronomy follows much larger fields like meteorology, which has had its own branch of the U.S. federal government since 1870 in the National Weather Service and its predecessors, and for which NASA still develops satellites and instrumentation.

The Changing Views on Near-Earth Objects

Though nearly ignored at the time, the early photographic near-Earth object discovery efforts of the 1970s and 1980s by the Shoemakers, Eleanor Helin, and Tom Gehrels began a discovery process that snowballed interest in these bodies. Each time a NEO was found, it often generated public (and sometimes congressional) interest—especially if the object made, or was about to make, a close Earth approach. Although NASA was initially slow to take advantage of this interest, it did eventually generate increased funding for these efforts. The advancements of efficient CCD detectors and computer-driven observations in the late 20th century dramatically accelerated the discovery rate, along with the number of known and upcoming close Earth approaches—thus generating more and more interest and more and more efforts to find the vast majority of those objects that could threaten Earth.

Once discovered, NEOs require so-called follow-up positional observations that allow an accurate orbit determination so that the object's future motion can be accurately predicted. In the last quarter of the 20th century, amateur astronomers were key to this effort, but the large-aperture, CCD-based surveys of today require professional astronomers with access to large telescopes to provide the follow-up observations that are most often too faint for amateur observers.

The characterization of NEOs has also made impressive advances since the efforts in the late 20th century to understand the physical and photometric characteristics of NEOs. The categorization of asteroid spectral responses has been extended from the visible into the near-infrared region of the spectrum, and thanks largely to the space-based NEOWISE satellite, there are now over 1300 NEOs and 39,475 solar system objects for which infrared data (and hence accurate diameters and albedos) are available.[11]

Radar observations of NEOs are now often capable of "imaging" NEOs to better than 4-meter resolution, thus achieving a level of spatial characterization that is only exceeded by some spacecraft observations. There have been nine comets and over a dozen asteroids visited by spacecraft and dust samples returned from both a comet and an asteroid. Two more near-Earth asteroids were added to the list when the Japanese Hayabusa2 mission reached the dark, kilometer-sized 162173 Ryugu in mid-2018 and when NASA's OSIRIS-REx reached the dark, half-kilometer-sized 101955 Bennu in late 2018. The Hayabusa2 return capsule brought samples to Earth in December 2020 when it landed in Australia.[12] There are plans to bring back surface samples from Bennu in September 2023.

Extensive ground-based and space-based observations have changed our views of these near-Earth objects. In the second half of the 20th century, comets and asteroids were considered distinct categories of solar system bodies. Comets were dirty snowballs that were remnants of the outer solar system formation process, while asteroids were thought to be rocky bodies that were left over from the inner solar system formation process. Extensive ground-based and space-based observations have altered this view considerably. Comets seem to be composed of at least some dust that formed in the inner solar system and ices that were acquired in the outer solar system, while some asteroids have both inner solar system dust and ices. Clearly, mixing of inner and outer solar system material has taken place. Some active comets have evolved into inactive asteroid-like bodies, and some asteroids have exhibited comet-like outgassing. Some, like the interstellar 'Oumuamua, defy classification. Comets and asteroids must now be considered members of the same solar system family of small bodies with widely diverse characteristics that run the

---

11. For up-to-date statistics, see JPL/CNEOS, "Discovery Statistics," *https://cneos.jpl.nasa. gov/stats/wise.html* and JPL, "The NEOWISE Project," *https://neowise.ipac.caltech.edu* (accessed 15 September 2021).

12. "Japan's Hayabusa2 Mission Brings Home Samples of Asteroid Dirt and Gas," *Planetary News*, 22 December 2020, *https://www.lpi.usra.edu/planetary_news/2020/12/22/ japans-hayabusa2-mission-brings-home-samples-of-asteroid-dirt-and-gas/* (accessed 15 September 2021).

gamut from outgassing cometary fluff balls, to inactive fragile rubble piles, to shattered rocks, to solid slabs of iron. NEOs can pose threats to Earth, serve as repositories of future space resources, and open windows into the early solar system formation process. The study of asteroids and especially near-Earth objects has evolved from their being considered nuisance objects ("vermin of the skies") by stellar astronomers in the late 19th century to objects deserving considerable attention and resources by today's research scientists. They have also become objects of policy, as we have seen. NASA's Near-Earth Objects Observations Program has been a major enabler for this important research.

Near-Earth objects are being integrated into our understanding of Earth's history. The cumulative changes due to gradualism between catastrophes appear to be less than the changes due to a single catastrophe. In the case of the Cretaceous–Paleogene impact, the cosmic catastrophe provided the conditions for the rapid repopulation of life after cosmic extinction. Occasional cosmic impacts transform planetary surfaces and serve as a reset button for evolution.

The near-Earth asteroid discovery rate, as well as the scientific and public interest in these objects, has grown dramatically in the last three decades. At the beginning of 1990, there were 134 known near-Earth asteroids. In mid-2019, there were well over 20,000—and the discovery rate, facilitated by modern detector and computer technologies, keeps increasing. As we have seen, the ability of the automated surveys to discover new objects exceeds the ability of follow-up observers to keep up, so even today, not all discovered objects have well-established orbits. For those objects with well-known orbits, JPL's Center for Near-Earth Objects web page lists over 47 future close-Earth approaches to within the distance of the Moon. Predicted future close-Earth approaches by near-Earth asteroids will enable extensive ground-based observing programs like those being planned for the passage of the 340-meter asteroid Apophis to within 5 Earth radii above Earth's surface on 13 April (Friday the 13th) 2029. Future observations will, no doubt, enable accurate predictions for actual Earth atmosphere impacts for an increasing number of small near-Earth asteroids. These predictions will allow the gathering up, and study, of meteorites from known objects—and perhaps a cottage industry for future tour groups that will witness impressive atmospheric impact events from a safe distance. Advance warnings for the much rarer future impacts by large, destructive near-Earth asteroids will facilitate the necessary mitigation efforts. In the past three decades, more than a dozen spacecraft have studied asteroids during flyby or rendezvous encounters. Observations of these ancient objects are providing clues to the conditions during the solar system formation process, and their minerals and water resources may one day provide the raw materials for space structures and rocket fuel.

The American public and Congress already consider planetary defense and the continued discovery and study of near-Earth asteroids to be one of NASA's top priorities. This book has outlined the reasons for this extraordinary interest.

APPENDIX 1

ASTEROID AND COMET DESIGNATIONS, NUMBERS, AND NAMES

Upon discovery, both comets and asteroids are first given temporary, or provisional, designations; then, when their orbits become secure (i.e., accurate), they can receive permanent numbers and names.

Table A1-1. Naming conventions for asteroids and comets.

	Provisional Designations	Permanent Numbers and Names
Asteroids	Discovery year and alphanumeric code providing the discovery year, half month, and order of discovery within the half month.	An asteroid number is normally assigned in sequential order when its orbit is secure. It is then eligible for naming—usually by the discoverer(s).
Asteroid Example	1992 AB is the second (B) asteroid discovered in the first half of January (A) 1992.*	8373 Stephengould = 1992 AB was named after Harvard scientist Stephen Jay Gould by the discoverers Gene and Carolyn Shoemaker.
Comets	Short periodic and long periodic comets are designated respectively by a leading "P" or "C" with the discovery year and half-month capital letter followed by the numerical order of the discovery during that half month. Comets designated with a "D" or "I" are defunct or interstellar.	All comets are normally named after the discoverer(s) or discovery program. Short periodic comets are numbered sequentially after they have been recovered or observed at a second perihelion passage.
Comet Examples	Non-short periodic Comet C/2015 A2 (Pan-STARRS) was the second comet discovery announcement made during the first half of January 2015. It was discovered by the Pan-STARRS discovery program.	1P/Halley was the first short periodic comet recognized as such—by Edmond Halley. This comet's return in 1682 is designated 1P/1682 Q1 since it was the first comet seen in the second half of August of that year (i.e., first seen on 24 August by Arthur Storer in Maryland).

* The letters "I" and "Z" are not utilized in designating asteroid or comet half months, and each month is divided into the first half month (days 1–15) and the second half (all the remaining days of that month).

A Historical Development of Asteroid and Comet Designations, Numbers, and Names

Asteroid vs. Minor Planet

In his 2014 Ph.D. thesis, Clifford Cunningham identified the true origin of the term "asteroid."[1] Although it is often attributed to William Herschel, due credit should actually go to the English music historian Dr. Charles Burney, Sr., and his son, Charles Jr. In May 1802, Herschel asked Charles Burney, Sr., if he would furnish a Latin or Greek name for the small "stars" that had lately been found. In a subsequent letter to his son, the senior Burney suggested the Greek word "aster" to denote "starlike," and his son suggested that "oid" be added to form "asteroid." Herschel first used the term on 6 May 1802 in his memoir presented to the Royal Society entitled "Observations of the Two Lately Discovered Celestial Bodies," but his choice was not immediately greeted with great enthusiasm. Indeed, Herschel was accused of applying a lesser designation for these objects so as not to detract from his own discovery of the planet Uranus in 1781, and the English author John Corry even accused Hershel of "philosophical quackery." Giuseppe Piazzi himself also rejected the term "asteroid," perhaps because his pride would not allow his discovery to be known as anything other than a primary planet. Even so, by 1830, the term "asteroid" was commonly used in England, and the American astronomer Benjamin Apthorp Gould, founder of the *Astronomical Journal* in 1849, gave the term his stamp of approval in 1848.[2] By the second half of the 19th century, the term "minor planet" was often used instead of "asteroid." Currently, the International Astronomical Union (IAU) and the Minor Planet Center (MPC) in Cambridge, Massachusetts, use the term "minor planet," but the two terms are interchangeable, and both are widely used.

By July 1851, there were only 15 discovered asteroids, few enough that they were usually listed along with the then-known planets. Fourteen of these asteroids had assigned symbols, but these symbols, meant to be shortcut designations, were getting complicated and difficult to draw. For example,

---

1.   Clifford J. Cunningham, "The First Four Asteroids: A History of Their Impact on English Astronomy in the Early Nineteenth Century" (Ph.D. thesis., University of Southern Queensland, 2014), pp. 46–97.

2.   Ibid., p. 96.

asteroid (14) Irene's symbol was a dove carrying an olive branch, with a star on its head. In 1851, the German astronomer Johann F. Encke introduced encircled numbers instead of symbols, but this short-lived scheme began with the asteroid Astraea, given the number (1), and went through (11) Eunomia. Ceres, Pallas, Juno, and Vesta continued to be denoted by their traditional symbols. In 1851, B. A. Gould was using a short-lived designation system consisting of a circle containing the number of the asteroid in the chronological order of discovery. For a time, this system began with (5) Astraea with Ceres, Pallas, Juno, and Vesta retaining their old symbols rather than their appropriate numbers; but by 1864, J. F. Encke, in the influential journal *Berliner Astronomisches Jahrbuch*, was numbering the first four asteroids as (1) Ceres, (2) Pallas, (3) Juno, and (4) Vesta.

Currently, the MPC is responsible for assigning designations, numbers, and names for asteroids and comets. Numbers (in parentheses) are assigned sequentially when the asteroid's orbit is secure such that the observational recovery of the asteroid at future returns is assured. However, when this numbering system was first introduced in the mid-19th century and for many years thereafter, a secure orbit was not required, and some numbered asteroids were subsequently lost. For example, when the numbered asteroid 878 Mildred was discovered in September 1916 by Seth Nicholson at Mount Wilson Observatory, there were not enough astrometric observations to render its orbit secure. As a result, 878 Mildred, named after the daughter of the American astronomer Harlow Shapley, was lost until 1991, when Gareth Williams at the Minor Planet Center identified its few 1916 observations with single-night observations made in 1985 and 1991.[3]

Asteroid numbers are sequentially assigned as their orbits are secured, and then they become eligible for naming. For the first few hundred asteroids discovered, classical names were often assigned; thereafter, at least for a time, nonclassical female names were often used. The first near-Earth object, 433 Eros, discovered in 1898, was an exception. Currently, asteroids are being named for well-known, and not-so-well-known, scientists, personalities, geographic locations, and science contest award winners and their mentors, along with a host of other entities. The asteroid discoverer, or discovery team,

---

3. *International Astronomical Union Circular* 5275, dated 25 May 1991. Mildred Shapley became Mildred Mathews, editor of the noted University of Arizona series of books on asteroids and other solar system objects.

normally has the option of naming their discovered asteroids—subject to the guidelines and approval of the IAU Committee on Small Body Nomenclature (CSBN).[4] However, by 2018, the asteroid discovery rate was well over 10,000 per year, so only a relatively small subset of the numbered asteroids discovered by major discovery teams currently receive names.

Periodic and non-periodic comets also get designations and are generally (but not always) named after the discoverer, the discoverers, or the discovery program. Because there are far fewer comet discoveries than asteroid discoveries, cometary designations use a numerical sequence to denote the order of discovery in a particular half month, rather than the more complex alphanumeric code used to designate asteroids. The rules for comet designations, numbers, and names are maintained by the MPC.[5]

4.   IAU MPC, "How Are Minor Planets Named?" *https://minorplanetcenter.net/iau/info/ HowNamed.html* (accessed 23 September 2021).

5.   IAU MPC, "Cometary Designation System," *https://www.minorplanetcenter.net/iau/ lists/CometResolution.html* (accessed 18 April 2018).

APPENDIX 2

HELIOCENTRIC ORBIT CLASSIFICATIONS FOR NEAR-EARTH OBJECTS

The term "near-Earth object" (NEO) applies to any asteroid or comet whose perihelion distance is less than 1.3 au, a condition that is a bit arbitrary but ensures that the object can come within 0.3 au of Earth's distance from the Sun. Those objects whose orbits can bring them much closer—to within 0.05 au of Earth's orbit—are termed Potentially Hazardous Objects (PHOs). That is, their Minimum Orbital Intersection Distance (MOID) from Earth is

Near Earth Object Orbit Classes	Orbit Criteria	
Amors	Earth-approaching asteroids with orbits exterior to Earth's but interior to the orbit of Mars. Hence their semimajor axes are larger than 1.0 au and their perihelia are between 1.017 and 1.3 au.	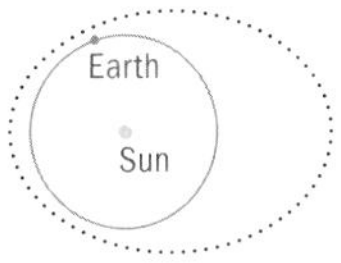
Apollos	Earth-orbit-crossing asteroids with semimajor axes larger than Earth's (a is larger than 1.0 au) and with perihelia less than 1.017 au.	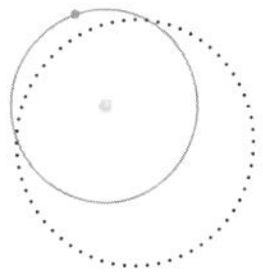
Atens	Earth-orbit-crossing asteroids with their semimajor axes smaller than Earth's and their aphelia larger than 0.983 au.	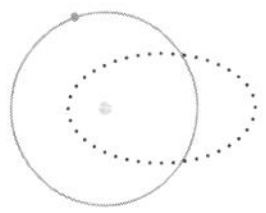
Atiras	Asteroids with their orbits contained entirely within that of Earth. Hence their semimajor axes are less than 1 au and their aphelia are less than 0.983 au.	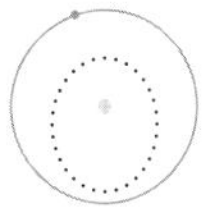

Figure A2-1. Orbit classes.

0.05 au. For a MOID of this size, planetary gravitational perturbations could further decrease the MOID over several centuries, rendering the object truly, rather than just potentially, hazardous. Often, the definition of a PHO (or PHA for a potentially hazardous asteroid) includes the further restriction that its absolute magnitude must be equal to or less than 22, which roughly limits its diameter to 140 meters or larger. Based upon their orbital characteristics, there are four groups of near-Earth objects: the Amors, the Apollos, the Atens, and the Atiras. This latter group is sometimes called the Inner-Earth objects, or Apoheles.

Amors: Asteroids in this group have heliocentric orbits exterior to Earth's but generally interior to the orbit of Mars. An Amor-class asteroid's perihelion distance (q) is greater than Earth's aphelion distance (1.017 au). Amors are also near-Earth objects, so their perihelia are less than 1.3 au (q < 1.3 au). Hence, an asteroid is in the Amor class if its perihelion distance is between the values of 1.017 and 1.3 au. The namesake for this group is asteroid 1221 Amor, which was discovered photographically on 12 March 1932 by Eugène Joseph Delporte in Uccle, Belgium. Amor is the Latin name for the Greek god of love. Asteroid 433 Eros is an Amor object.

Apollos: Apollo-class asteroids are so-called Earth-crossers because they cross over the heliocentric distance of Earth on their way to perihelion. Because of their inclinations with respect to the ecliptic plane, their orbits do not necessarily intersect Earth's orbit. The perihelia of Apollos are less than Earth's aphelion distance (q < 1.017 au), and their semimajor axes (a) are greater than 1.0 au. Apollos can cross the orbits of Mars, Earth, and Venus. The namesake of this group is 1862 Apollo, named for the Greek god of the Sun, which was discovered on 24 April 1932 during a photographic search for main-belt asteroids by the German astronomer Karl Reinmuth at Heidelberg Observatory. However, the 1932 observations covered less than a month, so the orbit was not well defined, or secured, and it was lost until it was observed again 41 years later on 28 March 1973 by the Oak Ridge Observatory in Massachusetts. As a result, 1862 Apollo has a higher number than some other so-called Apollo asteroids, like 1566 Icarus and 1620 Geographos, which were discovered after Apollo but received enough observations to secure their orbits before the orbit of Apollo could be secured in 1973.

Atens: The Aten-class asteroids are Earth-crossing, with their semimajor axes less than that of Earth's (a < 1.0 au) and their aphelia larger than Earth's perihelion (Q > 0.983 au). The first-discovered Aten, and the namesake for this group, was 2062 Aten, found on 7 January 1976 by Eleanor F. "Glo" Helin at the Palomar Mountain Observatory. Aten was the ancient Egyptian Sun god.

Atiras: The Atiras, or inner-Earth objects, have their entire orbits interior to Earth's orbit. Hence their semimajor axes are less than 1.0 au and their aphelia are less than Earth's perihelion (Q < 0.983 au). Atiras were named after the Pawnee Indian goddess of Earth and the evening star. This group is sometimes referred to as Apohele asteroids. The name Apohele, Hawaiian for "orbit," was suggested by Dave Tholen after his 1998 discovery of the first suspected member of this group, 1998 DK36. However, this object's orbit was not secure enough for it to receive a permanent number, whereas the first confirmed inner-Earth object with a secure orbit was 163693 Atira discovered by the LINEAR survey in 2003.

GLOSSARY

Absolute magnitude: The magnitude of an asteroid at zero phase angle and at unit heliocentric and geocentric distances.

Albedo: Geometric albedo is the ratio of a body's brightness at zero phase angle to the brightness of a perfectly diffusing disk with the same position and apparent size as the body.

Amor (class): Asteroids in this group have heliocentric orbits exterior to Earth's but generally interior to the orbit of Mars. An asteroid is in the Amor class if its perihelion distance is between the values of 1.017 and 1.3 astronomical units (au). (See appendix 2 for details.)

Anhydrous: A substance containing no water.

Aphelion: That point of a celestial body's orbit furthest from the Sun (plural: aphelia).

Apollo (class): The perihelia of Apollo asteroids are less than Earth's aphelion distance (q < 1.017 au), and their semimajor axes are greater than 1.0 (a > 1.0 au). They can be Venus-, Earth-, and/or Mars-crossing. (See appendix 2 for details.)

Apparent magnitude: The brightness of a celestial body as seen from Earth, with low numbers brighter. The Sun is apparent magnitude –26.7, while the full Moon is –12.7. The star Vega is the definition of zero on this scale.

Astrobleme: An impact crater's remnant on Earth.

Astronomical unit (au): The approximate mean distance between the centers of Earth and the Sun, 149.6 million kilometers or 93 million miles.

Aten (class): The Aten asteroids are Earth-crossing, with their semimajor axes less than that of Earth's (a < 1.0 au) and their aphelia larger than Earth's perihelion (Q > 0.983 au). (See appendix 2 for details.)

Atira (class): The Atiras, or Interior-Earth Objects, have their entire orbits interior to Earth's orbit. Hence their semimajor axes are less than 1.0 au and their aphelia are less than Earth's perihelion (Q < 0.983 au). (See appendix 2 for details.)

Bolide: A meteor that explodes in Earth's atmosphere.

Chondrite: Chondrite meteorites take their name from chondrules, the nearly spherical, silicate-rich particles they contain. They are the most abundant type of stony meteorites.

Earth-crossing asteroid: Any asteroid whose heliocentric orbit crosses Earth's orbit. The Apollo- and Aten-class asteroids can be Earth-crossing.

Ephemeris: The predicted positions of an object in the sky at given times.

Fireball: A brighter-than-usual meteor. The International Astronomical Union defines a fireball as "a meteor brighter than any of the planets" (apparent magnitude −4 or brighter).

Hazard: The population (sometimes called "flux") of objects that can closely approach Earth over time.

Hypering: A chemical process for increasing the sensitivity of film.

Keyhole: A relatively small region near Earth that allows a passing near-Earth object to modify its orbit just enough, and in the right direction, to set up an Earth impact at a subsequent Earth return.

Limiting magnitude: The faintest apparent magnitude a particular telescope can detect.

Long-period comet: A comet that has an orbit period greater than 200 years.

Lunation: A lunar month as measured from new Moon to the next new Moon (about 29.5 days).

Megaton: This unit is often used to describe the energy of a nuclear blast. A 1-megaton explosion has the energy of 1 million tons of TNT explosives.

Meteor: Known colloquially as a "shooting star," a meteor is the visible passage of a small particle from a comet or asteroid being heated to incandescence by collisions with air molecules in the upper atmosphere. Most meteors are associated with objects only the size of a sand particle.

Meteorite: A portion of an asteroid that survives its passage through the atmosphere (fireball phase) and hits the ground without being completely destroyed.

Micron: A unit of length equal to one-millionth of a meter.

MOID: Minimum Orbital Intersection Distance.

Near-Earth space: A region in space within 0.3 au (about 45 million kilometers) of Earth's orbit.

NEOOP: Near-Earth Objects Observations Program

Newtonian telescope: A reflector telescope composed of a concave primary mirror and a smaller, flat secondary mirror that feeds an eyepiece (or instruments).

Occultation: An event that occurs when one object is hidden by another object that passes between it and the observer.

Opposition: A celestial object is in opposition to Earth when it is on the opposite side of the Sun so that there is an Earth—Sun—object alignment.

Perihelion: The point of a celestial body's orbit closest to the Sun (plural: perihelia).

Polarization: Polarized light waves are light waves in which the vibrations occur in a single plane.

Potentially Hazardous Object: Objects whose orbit approaches within 0.05 au of Earth's orbit (e.g., have a MOID of 0.05 au or less).

Regolith: The layer of unconsolidated rocky material covering bedrock.

Risk: The negative effects that could occur from Earth encountering a "Threat." An impact of a relatively small object has the "risk" of wiping out a city from airburst and thermal effects, or it may be harmless if in the middle of the ocean.

Schmidt telescope: A reflector telescope composed of a spherical primary mirror and a correcting lens or "field flattener" at the primary mirror's prime focus point, where film or instruments are also mounted.

Short-period comet: A comet with an orbital period of less than 200 years.

Signal-to-Noise Ratio (SNR): The relationship between the amount of a signal received to the amount of noise also received during a measurement.

Spectrum: A chart or a graph that shows the intensity of light being emitted over a range of wavelengths.

Taxonomy: Asteroid taxonomy refers to the classification of these objects by their reflected light characteristics, including their spectra and albedo.

Threat: A detected and tracked object with a non-negligible Earth impact probability.

Trojan (class): Small bodies that share an orbit with another body, maintaining the same relative position to that body. The only known Earth trojan is 2010 TK7.

Yarkovsky effect: Named after the Polish engineer Ivan Yarkovsky, this effect refers to the small thrust introduced on a rotating asteroid due to the thermal reradiation of sunlight.

YORP effect: The Yarkovsky-O'Keefe-Radzievskii-Paddack effect is due to reradiation of sunlight from irregularly shaped asteroids where either the morning or evening edge of the rotating asteroid is more effective than the other in catching and re-emitting solar radiation—thus either spinning up (or down) the affected asteroid.

Zodiacal light: A band of faint light around the ecliptic caused by the scattering of sunlight by interplanetary dust.

ACRONYMS AND ABBREVIATIONS

^{26}Al	aluminum-26
AA	Associate Administrator
AAAS	American Association for the Advancement of Science
AFRL	Air Force Research Laboratory
AIAA	American Institute of Aeronautics and Astronautics
AIDA	Asteroid Impact and Deflection Assessment
AIM	Asteroid Impact Mission
AMOS	Air Force Maui Optical Station
AOA	Analysis of Alternatives
APL	Applied Physics Laboratory
ARM	Asteroid Redirect Mission
ARPA	Advanced Research Projects Agency
ARRM	Asteroid Redirect Robotic Mission
ASE	Association of Space Explorers
ASEE	American Society for Engineering Education
ASME	American Society of Mechanical Engineers
ATAP	Asteroid Threat Assessment Project
ATLAS	Asteroid Terrestrial-impact Last Alert System
au	astronomical unit
BMDO	Ballistic Missile Defense Organization
CAI	calcium-aluminum inclusion
Caltech	California Institute of Technology
CCD	charge-coupled detector
CEV	Crew Exploration Vehicle
CHON	carbon, hydrogen, oxygen, and nitrogen
CLOMON	Close Approach Monitoring System
CNEOS	Center for Near Earth Object Studies
CONTOUR	COmet Nucleus TOUR
COPUOS	Committee on the Peaceful Uses of Outer Space

CRAF	Comet Rendezvous Asteroid Flyby
CSBN	Committee on Small Body Nomenclature
CSS	Catalina Sky Survey
DAMIEN	Interagency Working Group for Detecting and Mitigating the Impact of Earth-bound Near-Earth Objects
DARPA	Defense Advanced Research Projects Agency
DART	Double Asteroid Redirection Test
DHS	Department of Homeland Security
DLR	German Aerospace Center
DOD	Department of Defense
DOE	Department of Energy
DOS	Department of State
DSI	Deep Space Industries
DSN	Deep Space Network
EPOXI	Extrasolar Planet Observation and Deep Impact Extended Investigation
ESA	European Space Agency
ESOC	European Space Operations Centre
ESRIN	European Space Research Institute
ETS	Experimental Test Site
FARCE	Far Away Robotic sandCastle Experiment
FEMA	Federal Emergency Management Agency
FROSST	Fast Resident Object Surveillance Simulation Tool
FY	fiscal year
GAO	Government Accountability Office
GEODSS	Ground-based Electro-Optical Deep-Space Surveillance
GRAIL	Gravity Recovery and Interior Laboratory
H	absolute magnitude of an asteroid
HCO	Harvard College Observatory
HED	howardites, eucrites, and diogenites
HRC	Historical Reference Collection
IAA	International Academy of Astronautics
IAU	International Astronomical Union
IAUC	*International Astronomical Union Circular*
IAWN	International Asteroid Warning Network
ICE	International Comet Explorer
IfA	Institute for Astronomy

IG	Inspector General
IPAC	Infrared Processing and Analysis Center
IR	infrared
IRAS	Infrared Astronomical Satellite
IRTF	Infrared Telescope Facility
ISEE	International Sun-Earth Explorer
ISS	International Space Station
JHU	Johns Hopkins University
JPL	Jet Propulsion Laboratory
KISS	Keck Institute for Space Studies
LANL	Los Alamos National Laboratory
LEO	low-Earth orbit
LINEAR	Lincoln Near-Earth Asteroid Research
LLNL	Lawrence Livermore National Laboratory
LONEOS	Lowell Observatory NEO Survey
LPL	Lunar and Planetary Laboratory
LSST	Large Synoptic Survey Telescope
Ma	mega annum (millions of years)
MESSENGER	MErcury Surface, Space ENvironment, GEochemistry, and Ranging
MICE	Megaton Ice-Contained Explosion
MIT	Massachusetts Institute of Technology
MMT	Multi-Mirror Telescope
MODP	Moving Object Detection Program
MOID	Minimum Orbital Intersection Distance
MPC	Minor Planet Center
MPEC	*MPC Daily Electronic Circular*
MSX	Mid-Course Space Experiment
NAC	NASA Advisory Committee
NASA	National Aeronautics and Space Administration
NATO	North Atlantic Treaty Organization
NEA	near-Earth asteroid
NEAP	Near-Earth Asteroid Prospector
NEAR	Near Earth Asteroid Rendezvous
NEAT	Near-Earth Asteroid Tracking
NEATM	Near-Earth Asteroid Thermal Model
NEO	near-Earth object

NEOCam	Near-Earth Object Camera
NEOCP	NEO Confirmation web page
NEODyS	Near Earth Objects Dynamic Site
NEOOP	Near-Earth Objects Observations Program
NEOSSat	Near-Earth Object Surveillance Satellite
NEOWISE	Near-Earth Object Wide-field Infrared Survey Explorer
NExT	New Exploration of Tempel 1
NHATS	Near-Earth Object Human Space Flight Accessible Targets Study
NIAC	NASA Innovative and Advanced Concepts
NOAA	National Oceanic and Atmospheric Administration
NRC	National Research Council
OPAE	Office of Program Analysis and Evaluation
OSIRIS-REx	Origins, Spectral Interpretation, Resource Identification, Security, and Regolith Explorer
OSTP	Office of Science and Technology Policy
P	percentage polarization
PACS	Palomar Asteroid and Comet Survey
PAIR	Probabilistic Asteroid Impact Risk
Pan-STARRS	Panoramic Survey Telescope and Rapid Response System
PCAS	Palomar Planet-Crossing Asteroid Survey
PDC	Planetary Defense Conference
PDCO	Planetary Defense Coordination Office
PHA	Potentially Hazardous Asteroid
PHO	Potentially Hazardous Object
PI	Principal Investigator
Pu-244	plutonium-244
Pv	albedo of an asteroid
RAP	Robotics Asteroid Prospector
ROTC	Reserve Officers' Training Corps
SBAG	Small Bodies Assessment Group
SDIO	Strategic Defense Initiative Organization
SDT	Science Definition Team
SEP	solar-electric propulsion
SETI	Search for Extraterrestrial Intelligence
SNR	Signal-to-Noise Ratio
SSA	Space Situational Awareness
SSI	Space Studies Institute

STM	standard thermal model
TOPEX/Poseidon	Topography Experiment/Poseidon
UBV	ultraviolet, blue, and visual
UCLA	University of California, Los Angeles
UFO	unidentified flying object
UMD	University of Maryland
UNISPACE	United Nations Conference on the Exploration and Peaceful Uses of Outer Space
USAF	U.S. Air Force
USGS	United States Geological Survey
VLA	Very Large Array
VLBA	Very Long Baseline Array
VSE	Vision for Space Exploration
WISE	Wide-field Infrared Survey Explorer
WWI	World War I or First World War
YORP	Yarkovsky-O'Keefe-Radzievskii-Paddack

SELECT BIBLIOGRAPHY

Published Books

Alvarez, Walter. *T-Rex and the Crater of Doom*. Princeton, NJ: Princeton University Press, 1997.

Badash, Lawrence. *A Nuclear Winter's Tale: Science and Politics in the 1980s*. Cambridge, MA: MIT Press, 2009.

Baldwin, Ralph Belknap. *The Face of the Moon*. Chicago: University of Chicago Press, 1949.

Baxter, John, and Thomas Atkins. *The Fire Came By: The Riddle of the Great Siberian Explosion*. New York: Doubleday & Company, 1976.

Binzel, Richard P., Tom Gehrels, and Mildred Shapley Matthews, eds. *Asteroids II*. Space Science Series. Tucson, AZ: University of Arizona Press, 1989.

Bell, Jim, and Jacqueline Mitton, eds. *Asteroid Rendezvous: NEAR Shoemaker's Adventures at Eros*. New York: Cambridge University Press, 2002.

Belton, Michael J. S., Thomas H. Morgan, Nalin H. Samarasinha, and Donald K. Yeomans. *Mitigation of Hazardous Comets and Asteroids*. Cambridge, U.K.: Cambridge University Press, 2004.

Bottke, W. F., Jr., A. Cellino, P. Paolicchi, and R. P. Binzel, eds. *Asteroids III*. Tucson, AZ: University of Arizona Press, 2002.

Brush, Stephen G. *A History of Modern Planetary Physics: Nebulous Earth*. Cambridge, U.K.: Cambridge University Press, 1996.

———. *A History of Modern Planetary Physics: Transmuted Past*. Cambridge, U.K.: Cambridge University Press, 1996.

———. *A History of Modern Planetary Physics: Fruitful Encounters*. Cambridge, U.K.: Cambridge University Press, 1996.

Burke, John G. *Cosmic Debris: Meteorites in History*. Berkeley, CA: University of California Press, 1991.

Burrows, William E. *The Asteroid Threat*. Amherst, NY: Prometheus Books, 2014.

Butrica, Andrew J. *To See the Unseen: A History of Planetary Radar Astronomy.* NASA History Series. Washington, DC: NASA, 1996.

Chapman, Clark R., and David Morrison. *Cosmic Catastrophes.* New York: Plenum Press, 1989.

Clerke, Agnes Mary. *A Popular History of Astronomy During the Nineteenth Century.* Edinburgh, U.K.: A. & C. Black, 1893.

Conway, Erik M. *Atmospheric Science at NASA: A History.* Baltimore: Johns Hopkins University Press, 2008.

Dick, Steven J., and James E. Strick. *The Living Universe: NASA and the Development of Astrobiology.* New Brunswick, NJ: Rutgers University Press, 2004.

Doel, Ronald Edmund. *Solar System Astronomy in America: Communities, Patronage, and Interdisciplinary Science, 1920–1960.* Cambridge, U.K.: Cambridge University Press, 2010.

Festou, M., H. U. Keller, and Harold A. Weaver, eds. *Comets II.* Tucson, AZ: University of Arizona Press, 2004.

Gehrels, Tom, Mildred Shapley Matthews, and A. M. Schumann, eds. *Hazards Due to Comets and Asteroids.* Tucson, AZ: University of Arizona Press, 1994.

Halley, Edmond. "A Synopsis of the Astronomy of Comets." 1705. *https://library. si.edu/digital-library/book/synopsisofastron00hall.* Accessed 7 July 2021.

Hoyt, William Graves. *Coon Mountain Controversies: Meteor Crater and the Development of Impact Theory.* Tucson, AZ: University of Arizona Press, 1987.

Jardine, N., J. A. Secord, and E. C. Spary. *Cultures of Natural History.* Cambridge, U.K.: Cambridge University Press, 1996.

Kleiman, Luis A., ed. *Project Icarus.* Cambridge, MA: MIT Press, 1968. *https:// mitpress.mit.edu/books/project-icarus-systems-engineering.*

Kronk, Gary W. *Cometography: A Catalog of Comets.* Vol. 1, *Ancient–1799.* Cambridge, U.K.: Cambridge University Press, 1999.

Levy, David H. *Shoemaker by Levy—The Man Who Made an Impact.* Princeton, NJ: Princeton University Press, 2002.

Logsdon, John M., et al., *Exploring the Unknown: Selected Documents in the History of the U.S. Civil Space Program.* Vol. 3, *Using Space.* Washington, DC: NASA, 1998.

Lunan, Duncan. *Incoming Asteroid!* New York: Springer New York, 2014.

Mark, Kathleen. *Meteorite Craters.* Tucson, AZ: University of Arizona Press, 1995.

Markley, Robert. *Dying Planet: Mars in Science and the Imagination.* Durham, NC: Duke University Press, 2005.

Matloff, Greg, C. Bangs, and Les Johnson. *Harvesting Space for a Greener Earth.* New York: Springer New York, 2014. *http://link.springer.com/10.1007/978-1-4614-9426-3.* Accessed 7 July 2021.

McCall, Gerald Joseph Home, A. J. Bowden, and Richard John Howarth. *The History of Meteoritics and Key Meteorite Collections: Fireballs, Falls and Finds.* Trowbridge, U.K.: Geological Society of London, 2006.

McCray, W. Patrick. *Giant Telescopes: Astronomical Ambition and the Promise of Technology.* Cambridge, MA: Harvard University Press, 2004.

Meadows, Donella H., Dennis L. Meadows, Jorgen Randers, and William W. Behrens III. *The Limits to Growth.* New York: Universe Books, 1972. *https://www.clubofrome.org/report/the-limits-to-growth/.* Accessed 7 July 2021.

Michel, Patrick, Francesca E. DeMeo, and William F. Bottke, eds. *Asteroids IV.* Tucson, AZ: University of Arizona Press, 2015.

Michaud, Michael A. G. *Reaching for the High Frontier.* Westport, CT: Praeger, 1986.

Mohun, Arwen. *Risk: Negotiating Safety in American Society.* Baltimore, MD: Johns Hopkins University Press, 2013.

Napolitano, L. G., ed. *Applications of Space Developments: Selected Papers from the XXXI International Astronautical Congress, Tokyo, 21–28 September 1980.* London: Elsevier Science, October 2013.

National Commission on Space. *Pioneering the Space Frontier.* New York: Bantam Books, 1986.

Noll, Keith S., Harold A. Weaver, and Paul D. Feldman. *The Collision of Comet Shoemaker-Levy 9 and Jupiter: IAU Colloquium 156.* Cambridge, U.K.: Cambridge University Press, 2006.

Oreskes, Naomi, and John Krige, eds. *Science and Technology in the Global Cold War.* Cambridge, MA: MIT Press, 2014.

Osterbrock, Donald E., and P. Kenneth Seidelmann. "Paul Herget." In *Biographical Memoirs*, vol. 57, p. 59. Washington, DC: National Academy Press, 1987.

Powell, James Lawrence. *Night Comes to the Cretaceous: Dinosaur Extinction and the Transformation of Modern Geology.* New York: W. H. Freeman and Co., 1998.

Porter, Theodore. *Trust in Numbers: The Pursuit of Objectivity in Science and Public Life.* Princeton, NJ: Princeton University Press, 1995.

Randall, Lisa. *Dark Matter and the Dinosaurs: The Astounding Interconnectedness of the Universe.* 1st ed. New York: Ecco, 2015.

Raup, David M. *The Nemesis Affair: A Story of the Death of Dinosaurs and the Ways of Science*. 2nd ed. New York: W. W. Norton, 1999.

Sagan, Carl. *Pale Blue Dot: A Vision of the Human Future in Space*. New York: Ballantine Books, 1994.

Schmadel, Lutz D. *Dictionary of Minor Planet Names*. Berlin: Springer, 1999. *http://adsabs.harvard.edu/abs/1999dmpn.book.....S*. Accessed 7 July 2021.

Serio, Giorgia Foderà, Alessandro Manara, Piero Sicoli, and William Frederick Bottke. *Giuseppe Piazzi and the Discovery of Ceres*. Tucson, AZ: University of Arizona Press, 2002.

Serviss, Garrett P. *Edison's Conquest of Mars*. Gutenberg e-book #19141, 2006. *https://www.gutenberg.org/files/19141/19141-h/19141-h.htm*. Accessed 7 July 2021.

Shapley, Harlow. *Flights from Chaos: A Survey of Material Systems from Atoms to Galaxies, Adapted from Lectures at the College of the City of New York, Class of 1872 Foundation*. New York: Whittlesey House, McGraw-Hill Book Company, Inc., 1930.

Sharpton, Virgil L., and Peter D. Ward, eds. *Global Catastrophes in Earth History: An Interdisciplinary Conference on Impacts, Volcanism, and Mass Mortality*. Geological Society of America Special Papers 247. Boulder, CO: Geological Society of America, 1990.

Sheehan, William, and Thomas A. Dobbins. *Epic Moon: A History of Lunar Exploration in the Age of the Telescope*. Richmond, VA: Willmann-Bell, 2001.

Silver, Leon T., and Peter H. Schultz. *Geological Implications of Impacts of Large Asteroids and Comets on the Earth*. Geological Society of America Special Papers 190. Boulder, CO: Geological Society of America, 1982.

Spencer, John R., and Jacqueline Mitton, eds. *The Great Comet Crash: The Collision of Comet Shoemaker-Levy 9 and Jupiter*. New York: Cambridge University Press Archive, 1995.

Watson, Fletcher G. *Between the Planets*. 2nd rev. ed. Cambridge, MA: Harvard University Press, 1956.

Whitaker, Ewen A. *Mapping and Naming the Moon: A History of Lunar Cartography and Nomenclature*. Cambridge, U.K.: Cambridge University Press, 2003.

Yeomans, Donald K. *Comets. A Chronological History of Observation, Science, Myth, and Folklore*. New York: Wiley, 1991.

———. *Near-Earth Objects: Finding Them Before They Find Us*. 1st ed. Princeton, NJ: Princeton University Press, 2012.

Reports, Studies, and Legislation

AIAA. "Protecting Earth from Asteroids and Comets: An AIAA Position Paper." AIAA, October 2004.

Brophy, John R., F. Culick, and L. Friedman. *Asteroid Retrieval Feasibility Study*. Pasadena: Keck Institute for Space Studies, 2 April 2012. *http://kiss.caltech.edu/ final_reports/Asteroid_final_report.pdf*. Accessed 7 July 2021.

Camacho, Sergio. "Report of the Action Team on Near-Earth Objects: Recommendations for an International Response to a NEO Threat." Presented at the 50th Session Scientific and Technical Subcommittee, Vienna, Austria, 11 February 2013. *http://www.unoosa.org/pdf/pres/stsc2013/2013neo-01E.pdf*. Accessed 7 July 2021.

Canavan, Gregory H., Johndale C. Solem, and John D. G. Rather, eds. *Proceedings of the Near-Earth Object Interception Workshop*. Held at the Los Alamos National Laboratory, 14–16 January 1992. Los Alamos, NM: Los Alamos National Laboratory, February 1993. *http://ntrs.nasa.gov/archive/nasa/ casi.ntrs.nasa.gov/19930019383.pdf*. Accessed 7 July 2021.

"Collision of Asteroids and Comets with the Earth: Physical and Human Consequences." Report of a Workshop Held at Snowmass, Colorado, 13–16 July 1981. (Unpublished typescript).

Committee on Near-Earth Object Observations in the Infrared and Visible Wavelengths. *Finding Hazardous Asteroids Using Infrared and Visible Wavelength Telescopes*. Washington, DC: National Academies Press, July 2019 (prepublication draft).

Defending Planet Earth: Near-Earth Object Surveys and Hazard Mitigation Strategies. Washington, DC: National Academies Press, 2010. *http://www.nap. edu/catalog/12842*. Accessed 7 July 2021.

Directorate of Strategic Planning, U.S. Air Force Headquarters (AF/A8X). "Natural Impact Hazard (Asteroid Strike) Interagency Deliberate Planning Exercise After Action Report." December 2008. *https://cneos.jpl.nasa.gov/doc/ Natural_Impact_After_Action_Report.pdf*. Accessed 22 July 2019.

Gehrels, Tom, ed. *Physical Studies of Minor Planets*. Washington, DC: NASA SP-267, 1971.

Hills, Jack G., and P. T. Leonard. *Proceedings of the Planetary Defense Workshop*. Livermore, CA: Lawrence Livermore National Laboratory, 1995.

House Appropriations Committee. Commerce, Justice, Science, and Related Agencies Appropriations Bill, 2020, H. Rep. 116-101, 3 June 2019.

https://www.congress.gov/116/crpt/hrpt101/CRPT-116hrpt101.pdf. Accessed 12 November 2019.

House Appropriations Committee. Consolidated Appropriations Act, 2008: Division B—Commerce, Justice, Science and Related Agencies Appropriations Act. PL 110-161. *https://www.govinfo.gov/content/pkg/ CPRT-110HPRT39564/pdf/CPRT-110HPRT39564-DivisionB.pdf.* Accessed 20 August 2019.

House Committee on Science. George E. Brown, Jr. Near-Earth Object Survey Act, H. Rept. 109-158, 27 June 2005. *https://www.congress.gov/congressional-report/109th-congress/house-report/158/1.* Accessed 11 August 2017.

House Committee on Science. National Aeronautics and Space Administration Authorization Act, FY 1996, H. Rept. 104-233, 4 August 1995. *https://www. congress.gov/congressional-report/104th-congress/house-report/233/1.* Accessed 23 December 2015.

House Committee on Science and Technology. National Aeronautics and Space Administration Authorization Act of 2008, H. Rep. 110-702, 9 June 2008. *https://www.congress.gov/congressional-report/110th-congress/house-report/702/1?q =%7B%22search%22%3A%5B%22NASA+Appropriations+2008%22%5D%7 D&s=3&r=165.* Accessed 20 August 2019.

House Committee on Science and Technology. *Near-Earth Objects: Status of the Survey Program and Review of NASA's 2007 Report to Congress.* 110th Congress, 1st session, 8 November 2007.

Jones, Thomas D., and Russell L. Schweickart. "Report of the NASA Advisory Council Ad Hoc Task Force on Planetary Defense." Washington, DC: NASA Advisory Committee, 6 October 2010. *http://www.boulder.swri.edu/clark/ tf10pub.pdf.* Accessed 7 July 2021.

Kennedy, Fred L., and Lindley N. Johnson, cochairs. "National Near-Earth Object Preparedness Strategy." Washington, DC: National Science and Technology Council, December 2016.

McKay, Mary Fae, David S. McKay, and Michael Duke, eds. "Space Resources." Houston, TX: NASA SP-509, 1 January 1992. *https://ntrs.nasa.gov/search. jsp?R=19930007680.* Accessed 7 July 2021.

Miles, Aron R., and Lindley N. Johnson, cochairs. "National Near-Earth Object Preparedness Strategy and Action Plan." Washington, DC: National Science and Technology Council, June 2018. *https://www.nasa.gov/sites/default/files/ atoms/files/national_near-earth_object_preparedness_strategy_tagged.pdf.*

Morrison, David, chair. "The Spaceguard Survey: Report of the NASA International Near-Earth-Object Detection Workshop." Pasadena, CA: Jet Propulsion Laboratory, 1992. *http://ntrs.nasa.gov/search.jsp?R=19920025001*. Accessed 7 July 2021.

Morrison, David, and Ethiraj Venkatapathy. "Asteroid Generated Tsunami: Summary of NASA/NOAA Workshop." Moffett Field, CA: NASA TM-219463, January 2017. *https://ntrs.nasa.gov/archive/nasa/casi.ntrs.nasa.gov/20170005214.pdf*. Accessed 20 August 2019.

NASA Office of Inspector General. "NASA's Efforts To Identify Near-Earth Objects and Mitigate Hazards." Washington, DC: NASA, 15 September 2014. *https://oig.nasa.gov/audits/reports/FY14/*. Accessed 7 July 2021.

National Aeronautics and Space Administration Authorization Act of 2005. PL 109-155. 109th Congress (2005–06), 30 December 2005. *https://www.gpo.gov/fdsys/pkg/PLAW-109publ155/pdf/PLAW-109publ155.pdf*. Accessed 7 July 2021.

National Aeronautics and Space Administration Authorization Act of 2008. H.R. 6063. 110th Congress (2007–08). Became PL 110-422, 15 October 2008. *https://www.congress.gov/bill/110th-congress/house-bill/6063/text*. Accessed 7 July 2021.

National Research Council. *Exploration of Near Earth Objects*. Washington, DC: National Academies Press, 1998. *https://www.nap.edu/catalog/6106/exploration-of-near-earth-objects*.

National Research Council. *Near-Earth Object Surveys and Hazard Mitigation Strategies: Interim Report*. Washington, DC: National Academies Press, 2009.

Near-Earth Object Science Definition Team. *Study To Determine the Feasibility of Extending the Search for Near-Earth Objects to Smaller Limiting Diameters*. Washington, DC: NASA, 22 August 2003. *https://cneos.jpl.nasa.gov/doc/neoreport030825.pdf*. Accessed 7 July 2021.

Near-Earth Object Science Definition Team. "Update to Determine the Feasibility of Enhancing the Search and Characterization of NEOs." Washington, DC: NASA, September 2017. *https://cneos.jpl.nasa.gov/doc/SDT_report_2017.html*. Accessed 7 July 2021.

Office of Program Analysis and Evaluation. *2006 Near-Earth Object Survey and Deflection Study* (DRAFT). Washington, DC: NASA, 28 December 2006.

O'Neill, Gerard K., John Billingham, William Gilbreath, Brian O'Leary, and Beulah Gossett, eds. *Space Resources and Space Settlements*. Washington, DC: NASA SP-428, 1979.

Ree, Sidney G., Richard H. Van Atta, and Seymour J. Deitchman. *DARPA Technical Accomplishments: An Historical Review of Selected DARPA Projects.* Vol. 1. Alexandria, VA: Institute for Defense Analyses, February 1990. *https:// fas.org/spp/military/program/track/amos-history.pdf.* Accessed 7 July 2021.

Review of U.S. Human Spaceflight Plans Committee. *Seeking a Human Spaceflight Program Worthy of a Great Nation.* Washington, DC: U.S. Human Spaceflight Plans Committee, October 2009. *https://www.nasa.gov/ pdf/396093main_HSF_Cmte_FinalReport.pdf.* Accessed 7 July 2021.

"Threats from Space: A Review of U.S. Government Efforts to Track and Mitigate Asteroids and Meteors (Part I & Part II)": hearing before the Committee on Science, Space, and Technology, 113th Congress, 1st session, 19 March and 10 April 2013.

United Nations Office of Outer Space Affairs, "Report of the Third United Nations Conference on the Exploration and Peaceful Uses of Outer Space," held in Vienna, Austria, 19–30 July 1999. Report A/CONF.184/6. *http://www.unoosa. org/oosa/en/ourwork/psa/schedule/1999/unispace-iii.html.* Accessed 7 July 2021.

ABOUT THE AUTHORS

Dr. Erik M. Conway has been the Jet Propulsion Laboratory's (JPL's) historian since 2004. He is the author of seven books, including *Exploration and Engineering: The Jet Propulsion Laboratory and the Quest for Mars; High Speed Dreams: NASA and the Technopolitics of Supersonic Transportation;* and *Atmospheric Science at NASA: A History.* He served as an officer in the United States Navy prior to completing a doctorate in the history of science and technology.

Dr. Donald K. Yeomans was a JPL fellow and senior research scientist prior to his retirement in 2015. From 1998 through early 2015, he was the manager of NASA's Near-Earth Object Program Office at JPL. He provided the accurate predictions that led to the telescopic recovery of Comet Halley at Palomar Observatory on 16 October 1982 and allowed the recognition of 164 BCE Babylonian observations of Comet Halley on clay tablets in the British Museum. He was a Science Team member for NASA's Near Earth Asteroid Rendezvous (NEAR) mission, for NASA's Deep Impact mission, and for the Japanese Hayabusa mission. Asteroid 2956 was renamed asteroid 2956 Yeomans to honor his professional achievements, and in 2013, he was named as one of the 100 most influential persons by *Time* magazine.

Dr. Meg Rosenburg received her Ph.D. in planetary geophysics, with additional specialization in the history of science, from the California Institute of Technology in 2014, focusing on the interpretation of cratered terrains, particularly the lunar surface. She is especially interested in the role of analogy in understanding the physics of impact and other extreme processes. She is currently a science communicator and educator at the Museum of Science in Boston.

THE NASA HISTORY SERIES

Reference Works, NASA SP-4000

Grimwood, James M. *Project Mercury: A Chronology*. NASA SP-4001, 1963.

Grimwood, James M., and Barton C. Hacker, with Peter J. Vorzimmer. *Project Gemini Technology and Operations: A Chronology*. NASA SP-4002, 1969.

Link, Mae Mills. *Space Medicine in Project Mercury*. NASA SP-4003, 1965.

Astronautics and Aeronautics, 1963: Chronology of Science, Technology, and Policy. NASA SP-4004, 1964.

Astronautics and Aeronautics, 1964: Chronology of Science, Technology, and Policy. NASA SP-4005, 1965.

Astronautics and Aeronautics, 1965: Chronology of Science, Technology, and Policy. NASA SP-4006, 1966.

Astronautics and Aeronautics, 1966: Chronology of Science, Technology, and Policy. NASA SP-4007, 1967.

Astronautics and Aeronautics, 1967: Chronology of Science, Technology, and Policy. NASA SP-4008, 1968.

Ertel, Ivan D., and Mary Louise Morse. *The Apollo Spacecraft: A Chronology, Volume I, Through November 7, 1962*. NASA SP-4009, 1969.

Morse, Mary Louise, and Jean Kernahan Bays. *The Apollo Spacecraft: A Chronology, Volume II, November 8, 1962–September 30, 1964*. NASA SP-4009, 1973.

Brooks, Courtney G., and Ivan D. Ertel. *The Apollo Spacecraft: A Chronology, Volume III, October 1, 1964–January 20, 1966*. NASA SP-4009, 1973.

Ertel, Ivan D., and Roland W. Newkirk, with Courtney G. Brooks. *The Apollo Spacecraft: A Chronology, Volume IV, January 21, 1966–July 13, 1974*. NASA SP-4009, 1978.

Astronautics and Aeronautics, 1968: Chronology of Science, Technology, and Policy. NASA SP-4010, 1969.

Newkirk, Roland W., and Ivan D. Ertel, with Courtney G. Brooks. *Skylab: A Chronology*. NASA SP-4011, 1977.

Van Nimmen, Jane, and Leonard C. Bruno, with Robert L. Rosholt. *NASA Historical Data Book, Volume I: NASA Resources, 1958–1968*. NASA SP-4012, 1976; rep. ed. 1988.

Ezell, Linda Neuman. *NASA Historical Data Book, Volume II: Programs and Projects, 1958–1968*. NASA SP-4012, 1988.

Ezell, Linda Neuman. *NASA Historical Data Book, Volume III: Programs and Projects, 1969–1978*. NASA SP-4012, 1988.

Gawdiak, Ihor, with Helen Fedor. *NASA Historical Data Book, Volume IV: NASA Resources, 1969–1978*. NASA SP-4012, 1994.

Rumerman, Judy A. *NASA Historical Data Book, Volume V: NASA Launch Systems, Space Transportation, Human Spaceflight, and Space Science, 1979–1988*. NASA SP-4012, 1999.

Rumerman, Judy A. *NASA Historical Data Book, Volume VI: NASA Space Applications, Aeronautics and Space Research and Technology, Tracking and Data Acquisition/Support Operations, Commercial Programs, and Resources, 1979–1988*. NASA SP-4012, 1999.

Rumerman, Judy A. *NASA Historical Data Book, Volume VII: NASA Launch Systems, Space Transportation, Human Spaceflight, and Space Science, 1989–1998*. NASA SP-2009-4012, 2009.

Rumerman, Judy A. *NASA Historical Data Book, Volume VIII: NASA Earth Science and Space Applications, Aeronautics, Technology, and Exploration, Tracking and Data Acquisition/Space Operations, Facilities and Resources, 1989–1998*. NASA SP-2012-4012, 2012.

No SP-4013.

Astronautics and Aeronautics, 1969: Chronology of Science, Technology, and Policy. NASA SP-4014, 1970.

Astronautics and Aeronautics, 1970: Chronology of Science, Technology, and Policy. NASA SP-4015, 1972.

Astronautics and Aeronautics, 1971: Chronology of Science, Technology, and Policy. NASA SP-4016, 1972.

Astronautics and Aeronautics, 1972: Chronology of Science, Technology, and Policy. NASA SP-4017, 1974.

Astronautics and Aeronautics, 1973: Chronology of Science, Technology, and Policy. NASA SP-4018, 1975.

Astronautics and Aeronautics, 1974: Chronology of Science, Technology, and Policy. NASA SP-4019, 1977.

Astronautics and Aeronautics, 1975: Chronology of Science, Technology, and Policy. NASA SP-4020, 1979.

Astronautics and Aeronautics, 1976: Chronology of Science, Technology, and Policy. NASA SP-4021, 1984.

Astronautics and Aeronautics, 1977: Chronology of Science, Technology, and Policy. NASA SP-4022, 1986.

Astronautics and Aeronautics, 1978: Chronology of Science, Technology, and Policy. NASA SP-4023, 1986.

Astronautics and Aeronautics, 1979–1984: Chronology of Science, Technology, and Policy. NASA SP-4024, 1988.

Astronautics and Aeronautics, 1985: Chronology of Science, Technology, and Policy. NASA SP-4025, 1990.

Noordung, Hermann. *The Problem of Space Travel: The Rocket Motor.* Edited by Ernst Stuhlinger and J. D. Hunley, with Jennifer Garland. NASA SP-4026, 1995.

Gawdiak, Ihor Y., Ramon J. Miro, and Sam Stueland. *Astronautics and Aeronautics, 1986–1990: A Chronology.* NASA SP-4027, 1997.

Gawdiak, Ihor Y., and Charles Shetland. *Astronautics and Aeronautics, 1991–1995: A Chronology.* NASA SP-2000-4028, 2000.

Orloff, Richard W. *Apollo by the Numbers: A Statistical Reference.* NASA SP-2000-4029, 2000.

Lewis, Marieke, and Ryan Swanson. *Astronautics and Aeronautics: A Chronology, 1996–2000.* NASA SP-2009-4030, 2009.

Ivey, William Noel, and Marieke Lewis. *Astronautics and Aeronautics: A Chronology, 2001–2005.* NASA SP-2010-4031, 2010.

Buchalter, Alice R., and William Noel Ivey. *Astronautics and Aeronautics: A Chronology, 2006.* NASA SP-2011-4032, 2010.

Lewis, Marieke. *Astronautics and Aeronautics: A Chronology, 2007.* NASA SP-2011-4033, 2011.

Lewis, Marieke. *Astronautics and Aeronautics: A Chronology, 2008.* NASA SP-2012-4034, 2012.

Lewis, Marieke. *Astronautics and Aeronautics: A Chronology, 2009.* NASA SP-2012-4035, 2012.

Flattery, Meaghan. *Astronautics and Aeronautics: A Chronology, 2010.* NASA SP-2013-4037, 2014.

Siddiqi, Asif A. *Beyond Earth: A Chronicle of Deep Space Exploration, 1958–2016.* NASA SP-2018-4041, 2018.

Management Histories, NASA SP-4100

Rosholt, Robert L. *An Administrative History of NASA, 1958–1963.* NASA SP-4101, 1966.

Levine, Arnold S. *Managing NASA in the Apollo Era.* NASA SP-4102, 1982.

Roland, Alex. *Model Research: The National Advisory Committee for Aeronautics, 1915–1958.* NASA SP-4103, 1985.

Fries, Sylvia D. *NASA Engineers and the Age of Apollo.* NASA SP-4104, 1992.

Glennan, T. Keith. *The Birth of NASA: The Diary of T. Keith Glennan.* Edited by J. D. Hunley. NASA SP-4105, 1993.

Seamans, Robert C. *Aiming at Targets: The Autobiography of Robert C. Seamans.* NASA SP-4106, 1996.

Garber, Stephen J., ed. *Looking Backward, Looking Forward: Forty Years of Human Spaceflight Symposium.* NASA SP-2002-4107, 2002.

Mallick, Donald L., with Peter W. Merlin. *The Smell of Kerosene: A Test Pilot's Odyssey.* NASA SP-4108, 2003.

Iliff, Kenneth W., and Curtis L. Peebles. *From Runway to Orbit: Reflections of a NASA Engineer.* NASA SP-2004-4109, 2004.

Chertok, Boris. *Rockets and People, Volume I.* NASA SP-2005-4110, 2005.

Chertok, Boris. *Rockets and People: Creating a Rocket Industry, Volume II.* NASA SP-2006-4110, 2006.

Chertok, Boris. *Rockets and People: Hot Days of the Cold War, Volume III.* NASA SP-2009-4110, 2009.

Chertok, Boris. *Rockets and People: The Moon Race, Volume IV.* NASA SP-2011-4110, 2011.

Laufer, Alexander, Todd Post, and Edward Hoffman. *Shared Voyage: Learning and Unlearning from Remarkable Projects.* NASA SP-2005-4111, 2005.

Dawson, Virginia P., and Mark D. Bowles. *Realizing the Dream of Flight: Biographical Essays in Honor of the Centennial of Flight, 1903–2003.* NASA SP-2005-4112, 2005.

Mudgway, Douglas J. *William H. Pickering: America's Deep Space Pioneer.* NASA SP-2008-4113, 2008.

Wright, Rebecca, Sandra Johnson, and Steven J. Dick. *NASA at 50: Interviews with NASA's Senior Leadership.* NASA SP-2012-4114, 2012.

Project Histories, NASA SP-4200

Swenson, Loyd S., Jr., James M. Grimwood, and Charles C. Alexander. *This New Ocean: A History of Project Mercury*. NASA SP-4201, 1966; rep. ed. 1999.

Green, Constance McLaughlin, and Milton Lomask. *Vanguard: A History*. NASA SP-4202, 1970; rep. ed. Smithsonian Institution Press, 1971.

Hacker, Barton C., and James M. Grimwood. *On the Shoulders of Titans: A History of Project Gemini*. NASA SP-4203, 1977; rep. ed. 2002.

Benson, Charles D., and William Barnaby Faherty. *Moonport: A History of Apollo Launch Facilities and Operations*. NASA SP-4204, 1978.

Brooks, Courtney G., James M. Grimwood, and Loyd S. Swenson, Jr. *Chariots for Apollo: A History of Manned Lunar Spacecraft*. NASA SP-4205, 1979.

Bilstein, Roger E. *Stages to Saturn: A Technological History of the Apollo/Saturn Launch Vehicles*. NASA SP-4206, 1980 and 1996.

No SP-4207.

Compton, W. David, and Charles D. Benson. *Living and Working in Space: A History of Skylab*. NASA SP-4208, 1983.

Ezell, Edward Clinton, and Linda Neuman Ezell. *The Partnership: A History of the Apollo-Soyuz Test Project*. NASA SP-4209, 1978.

Hall, R. Cargill. *Lunar Impact: A History of Project Ranger*. NASA SP-4210, 1977.

Newell, Homer E. *Beyond the Atmosphere: Early Years of Space Science*. NASA SP-4211, 1980.

Ezell, Edward Clinton, and Linda Neuman Ezell. *On Mars: Exploration of the Red Planet, 1958–1978*. NASA SP-4212, 1984.

Pitts, John A. *The Human Factor: Biomedicine in the Manned Space Program to 1980*. NASA SP-4213, 1985.

Compton, W. David. *Where No Man Has Gone Before: A History of Apollo Lunar Exploration Missions*. NASA SP-4214, 1989.

Naugle, John E. *First Among Equals: The Selection of NASA Space Science Experiments*. NASA SP-4215, 1991.

Wallace, Lane E. *Airborne Trailblazer: Two Decades with NASA Langley's 737 Flying Laboratory*. NASA SP-4216, 1994.

Butrica, Andrew J., ed. *Beyond the Ionosphere: Fifty Years of Satellite Communications*. NASA SP-4217, 1997.

Butrica, Andrew J. *To See the Unseen: A History of Planetary Radar Astronomy*. NASA SP-4218, 1996.

Mack, Pamela E., ed. *From Engineering Science to Big Science: The NACA and NASA Collier Trophy Research Project Winners.* NASA SP-4219, 1998.

Reed, R. Dale. *Wingless Flight: The Lifting Body Story.* NASA SP-4220, 1998.

Heppenheimer, T. A. *The Space Shuttle Decision: NASA's Search for a Reusable Space Vehicle.* NASA SP-4221, 1999.

Hunley, J. D., ed. *Toward Mach 2: The Douglas D-558 Program.* NASA SP-4222, 1999.

Swanson, Glen E., ed. *"Before This Decade Is Out…" Personal Reflections on the Apollo Program.* NASA SP-4223, 1999.

Tomayko, James E. *Computers Take Flight: A History of NASA's Pioneering Digital Fly-By-Wire Project.* NASA SP-4224, 2000.

Morgan, Clay. *Shuttle-Mir: The United States and Russia Share History's Highest Stage.* NASA SP-2001-4225, 2001.

Leary, William M. *"We Freeze to Please": A History of NASA's Icing Research Tunnel and the Quest for Safety.* NASA SP-2002-4226, 2002.

Mudgway, Douglas J. *Uplink-Downlink: A History of the Deep Space Network, 1957–1997.* NASA SP-2001-4227, 2001.

No SP-4228 or SP-4229.

Dawson, Virginia P., and Mark D. Bowles. *Taming Liquid Hydrogen: The Centaur Upper Stage Rocket, 1958–2002.* NASA SP-2004-4230, 2004.

Meltzer, Michael. *Mission to Jupiter: A History of the Galileo Project.* NASA SP-2007-4231, 2007.

Heppenheimer, T. A. *Facing the Heat Barrier: A History of Hypersonics.* NASA SP-2007-4232, 2007.

Tsiao, Sunny. *"Read You Loud and Clear!" The Story of NASA's Spaceflight Tracking and Data Network.* NASA SP-2007-4233, 2007.

Meltzer, Michael. *When Biospheres Collide: A History of NASA's Planetary Protection Programs.* NASA SP-2011-4234, 2011.

Conway, Erik M., Donald K. Yeomans, and Meg Rosenburg. *A History of Near-Earth Objects Research.* NASA SP-2022-4235, 2022.

Gainor, Christopher. *Not Yet Imagined: A Study of Hubble Space Telescope Operations.* NASA SP-2020-4237, 2020.

Center Histories, NASA SP-4300

Rosenthal, Alfred. *Venture into Space: Early Years of Goddard Space Flight Center.* NASA SP-4301, 1985.

Hartman, Edwin P. *Adventures in Research: A History of Ames Research Center, 1940–1965.* NASA SP-4302, 1970.

Hallion, Richard P. *On the Frontier: Flight Research at Dryden, 1946–1981.* NASA SP-4303, 1984.

Muenger, Elizabeth A. *Searching the Horizon: A History of Ames Research Center, 1940–1976.* NASA SP-4304, 1985.

Hansen, James R. *Engineer in Charge: A History of the Langley Aeronautical Laboratory, 1917–1958.* NASA SP-4305, 1987.

Dawson, Virginia P. *Engines and Innovation: Lewis Laboratory and American Propulsion Technology.* NASA SP-4306, 1991.

Dethloff, Henry C. *"Suddenly Tomorrow Came… ": A History of the Johnson Space Center, 1957–1990.* NASA SP-4307, 1993.

Hansen, James R. *Spaceflight Revolution: NASA Langley Research Center from Sputnik to Apollo.* NASA SP-4308, 1995.

Wallace, Lane E. *Flights of Discovery: An Illustrated History of the Dryden Flight Research Center.* NASA SP-4309, 1996.

Herring, Mack R. *Way Station to Space: A History of the John C. Stennis Space Center.* NASA SP-4310, 1997.

Wallace, Harold D., Jr. *Wallops Station and the Creation of an American Space Program.* NASA SP-4311, 1997.

Wallace, Lane E. *Dreams, Hopes, Realities. NASA's Goddard Space Flight Center: The First Forty Years.* NASA SP-4312, 1999.

Dunar, Andrew J., and Stephen P. Waring. *Power to Explore: A History of Marshall Space Flight Center, 1960–1990.* NASA SP-4313, 1999.

Bugos, Glenn E. *Atmosphere of Freedom: Sixty Years at the NASA Ames Research Center.* NASA SP-2000-4314, 2000.

Bugos, Glenn E. *Atmosphere of Freedom: Seventy Years at the NASA Ames Research Center.* NASA SP-2010-4314, 2010. Revised version of NASA SP-2000-4314.

Bugos, Glenn E. *Atmosphere of Freedom: Seventy Five Years at the NASA Ames Research Center.* NASA SP-2014-4314, 2014. Revised version of NASA SP-2000-4314.

No SP-4315.

Schultz, James. *Crafting Flight: Aircraft Pioneers and the Contributions of the Men and Women of NASA Langley Research Center*. NASA SP-2003-4316, 2003.

Bowles, Mark D. *Science in Flux: NASA's Nuclear Program at Plum Brook Station, 1955–2005*. NASA SP-2006-4317, 2006.

Wallace, Lane E. *Flights of Discovery: An Illustrated History of the Dryden Flight Research Center*. NASA SP-2007-4318, 2007. Revised version of NASA SP-4309.

Wallace, Lane E,. and Christian Gelzer. *Flights of Discovery: 75 Years of Flight Research at NASA Armstrong Flight Research Center*. NASA SP-2021-4309. Revised version of NASA SP-2007-4318.

Arrighi, Robert S. *Revolutionary Atmosphere: The Story of the Altitude Wind Tunnel and the Space Power Chambers*. NASA SP-2010-4319, 2010.

Gelzer, Christian, ed. *NASA Armstrong Flight Research Center's Contributions to the Space Shuttle Program*. NASA SP-2020-4322, 2022.

General Histories, NASA SP-4400

Corliss, William R. *NASA Sounding Rockets, 1958–1968: A Historical Summary*. NASA SP-4401, 1971.

Wells, Helen T., Susan H. Whiteley, and Carrie Karegeannes. *Origins of NASA Names*. NASA SP-4402, 1976.

Anderson, Frank W., Jr. *Orders of Magnitude: A History of NACA and NASA, 1915–1980*. NASA SP-4403, 1981.

Sloop, John L. *Liquid Hydrogen as a Propulsion Fuel, 1945–1959*. NASA SP-4404, 1978.

Roland, Alex. *A Spacefaring People: Perspectives on Early Spaceflight*. NASA SP-4405, 1985.

Bilstein, Roger E. *Orders of Magnitude: A History of the NACA and NASA, 1915–1990*. NASA SP-4406, 1989.

Logsdon, John M., ed., with Linda J. Lear, Jannelle Warren Findley, Ray A. Williamson, and Dwayne A. Day. *Exploring the Unknown: Selected Documents in the History of the U.S. Civil Space Program, Volume I: Organizing for Exploration*. NASA SP-4407, 1995.

Logsdon, John M., ed., with Dwayne A. Day and Roger D. Launius. *Exploring the Unknown: Selected Documents in the History of the U.S. Civil Space Program, Volume II: External Relationships*. NASA SP-4407, 1996.

Logsdon, John M., ed., with Roger D. Launius, David H. Onkst, and Stephen J. Garber. *Exploring the Unknown: Selected Documents in the History of the U.S. Civil Space Program, Volume III: Using Space.* NASA SP-4407, 1998.

Logsdon, John M., ed., with Ray A. Williamson, Roger D. Launius, Russell J. Acker, Stephen J. Garber, and Jonathan L. Friedman. *Exploring the Unknown: Selected Documents in the History of the U.S. Civil Space Program, Volume IV: Accessing Space.* NASA SP-4407, 1999.

Logsdon, John M., ed., with Amy Paige Snyder, Roger D. Launius, Stephen J. Garber, and Regan Anne Newport. *Exploring the Unknown: Selected Documents in the History of the U.S. Civil Space Program, Volume V: Exploring the Cosmos.* NASA SP-2001-4407, 2001.

Logsdon, John M., ed., with Stephen J. Garber, Roger D. Launius, and Ray A. Williamson. *Exploring the Unknown: Selected Documents in the History of the U.S. Civil Space Program, Volume VI: Space and Earth Science.* NASA SP-2004-4407, 2004.

Logsdon, John M., ed., with Roger D. Launius. *Exploring the Unknown: Selected Documents in the History of the U.S. Civil Space Program, Volume VII: Human Spaceflight: Projects Mercury, Gemini, and Apollo.* NASA SP-2008-4407, 2008.

Siddiqi, Asif A., *Challenge to Apollo: The Soviet Union and the Space Race, 1945–1974.* NASA SP-2000-4408, 2000.

Hansen, James R., ed. *The Wind and Beyond: Journey into the History of Aerodynamics in America, Volume 1: The Ascent of the Airplane.* NASA SP-2003-4409, 2003.

Hansen, James R., ed. *The Wind and Beyond: Journey into the History of Aerodynamics in America, Volume 2: Reinventing the Airplane.* NASA SP-2007-4409, 2007.

Hogan, Thor. *Mars Wars: The Rise and Fall of the Space Exploration Initiative.* NASA SP-2007-4410, 2007.

Vakoch, Douglas A., ed. *Psychology of Space Exploration: Contemporary Research in Historical Perspective.* NASA SP-2011-4411, 2011.

Ferguson, Robert G. *NASA's First A: Aeronautics from 1958 to 2008.* NASA SP-2012-4412, 2013.

Vakoch, Douglas A., ed. *Archaeology, Anthropology, and Interstellar Communication.* NASA SP-2013-4413, 2014.

Asner, Glen R., and Stephen J. Garber. *Origins of 21st-Century Space Travel: A History of NASA's Decadal Planning Team and the Vision for Space Exploration, 1999–2004.* NASA SP-2019-4415, 2019.

Launius, Roger D. *NACA to NASA to Now: The Frontiers of Air and Space in the American Century*. NASA SP-2022-4419, 2022.

Monographs in Aerospace History, NASA SP-4500

Launius, Roger D., and Aaron K. Gillette, comps. *Toward a History of the Space Shuttle: An Annotated Bibliography*. Monographs in Aerospace History, No. 1, 1992.

Launius, Roger D., and J. D. Hunley, comps. *An Annotated Bibliography of the Apollo Program*. Monographs in Aerospace History, No. 2, 1994.

Launius, Roger D. *Apollo: A Retrospective Analysis*. Monographs in Aerospace History, No. 3, 1994.

Hansen, James R. *Enchanted Rendezvous: John C. Houbolt and the Genesis of the Lunar-Orbit Rendezvous Concept*. Monographs in Aerospace History, No. 4, 1995.

Gorn, Michael H. *Hugh L. Dryden's Career in Aviation and Space*. Monographs in Aerospace History, No. 5, 1996.

Powers, Sheryll Goecke. *Women in Flight Research at NASA Dryden Flight Research Center from 1946 to 1995*. Monographs in Aerospace History, No. 6, 1997.

Portree, David S. F., and Robert C. Trevino. *Walking to Olympus: An EVA Chronology*. Monographs in Aerospace History, No. 7, 1997.

Logsdon, John M., moderator. *Legislative Origins of the National Aeronautics and Space Act of 1958: Proceedings of an Oral History Workshop*. Monographs in Aerospace History, No. 8, 1998.

Rumerman, Judy A., comp. *U.S. Human Spaceflight: A Record of Achievement, 1961–1998*. Monographs in Aerospace History, No. 9, 1998.

Portree, David S. F. *NASA's Origins and the Dawn of the Space Age*. Monographs in Aerospace History, No. 10, 1998.

Logsdon, John M. *Together in Orbit: The Origins of International Cooperation in the Space Station*. Monographs in Aerospace History, No. 11, 1998.

Phillips, W. Hewitt. *Journey in Aeronautical Research: A Career at NASA Langley Research Center*. Monographs in Aerospace History, No. 12, 1998.

Braslow, Albert L. *A History of Suction-Type Laminar-Flow Control with Emphasis on Flight Research*. Monographs in Aerospace History, No. 13, 1999.

Logsdon, John M., moderator. *Managing the Moon Program: Lessons Learned from Apollo*. Monographs in Aerospace History, No. 14, 1999.

Perminov, V. G. *The Difficult Road to Mars: A Brief History of Mars Exploration in the Soviet Union*. Monographs in Aerospace History, No. 15, 1999.

Tucker, Tom. *Touchdown: The Development of Propulsion Controlled Aircraft at NASA Dryden*. Monographs in Aerospace History, No. 16, 1999.

Maisel, Martin, Demo J. Giulanetti, and Daniel C. Dugan. *The History of the XV-15 Tilt Rotor Research Aircraft: From Concept to Flight*. Monographs in Aerospace History, No. 17, 2000. NASA SP-2000-4517.

Jenkins, Dennis R. *Hypersonics Before the Shuttle: A Concise History of the X-15 Research Airplane*. Monographs in Aerospace History, No. 18, 2000. NASA SP-2000-4518.

Chambers, Joseph R. *Partners in Freedom: Contributions of the Langley Research Center to U.S. Military Aircraft of the 1990s*. Monographs in Aerospace History, No. 19, 2000. NASA SP-2000-4519.

Waltman, Gene L. *Black Magic and Gremlins: Analog Flight Simulations at NASA's Flight Research Center*. Monographs in Aerospace History, No. 20, 2000. NASA SP-2000-4520.

Portree, David S. F. *Humans to Mars: Fifty Years of Mission Planning, 1950–2000*. Monographs in Aerospace History, No. 21, 2001. NASA SP-2001-4521.

Thompson, Milton O., with J. D. Hunley. *Flight Research: Problems Encountered and What They Should Teach Us*. Monographs in Aerospace History, No. 22, 2001. NASA SP-2001-4522.

Tucker, Tom. *The Eclipse Project*. Monographs in Aerospace History, No. 23, 2001. NASA SP-2001-4523.

Siddiqi, Asif A. *Deep Space Chronicle: A Chronology of Deep Space and Planetary Probes, 1958–2000*. Monographs in Aerospace History, No. 24, 2002. NASA SP-2002-4524.

Merlin, Peter W. *Mach 3+: NASA/USAF YF-12 Flight Research, 1969–1979*. Monographs in Aerospace History, No. 25, 2001. NASA SP-2001-4525.

Anderson, Seth B. *Memoirs of an Aeronautical Engineer: Flight Tests at Ames Research Center: 1940–1970*. Monographs in Aerospace History, No. 26, 2002. NASA SP-2002-4526.

Renstrom, Arthur G. *Wilbur and Orville Wright: A Bibliography Commemorating the One-Hundredth Anniversary of the First Powered Flight on December 17, 1903*. Monographs in Aerospace History, No. 27, 2002. NASA SP-2002-4527.

No monograph 28.

Chambers, Joseph R. *Concept to Reality: Contributions of the NASA Langley Research Center to U.S. Civil Aircraft of the 1990s.* Monographs in Aerospace History, No. 29, 2003. NASA SP-2003-4529.

Peebles, Curtis, ed. *The Spoken Word: Recollections of Dryden History, The Early Years.* Monographs in Aerospace History, No. 30, 2003. NASA SP-2003-4530.

Jenkins, Dennis R., Tony Landis, and Jay Miller. *American X-Vehicles: An Inventory—X-1 to X-50.* Monographs in Aerospace History, No. 31, 2003. NASA SP-2003-4531.

Renstrom, Arthur G. *Wilbur and Orville Wright: A Chronology Commemorating the One-Hundredth Anniversary of the First Powered Flight on December 17, 1903.* Monographs in Aerospace History, No. 32, 2003. NASA SP-2003-4532.

Bowles, Mark D., and Robert S. Arrighi. *NASA's Nuclear Frontier: The Plum Brook Research Reactor.* Monographs in Aerospace History, No. 33, 2004. NASA SP-2004-4533.

Wallace, Lane, and Christian Gelzer. *Nose Up: High Angle-of-Attack and Thrust Vectoring Research at NASA Dryden, 1979–2001.* Monographs in Aerospace History, No. 34, 2009. NASA SP-2009-4534.

Matranga, Gene J., C. Wayne Ottinger, Calvin R. Jarvis, and D. Christian Gelzer. *Unconventional, Contrary, and Ugly: The Lunar Landing Research Vehicle.* Monographs in Aerospace History, No. 35, 2006. NASA SP-2004-4535.

McCurdy, Howard E. *Low-Cost Innovation in Spaceflight: The History of the Near Earth Asteroid Rendezvous (NEAR) Mission.* Monographs in Aerospace History, No. 36, 2005. NASA SP-2005-4536.

Seamans, Robert C., Jr. *Project Apollo: The Tough Decisions.* Monographs in Aerospace History, No. 37, 2005. NASA SP-2005-4537.

Lambright, W. Henry. *NASA and the Environment: The Case of Ozone Depletion.* Monographs in Aerospace History, No. 38, 2005. NASA SP-2005-4538.

Chambers, Joseph R. *Innovation in Flight: Research of the NASA Langley Research Center on Revolutionary Advanced Concepts for Aeronautics.* Monographs in Aerospace History, No. 39, 2005. NASA SP-2005-4539.

Phillips, W. Hewitt. *Journey into Space Research: Continuation of a Career at NASA Langley Research Center.* Monographs in Aerospace History, No. 40, 2005. NASA SP-2005-4540.

Rumerman, Judy A., Chris Gamble, and Gabriel Okolski, comps. *U.S. Human Spaceflight: A Record of Achievement, 1961–2006.* Monographs in Aerospace History, No. 41, 2007. NASA SP-2007-4541.

Peebles, Curtis. *The Spoken Word: Recollections of Dryden History Beyond the Sky.* Monographs in Aerospace History, No. 42, 2011. NASA SP-2011-4542.

Dick, Steven J., Stephen J. Garber, and Jane H. Odom. *Research in NASA History.* Monographs in Aerospace History, No. 43, 2009. NASA SP-2009-4543.

Merlin, Peter W. *Ikhana: Unmanned Aircraft System Western States Fire Missions.* Monographs in Aerospace History, No. 44, 2009. NASA SP-2009-4544.

Fisher, Steven C., and Shamim A. Rahman. *Remembering the Giants: Apollo Rocket Propulsion Development.* Monographs in Aerospace History, No. 45, 2009. NASA SP-2009-4545.

Gelzer, Christian. *Fairing Well: From Shoebox to Bat Truck and Beyond, Aerodynamic Truck Research at NASA's Dryden Flight Research Center.* Monographs in Aerospace History, No. 46, 2011. NASA SP-2011-4546.

Arrighi, Robert. *Pursuit of Power: NASA's Propulsion Systems Laboratory No. 1 and 2.* Monographs in Aerospace History, No. 48, 2012. NASA SP-2012-4548.

Renee M. Rottner. *Making the Invisible Visible: A History of the Spitzer Infrared Telescope Facility (1971–2003).* Monographs in Aerospace History, No. 47, 2017. NASA SP-2017-4547.

Goodrich, Malinda K., Alice R. Buchalter, and Patrick M. Miller, comps. *Toward a History of the Space Shuttle: An Annotated Bibliography, Part 2 (1992–2011).* Monographs in Aerospace History, No. 49, 2012. NASA SP-2012-4549.

Ta, Julie B., and Robert C. Treviño. *Walking to Olympus: An EVA Chronology, 1997–2011*, Vol. 2. Monographs in Aerospace History, No. 50, 2016. NASA SP-2016-4550.

Gelzer, Christian. *The Spoken Word III: Recollections of Dryden History; The Shuttle Years.* Monographs in Aerospace History, No. 52, 2013. NASA SP-2013-4552.

Ross, James C. *NASA Photo One.* Monographs in Aerospace History, No. 53, 2013. NASA SP-2013-4553.

Launius, Roger D. *Historical Analogs for the Stimulation of Space Commerce.* Monographs in Aerospace History, No. 54, 2014. NASA SP-2014-4554.

Buchalter, Alice R., and Patrick M. Miller, comps. *The National Advisory Committee for Aeronautics: An Annotated Bibliography.* Monographs in Aerospace History, No. 55, 2014. NASA SP-2014-4555.

Chambers, Joseph R., and Mark A. Chambers. *Emblems of Exploration: Logos of the NACA and NASA.* Monographs in Aerospace History, No. 56, 2015. NASA SP-2015-4556.

Alexander, Joseph K. *Science Advice to NASA: Conflict, Consensus, Partnership, Leadership.* Monographs in Aerospace History, No. 57, 2017. NASA SP-2017-4557.

Electronic Media, NASA SP-4600

Remembering Apollo 11: The 30th Anniversary Data Archive CD-ROM. NASA SP-4601, 1999.

Remembering Apollo 11: The 35th Anniversary Data Archive CD-ROM. NASA SP-2004-4601, 2004. This is an update of the 1999 edition.

The Mission Transcript Collection: U.S. Human Spaceflight Missions from Mercury Redstone 3 to Apollo 17. NASA SP-2000-4602, 2001.

Shuttle-Mir: The United States and Russia Share History's Highest Stage. NASA SP-2001-4603, 2002.

U.S. Centennial of Flight Commission Presents Born of Dreams—Inspired by Freedom. NASA SP-2004-4604, 2004.

Of Ashes and Atoms: A Documentary on the NASA Plum Brook Reactor Facility. NASA SP-2005-4605, 2005.

Taming Liquid Hydrogen: The Centaur Upper Stage Rocket Interactive CD-ROM. NASA SP-2004-4606, 2004.

Fueling Space Exploration: The History of NASA's Rocket Engine Test Facility DVD. NASA SP-2005-4607, 2005.

Altitude Wind Tunnel at NASA Glenn Research Center: An Interactive History CD-ROM. NASA SP-2008-4608, 2008.

A Tunnel Through Time: The History of NASA's Altitude Wind Tunnel. NASA SP-2010-4609, 2010.

Conference Proceedings, NASA SP-4700

Dick, Steven J., and Keith Cowing, eds. *Risk and Exploration: Earth, Sea and the Stars.* NASA SP-2005-4701, 2005.

Dick, Steven J., and Roger D. Launius. *Critical Issues in the History of Spaceflight.* NASA SP-2006-4702, 2006.

Dick, Steven J., ed. *Remembering the Space Age: Proceedings of the 50th Anniversary Conference.* NASA SP-2008-4703, 2008.

Dick, Steven J., ed. *NASA's First 50 Years: Historical Perspectives.* NASA SP-2010-4704, 2010.

Billings, Linda, ed. *50 Years of Solar System Exploration: Historical Perspectives.* NASA SP-2021-4705, 2021.

Societal Impact, NASA SP-4800

Dick, Steven J., and Roger D. Launius. *Societal Impact of Spaceflight.* NASA SP-2007-4801, 2007.

Dick, Steven J., and Mark L. Lupisella. *Cosmos and Culture: Cultural Evolution in a Cosmic Context.* NASA SP-2009-4802, 2009.

Dick, Steven J. *Historical Studies in the Societal Impact of Spaceflight.* NASA SP-2015-4803, 2015.

INDEX

Page numbers in **bold** text indicate figures and tables

Numerals

1I/2017 U1 interstellar object ('Oumuamua), 307–8, 313

3π Steradian Survey, 237–38

1770 I Lexell, Comet, 112n16

1989 CJ1 asteroid, 151

1989 FC (4581 Asclepius) asteroid, 9, 111–12nn12–14, 111–13, 273

1989 UP asteroid, 63

1990 SS asteroid, 63–64

1995 HM asteroid, 191

1996 GT asteroid, 299

1996 MQ asteroid, 153, 153n29

1997 XF11 asteroid (35396)
    IAU circular on, 133–34, 134n78, 137
    impact and orbit calculations for, 105, 132–38, **133**, 136nn89–90, 166, 167
    press information sheet release about and media attention given to, 133–35, 134n78, 135n84, 135n86, 137, 137–38n93, 138

1998 DK36 asteroid, 323

1998 KY26 asteroid, 191

1998 OX4 asteroid, 166–67

1999 AN10 asteroid, 167–68, 168n64

1999 JV6 asteroid (85990), 161

1999 KW4 asteroid (66391), 189

2000 DP107 asteroid satellite, 194–95

2000 PH5 asteroid (54509 YORP), 162, 162n51

2000 SG344 asteroid, 168–69

2004 JG asteroid, 154

2004 MN4 asteroid (99942 Apophis), 213, 224–37, 238, 278, 280

2004 VD17 asteroid (144898), 234

2008 TC3 asteroid, 275–77, **275**, 276n7, 279, 279n20, 282

2009 BD asteroid, 260–61

2010 AL30 asteroid, 190

2011 AG5 asteroid, 277–79, 279n19, 296

2012 DA14 asteroid, 289n50, **290**

2012 TC4 asteroid, 297, 297n70

2014 AA asteroid, 279, 279n20

A

achondrite/basaltic achondrite meteorites, 66–67, 67n37, 163, 163n52, 209

Ackerman, Thomas P., 85, 87

Advanced Research Projects Agency (ARPA), 71

Agreement Governing the Activities of States on the Moon and Other Celestial Bodies (Moon Treaty), 250, 263–64, 264n71

A'Hearn, Michael, 199–200, 201

Aherns, Thomas "Tom," 86–87, 116–18

AIDA (Asteroid Impact and Deflection Assessment), 298–300

Ailor, William, 300n79

Air Force, U.S.
    CCD development role of, 130
    Future Concepts and Transformation Division, 285
    GEODSS systems development for and use by, 130–31, 148–52, 153, 224

mirror for Gehrels's telescope project from, 146–47

Near-Earth Objects Survey Working Group coordination with, 129

NEO discovery role of, 114–15

Phillips Laboratory, 130

planetary defense role of, 129–32, 132n77, 223–24, 273, 285–86, 291–92

Science Definition Team role of, 215

SpaceCast 2020 study of, 131–32, 131n75, 223–24

Space Command GEODSS research, 150–52

Space Command mission, 130–31

space surveillance operations of, 130–31

space surveillance telescope in Hawai'i, 148–52

space surveillance telescope in New Mexico, 148, 150

Spacewatch funding from, 148

telescopes from for survey network, 141

Air Force Maui Optical Station (AMOS) telescope, 150, 152

Air Force Research Laboratory (AFRL), 149–50, 152

AKARI telescope, 183, 185

albedo/reflectivity
asteroid classification system for spectra and, 67–68, 67nn38–39
asteroid size estimates from, 143–44, 143n3, 175–87, 177n7
color research and, 65–66, 65n31, 66n33
comet nucleus feature, 76
geometric albedo, 177, 177n7
viewing geometry and, 175–76, 176n1

Albritton, Claude C., 44, 44n37, 45

Alexander, George, 99–100

Alfvén, Hannes, 191

Alinda asteroid (887), 178

Allen, David A., 178

Almagest (Ptolemy), 2–3

Althaea asteroid (119), 22

aluminum-26 (^{26}Al), 208–9, 209n106

Alvarez, Luis, 9, 79–80, 82–84, 250

Alvarez, Walter, 9, 79–80, 81–84, 91, 92, 95–96, 98

Alvarez hypothesis, 9, 79–80, 103, 251

American Institute of Aeronautics and Astronautics (AIAA)
asteroid impact research by Space Systems Technical Committee, 112–13
white paper review and giggle factor about rocks falling from the sky, 113, 123, 276

Ames Research Center
evolution of life workshops at, 99
nuclear warfare, research of atmospheric effects at, 87–88
Search for Extraterrestrial Intelligence (SETI) office, 62
search program workshops, role of, 114
Vertical Gun Range at, 49
volcanic eruptions and atmospheric chemistry research at, 84–86
Yucatán Peninsula crater research at, 95

Amor-class asteroids
Amor (1221), 322
naming of, 322
orbital characteristics of, 321, **321**, 322
See also Eros asteroid (433)

Analysis of Alternative (AOA) study/ Office of Program Analysis and Evaluation (OPAE) study, 231–35, 274, 280–81

Anderson, Peter, 266–67

André, C., 193

Anglo-Australian Near-Earth Object Survey, 155

anhydrite (anhydrous calcium sulfate) layers, 97

Annefrank asteroid (5535), **184**

Antarctica meteorites, 103

Apohele asteroids, 323. *See also* Atira-class asteroids

Apollo 11 mission, 66

Apollo asteroid (5496) 1973 NA, 53

Apollo-class asteroids
1990 SS, 63–64
Apollo (1862), 161, 196, 322

discovery of, 142

Geographos (1620), 322

Icarus (1566), 322

naming of, 227, 322

orbital characteristics of, 142, 321, **321**, 322

Apophis (2004 MN4) asteroid (99942), 213, 224–37, 238, 278, 280

Applied Physics Laboratory, Johns Hopkins University, 45, 206, 298–300

Arago, François, 19

Arecibo radio observatory, 71–74, 189–90, 194, 229–30, 236, 281, 281n23, 297

Aristotle, 3

Arkyd Astronautics (Planetary Resources), 266–67, 268, 269–70

Armageddon, 135n84

Armstrong, Neil, 250

Artemis mission, 270, 270n94

Asaro, Frank, 83

Asclepius (1989 FC) asteroid (4581), 9, 111–12nn12–14, 111–13, 273

Ashanti crater, Ghana, 44

Association of Space Explorers (ASE), 228, 295

"Asteroid Arrester" cartoon (Radebaugh), 245, 247

Asteroid Impact Mission (AIM), 298–300

asteroid mining
    legal and ethical concerns related to, 247, 269
    private interest in, 262–71
    science fiction story about, 245
    value of asteroid resources, 268–69

Asteroid Redirect Mission (ARM), 257–62, **258**, 270, 270–71, 293

asteroids
    attitude toward and priority of hunting for, 138–39
    classification system for spectra and albedo, 67–68, 67nn38–39
    collisional history and family members of, 182–83, 192–93
    collision frequency of Earth-crossing, estimate of, 43
    computation of orbit and ephemeris through numerical modeling, 158–62, 160n46
    cratering on, 206
    dark time of new Moon to hunt for, 60
    debris tails of, 2
    designations, numbers, and names for, 157, 227–28, 317–20
    directive to discover and track, 9, 113, 113n21, 128–29, 129n70, 138–39, 213–14, 230, 236–37, 273, 303–4
    discovery of, 8–9, 10, 13–19, 317–20
    distinction between comets and, 210–11, 213
    effects of impacts on Earth's surface and life on Earth, 8–9, 30, 31, 77, 79–81, 99–104
    energy deposited in atmosphere by entry of, 1–2, 41–42, 42n30, 86–87
    as fragments of a planet, 18–19
    human exploration of, 245
    interdisciplinary research on, 43
    magnitude/brightness of, 18n13
    main-belt asteroids, 8, 163–64, 175, 179
    minor planet term for, 13, 318
    missions to and flybys of, 183–87, **184**, 198, 204–11, 247–62, 313
    nongravitational motion of and forces on, 160–62, **161**, **162**
    number known and number found per year, 43, 53, 57, 105, 142, 156, 173–74, **239**, 314, 318–19, 320
    orbits and ephemerides of, 8, 16–17, 16n7, 319
    orbits and ephemerides of and PCAS, 51–57, 54n3, 54n5
    origin and use of term, 17–18, 318
    origination and belt structure of, 19
    prevalence of events related to, 41
    recovery and precovery of, 20–22, 61–62, 319
    retrieval of, 247–48, 256–62, 257n43
    rotation and rotation rates of, 64–65, 68–69, 190–92, 206

shape modeling with radar, 70–71,
   177, 187–90, 188–89n40,
   194–95, 194–95nn58–59
size and size estimates of, 143–44,
   175–87, 177n7
structure of, 190–95, 191n49, 204–5,
   210–11, 313–14
surface texture of, 70–71, 179–80,
   179n13
Tunguska event attribution to, 41–42
visits and missions to, 77
See also near-Earth asteroids (NEAs);
   specific asteroids
Asteroid Terrestrial-impact Last Alert
   System (ATLAS), 237, 239–41, **239**,
   302
Asteroid Threat Assessment Project
   (ATAP), 301
Astraea asteroid, 19, 319
astroblemes, 9, 103. *See also* craters/
   impact craters
astrology
   characteristics of comets related to
      coming disaster, 3
   focus of, 3
   simultaneous practice of with astron-
      omy, 3
Astronomical Journal, 18, 318
astronomy
   focus of, 3, 13
   Ptolemy theory of epicycles, 2–3
   simultaneous practice of with astrol-
      ogy, 3
   time domain astronomy, 215
Aten-class asteroids
   1999 KW4 (66391), 189
   Aten (2062), 53, 161, 323
   discovery of, 53, 142
   naming of, 227, 323
   orbital characteristics of, 142, 321,
      321, 323
Atira-class asteroids
   1998 DK36, 323
   2004 JG, 154
   Atira (163693), 323
   naming of, 323

orbital characteristics of, 321, **321**,
   323
atmospheric science research, 303n88
Atomic Energy Commission, Megaton
   Ice-Contained Explosion (MICE)
   project, 46–49
Augustine Commission and Norman
   Augustine, 241, 253–54, 255
Australia
   Anglo-Australian Near-Earth Object
      Survey, 155
   Siding Spring Observatory, 63, 155,
      226, 275
   Uppsala Schmidt telescope in, 155

B

B612 Foundation, 214, 228, 234,
   242–43, 267
Baade, Walter, 43
Baalke, Ron, 128n68
Baker, Marcus, 24
Baldwin, Ralph B., 45–46
Ball Aerospace, 242
Ballistic Missile Defense Organization
   (BMDO), 119n39, 148, 183
Bamberg, Carl, 21–22n20
Bamberga asteroid (324), 178
Bambery, Raymond "Ray," 148–49, 151
Barbee, Brent, 300n79
Bardwell, Conrad, 157
Barnard, E. E., 178
Barringer, Daniel Moreau, 31–35, 33n4,
   38–39, 47
Barringer Meteorite Crater. *See* Meteor
   Crater (Barringer Meteorite Crater)
basalt, 98
basaltic achondrite meteorites, 66–67,
   67n37, 163, 163n52
Bauschinger, Julius, 22
Beals, Carlyle, 45–46
beaming parameter, 179–80, 179n13
Belize, 95–96
Bell Labs, 58
Beloc formation, Haiti, 93, 96–97
Belton, Michael, 200
Benner, Lance, 127, 189

Bennu asteroid (101955), 161, **184**, 185, 242, 313
Benson, Jim, 265
Berger, Eric, 260
Bernardi, Fabricio, 225
Berolina asteroid (422), 21
Beshore, Edward, 155–56
Bessel, Friedrich, 159
Betulia asteroid (1580), 71, 180
Bezos, Jeff, 265–66
Bickerton, A. W., 37n18
Billings, Linda, 296
binary objects and system, 68–69, 68–69n43, 193–94, 195, 195n60, 285–86, 299–300
Binzel, Richard "Dick," 67n39, 164, 170, 170n70, 181–82, 258–59
Binzel index, 170, 170n70
Biot, Jean-Baptiste, 7
Birks, John, 88
Blue Origin, 265–66
Bobrovnikoff, N. T., 65–66
Bode, Johann Elert, 15–16
Bode's law (Titius-Bode law), 8, 14, 16, 18
Bolden, Charles, 259, 278, 286, 291, 292
Boon, John D., 44, 44n37, 45
Borrelly, Comet, **184**, 197
Bosler, Jean, 36–37, 37n16, 38
Boslough, Mark, 285n38, 291, 300n79
Bottke, William, 164–65, 164n58, 216
Bourgeouis, Joanne, 92
Bowell, Edward "Ted," 67–68, 143n3, 144–45, 148, 154
Boynton, William, 92, 93
Bradford Space, 270
Braille asteroid (9969), **184**
Branca, W., 44n38
Brazon River deposit, Texas, 92
Briggs, Geoffrey, 59
Brilliant Pebbles concept, 115–16
British National Space Centre, 156
Broomfield Hazard Scale, 296–97
Brophy, John, 256
Brown, George E., 113, 115, 230, 273
Brown, Peter G., 275, 276, 279n20

Brown Act (George E. Brown Near-Earth Object Survey Act), 230–31, 240, 242, 273–74, 273n1, 280, 281, 286–87, 302
Brownlee, Donald, 198
Bruce, Catherine Wolfe, 20n18
Brucia asteroid (323), 20, 20n18
Brunk, William E., 59, 108
Bucher, Walter H., 44nn37–38
Buratti, Bonnie, 197
Burney, Charles, Jr., 17–18, 318
Burney, Charles, Sr., 17–18, 318
Burnt Frost Joint Task Force, 284, 284nn34–35
Bus, Schelte "Bobby," 66–67, 180–82
Bush administration and George H. W. Bush, 251
Byers, Carlos, 93–94

C

Cacciatore, Niccolò, 15n4
Calar Alto Observatory, 126, 288
calcium-aluminum inclusion (CAI) fragments, 198
California Institute of Technology (Caltech)
　CCD development role of, 58
　Infrared Processing and Analysis Center (IPAC), 183
　Planet 9 search by, 104
　Seismology Laboratory research on impact events, 86–87
Camacho, Sergio, 295
Camargo Zanoguera, Antonio, 93–94
Campbell, Don, 71, 74
Campins, Humberto, 182
Campo del Cielo, Argentina, craters, 6, 44
Canada
　cratering research in, 45–46
　NEOSSat launch by, 289
　Tagish Lake, Yukon Territory, impact event, 274–75
Canavan, Gregory, 115
Carnegie Committee for the Study of the Surface Features of the Moon, 42–43

Catalina Sky Survey, 155–56, 158, 169, 226, 238, 239, **239**, 275, 277, 279n20

Celestial Police (Himmelspolizei, Lilienthal Society), 15–16

cenotes, 95

Center for Near-Earth Object Studies (Near-Earth Object Program Office), 139, 141, 256n38, 280n21, 314

Ceres
composition, color, and structure of, 68, 210
discovery of, 8, 13–19, 15n4
dwarf planet status of, 8, 18
mission to, **184**, 209, 210
naming of, 16
orbit and ephemeris of, 16–17, 16n7
polarization curve for, 177, 177n5
reflected sunlight from, 66n33
size estimates of, 177, 180
symbol designation for, 319

Cerro Tololo telescope, 66

Chamberlin, Alan, 280n20

Chambers, Kenneth, 237–38

Chang'e-2 mission, **184**

Chao, Edward, 47, 49

Chapman, Clark
asteroid 1997 XF11 affair, case study by, 135n86
asteroid classification system development by, 67
B612 Foundation role of, 214, 267
Cosmic Catastrophes, 109
credibility of impact warnings, concern about, 135, 137, 137–38n93
Gaspra structure, opinion about, 205
Ida structure, opinion about, 205
Interception Workshop nuclear-focused objections of, 119, 119n41
international conference on Earth-approaching bodies role in, 114
nuclear device for deflection, opinion about, 234–35, 280–81
on opinions about impact hazards after Shoemaker-Levy–Jupiter collision, 128

press coverage of close call with asteroid, 112

press information sheet release about 1997 XF11, concerns about, 137

Snowmass workshop on hazard aspect of NEOs, role in, 108–9

S-type asteroid research by, 196n64

Swift-Tuttle comet impact probability correspondence with, 122

charge-coupled detectors (CCDs)
advantages over film, 57, 58
Air Force interest in, 130
computer-assisted analysis of images from, 11
development of, 58, 130
first fully automated discovery with, 63–64
improvement of, 10–11
LINEAR program use of, 130, 130n72
Palomar Observatory CCD–based camera, 10
replacement of measuring engine with, 54n5
as revolution for NEO discovery surveys, 57
See also Spacewatch

Charlois, Auguste, 22

Chelyabinsk, Russia, fireball and airburst, 1–2, 9, 274, 289–91, **290**

Chesley, Steven
Apophis (2004 MN4) observations, concerns about, 226
Apophis (2004 MN4) radio transponder tracking mission, discussions about, 228–29
asteroid 1999 AN10 impact risk calculation by, 167
on asteroid 2000 SG344 impact risk process, 168–69
CLOMON and NEODyS development by, 166, 167
computation of impact probabilities by, 166–67
JPL move by, 167
Palermo Scale development role of, 170, 173

Sentry system development by, 167

Sudan as impact area for 2008 TC3, projection by, 275

tsunamis-risk analysis by, 217

Yarkovsky drift verification by, 161

Chicxulub crater, Yucatán Peninsula, 80, 93–95, **94**, 97–98, 97n50, 101, 104, 128

China

astrological advice from astronomers in, 3, 3n6

Chang'e-2 mission, **184**

Chladni, E. E. F., 7

Chodas, Paul

asteroid 1997 XF11 impact and orbit calculations by, **133**, 134, 136–37, 136n90

asteroid 1999 AN10 impact risk verification by, 167

natural bodies and impact probability techniques use by, 166n60

Near-Earth Object Program Office role of, 139

Planetary Defense Conference scenario planning by, 300, 300n79

Shoemaker-Levy–Jupiter impact and orbital computations by, 125, 125–26nn61–62, 127

chondrite/ordinary chondrite meteorites, 163n52, 195–96, 205, 207, 208, 290

Churyumov-Gerasimenko (C-G), Comet, **184**, 198n68, 201, **202**, 204

Cincinnati Observatory, 54, 157

civil defense and civil defense exercises, 10, 274, 282, 300

clay layer composition and properties, 79–80, 82–83, 86, 88–89, 91, 92, 103

climate

dust-induced climatic cooling, 79–80, 83–84

effects of volcanic eruptions on, 84–86, 98

"impact winter" research, 88–90

Krakatoa explosion and global cooling, 85

mass extinction events related to, 98

nuclear winter and atmospheric effects of nuclear warfare, 87–88

CLOMON (Close Approach Monitoring System), 166

Closer Than We Think (Radebaugh), 245

Club of Rome, 246, 248–49

coesite, 47, 49

colonization of Moon and space, 248–52, 263–64

color of asteroids

composition and, 66–69, 67nn38–39, 195–96

rotation period and color variations, 65, 65n32

space weathering (solar wind particles) and, 195–96

spectral observations and, 64–69, 66n34, 68n42, 180–83, 182n22

COLT, Project, 146–47, 146n12

COmet Nucleus TOUR (CONTOUR) mission, 205–6n91

Comet Rendezvous Asteroid Flyby (CRAF) mission, 149

comets

albedo feature of nucleus, 76

anti-solar direction of tails of, 3n6

characteristics of related to coming disaster, 3

composition of, 198, 210–11

computation of orbit and ephemeris through numerical modeling, 158–62, 160n46

debris tails of, 2

designations, numbers, and names for, 157, 317–20

directive to discover and track, 128–29, 129n70

distinction between asteroids and, 210–11, 213

dust-to-gas ratios, 76–77, 76n67

effects of impacts on Earth's surface and life on Earth, 8–9, 30, 31, 77, 79–81, 99–104

energy deposited in atmosphere by entry of, 1–2, 41–42, 42n30, 86–87

Halley research on, 4
impact risk from, 217
malign intent and warning associated
    with, 2, 3–4, 5
missions to and flybys of, 75–77, **184**,
    197–204
nongravitational motion of and forces
    on, 159–60
nucleus observations and characteriza-
    tions, 75–77, 159–60, 198–99,
    200, 200n76, 203–4
observation and understanding of, 2–5,
    13
orbit and ephemeris of, 16n7
orbital characteristics of, 4, 159–60,
    160n46
political impacts of, 121–28
Ptolemy theory of epicycles, 2–3
structure of, 199–200, 210–11,
    313–14
Tunguska event attribution to, 41–42,
    41n27
visits and missions to, 74–77
See also specific comets
Commercial Space Launch
    Competitiveness Act, U.S. (SPACE
    Act), 269
Committee on the Peaceful Uses of
    Outer Space (COPUOS), 294–95,
    295n63
communications
    change in release of impact predictions,
        169, 296–97
    data release and public announcement
        guidelines, 138, 138n94, 167,
        168–69
    International Asteroid Warning
        Network (IAWN), 10, 295–97,
        298, 300
    notification about 2008 TC3 from,
        276, 276n7
    notification of Federal agencies and
        emergency response institutions
        of impending NEO threats, 274,
        276, 283–85
    notification protocols, 276, 276n6

press information sheet release about
    and media attention given to 1997
    XF11, 133–35, 134n78, 135n84,
    135n86, 137, 137–38n93, 138
risk assessment scales for communi-
    cating impact hazards, 169–74,
    171–72, 296–97
warning for predictions for short-term
    impacts and close approaches,
    279–80, 280n21
Congress, U.S.
    ARM hearings by, 259–60
    Commercial Space Launch
        Competitiveness Act, U.S., 269
    directive for a study of NEO survey
        capabilities, 213, 230–34, 274
    directive to NASA on NEO research,
        9, 113, 113n21, 128–29, 138–39,
        230, 231, 236–37, 273
    directive to track and deflect hazardous
        asteroids, 213–14, 303–4
    hearing on threats from space, 291
Conners, Martin, 187
ConsenSys, Inc., 269–70
Consolidated Appropriations Act, 274n3
Constellation program, 241, 241n64,
    253–55
Cook, A. F., 68
Cooper, Bob, 63
Cornell University, 71
Corry, John, 18, 318
Cosmic Catastrophes (Chapman and
    Morrison), 109
cratering mechanics theory and modern
    impact physics, 47
craters/impact craters
    astrobleme term for, 9, 103
    depth-to-diameter ratios, 45, 46
    destruction of, 91
    discovery of end-Cretaceous impactor,
        9
    excavation of, 40
    explosive event with no crater, 41–42,
        41–42nn29–30
    explosive processes and creation of, 24,
        31, 35–36n14, 35–38

geological markers for terrestrial impact sites, 44, 44nn37–38, 46

geological process, impact cratering as, 36, 42–46, 51

hypotheses on formation of, 24–25, 26–27, 26n37

identification of from extraterrestrial impacts, 8–9

impact origin of, 24, 31–39, 34–35nn10–11, 35–36n14, 37n18, 38–39n21

institutional backing of research on, 45–46

interdisciplinary approach to study of, 42–43, 45

lunar craters research, 23, 25–26n35, 25–30, 26n37, 27–28nn39–41, 33–38, 42–43, 42n31, 44–45, 49–50

magnetic surveys of, 24–25

Meteor Crater surveys, 22–26, 27, 30

moonlet theory for formation of, 27–28, 28n41, 29

number identified, 90–91

number of suspected impact structures, 43–44

shapes of and impacts angles for formation of, 27–28, 28n41, 35–36

space exploration and research on, 31

volcanic origin associated with, 23, 23n26, 26, 26n37, 28–29, 35, 37n18, 38

Cretaceous–Paleogene impactor and extinction (K–Pg)

Alvarez hypothesis on, 9, 79–80, 103

clay layer composition and properties, 79–80, 82–83, 86, 88–89, 91, 92, 103

fossil records and, 90

impactor crater search and identification, 80, 84, 90–98, **94**, 97n50, 101, 103–4, 128

iridium identification, 79–80, 81–84, 92, 103

mechanisms for mass extinction, 80–90, 96–98

timing of, 79

Cretaceous period, 9

Cretaceous–Tertiary extinction (K–T), 79n1

Crutzen, Paul, 88

cryptocurrency, 269–70

cryptovolcanic structures, 42, 44–45, 44nn37–38

C-type asteroids, 67–68, 180, 182–83, 204, 205–6, 209n106

Culberson, John, 260

Cunningham, Clifford, 318

D

D'Abramo, Germano, 301

Dactyl/Ida asteroid (243), **184**, 194, 196, 204, 205, 207

Daly, Reginald V., 45

DAMIEN: the Interagency Working Group for Detecting and Mitigating the Impact of Earth-bound Near-Earth Objects, 294

Dark Matter Telescope, 215

DART (Double Asteroid Redirection Test), 298–300, 304

David, Leonard, 296

Davis, Bob, 119n42

Dawn spacecraft mission, 66n33, 178, **184**, 205–6n91, 209–10

Dawson, Terry, 113, 113n21

death plunge, finding asteroids during, 239–41

Deccan Traps volcanism, 84, 98

Deep Bay Crater, 46

Deep Impact, 135n84

Deep Impact mission, 132n76, **184**, 199–201, 199n71, **199**, 205–6n91, 235–36, 283, 283n29, 288–89

Deep Sea Drilling Project, 93

Deep Space 1 (DS1) mission, **184**, 197, 209, 235

Deep Space Industries (DSI), 267–68, 269, 270

Deep Space Network (DSN), 69–74, 242

Defending Planet Earth (National Research Council), 281–82, 283

Defense, U.S. Department of (DOD)
congressional directive to NASA to
coordinate with on comet and
asteroid identification, 128–29
nuclear policy of and interest in atmospheric effects of nuclear warfare, 87
planetary defense role of, 273
Science Definition Team role of, 214
Spacewatch funding from, 62–63
Strategic Defense Initiative
Organization (SDIO), 115–16,
119, 119n39
Defense Advanced Research Projects
Agency (DARPA), 63, 71
defense against asteroid impacts. See planetary defense
Defense Nuclear Agency, 87
deflection. See mitigation/deflection
Delporte, Eugène Joseph, 322
DeMeo, Francesca, 181–82
Denneau, Larry, 186
deuterium-to-hydrogen (D/H) ratio, 201,
201n77
Diamandis, Peter, 266–67
Didymos asteroid system (65803) and
Didymoon, 299–300
Dietz, Robert S., 44–45, 44n38
dinosaur extinction, 8–9, 80, 90
diogenites, 67n37
"dirty snowball" model for cometary
nucleus, 77, 159–60, 204, 211, 313
discovery of near-Earth objects (NEOs)
advances in technology and increase in,
10–11, 173–74, 307–8, 312–15
automation and increase in discoveries,
11, 141–42, 174
computer-assisted analysis for, 11
congressional directive to discover and
track asteroids, 9, 113, 113n21,
128–29, 129n70, 138–39, 230,
236–37, 273
early discoveries, 8–9, 19–22,
20–22nn19–20, 21
follow-up observations of discoveries,
153, 153n30, 156, 220, 238–39,
239n59, 312

increase in and automated warning
systems for PHAs, 142
photograph comparison to find NEOs,
10
search and research efforts, 1
See also surveys/discovery surveys
Discovery Program, NASA, 198, 205–
6n91, 243–44, 303, 311
Dominion Observatory, Ottawa, 46
Don Quijote deflection demonstration
mission, 288–89
Double Asteroid Redirection Test
(DART), 298–300, 304
drift scanning, 61
drift scanning technique, 145
D-type asteroids, 68–69, 209n106
Duende asteroid (367943), 70
Duller, Charles, 95
dwarf planets, 8, 18, 158
Dying Planet (Markley), 246

E

Earth
distance from Sun, 14
effects of cosmic impacts on life on,
8–9, 30, 31, 77, 79–81, 99–104
evolution of life on, 99–104
formation from meteoric accretion, 45
geological changes to, 4–5
Lagrange points, 221–22, 221
sterilization from impact events, 101–2
transfer of life between Mars and, 102–3
water on, sources of, 201, 201n77
Earth Impact Database, University of
New Brunswick, 103–4
"Edison's Conquest of Mars" (Serviss),
245
Effelsberg antenna, 72
electronic asteroid-hunting telescope,
240–41
El Kef formation, Tunisia, 96
Elst-Pizarro, Comet (133P), 183n25
Elst-Pizarro asteroid (7968), 182–83,
183n25
Emery, Joshua, 182
Encke, Comet, 73

Encke, Johann F., 159, 319
energy
  Chelyabinsk airburst, 1–2
  deposited in atmosphere as small
      bodies enter, 1–2, 41–42, 42n30,
      86–87
  explosive force and shape of craters,
      35–36n14, 35–37
  heat from impact events, 101–2
  melting from energy associated with
      velocities of meteors, 27, 27n39
  Meteor Crater event, 36–37, 42n30
  quantitative energy scaling develop-
      ment, 45
  Shoemaker-Levy–Jupiter collision,
      126–27
  Tunguska event, 2, 41–42, 42n30
  wartime innovation and theories on
      crater formation, 31, 36–37
environmental objects, NEOs as, 10, 141
epicycles, theory of, 2–3
Erice, Sicily, retreat, 119–20, 121
Eros asteroid (433)
  Amor classification of, 322
  brightness variations and rotation of,
      64
  composition and structure of, 196, 207
  discovery of, 8, 19–22, 20–22nn19–20,
      21
  image of, **197**
  mission to, 173, **184**, 193–94, 196,
      197, 205, 205–6n91, 207
  naming of, 319
  radar observations of, 70–71
  shape and surface texture of, 70–71
  size estimate of, 179
  structure of, 193–94
eucrites, 67n37
Eunike asteroid (185), 21–22
Eunomia asteroid (15), 319
European Space Agency (ESA)
  deflection demonstration missions of,
      10, 288–89, 298–300
  Herschel observatory, 185
  Herschel Space Telescope, 201,
      201n77, 210

NEODyS system and web page, 173,
    173n73, 279–80, 288n46
planetary defense role of, 287–89
Space Situational Awareness (SSA) pro-
    gram, 156, 287–88, 288n46
European Space Research Institute
    (ESRIN), 288
Evans, Jenifer, 222
Evpatoria antenna, 72
Explorer program, 185
exposure gating technique, 56
extinction/mass extinction
  Alvarez hypothesis on, 9, 79–80, 103,
      251
  climatic catastrophes as mechanism, 98
  dinosaur extinction, 8–9, 80, 90
  extraterrestrial mechanisms, 99–103
  fossil records and, 90
  heat as mechanism for, 86–87
  impact event as mechanism, 9, 79–90,
      79n1, 80n3, 81n5, 88–90, 96–98
  periodicity in extinction events,
      99–100, 103–4
  Phanerozoic extinction events, 98
  risk of extinction-scale events, 215–16
  volcanic eruptions as mechanism,
      84–86, 98
  See also Cretaceous–Paleogene impactor
      and extinction (K–Pg)
Extrasolar Planet Observation and Deep
    Impact Extended Investigation
    (EPOXI), **184**, 201
extraterrestrial extinction mechanisms,
    99–103

F
Face of the Moon, The (Baldwin), 45
FARCE (Far Away Robotic sandCastle
    Experiment), 258–59
Farnocchia, Davide, 280
Farquhar, Robert, 74–75, 205–6n91
Fast Resident Object Surveillance
    Simulation Tool (FROSST), 222
Federal Emergency Management Agency
    (FEMA), Department of Homeland
    Security, 284, 285, 286, 300

Fender, Janet, 149–50
fires/wildfires, 2, 88, 89
Fomenkova, Marina, 123
forests and trees
    fires and wildfires, 2, 88, 89
    uprooting and flattening by impact
        events, 2, 40
fossil records, 90
Fraas, E., 44n38
fragmentation hypothesis, 18–19
fragmentation strategy, 117
Frasnian–Famennian boundary, 81
Frecker, Jack, 60
Friedlander, Alan, 108
Friedman, George, 253
Friedman, Louis D., 256–57, 259, 270
Frosch, Robert A., 250

G

Gaffey, Michael, 249–50
Galileo mission
    asteroid color observations by, 196
    asteroid satellite discovery by, 194
    asteroid visits, **184**, 204–5
    CCDs use on, 58
    Shoemaker-Levy–Jupiter collision wit-
        nessed during, 9, 125, 204
Gamma Ray Spectrometer, 207
Garradd, Gordon, 155, 275
Garretson, Peter, 285
Gartner, Stephan, 92
Garver, Lori, 270
Gaspra asteroid (951), **184**, 196, 204–5
Gates, Michele, 260–61
Gauss, Carl Friedrich, 17
Gehrels, Tom
    asteroid-hunting telescope, advocacy
        for, 240
    career of, 57
    hazard aspect of NEOs and efforts to
        forecast collisions, 59
    Hazards Due to Comets and Asteroids
        role of and efforts for balanced
        views by, 120–21
    Johnson interaction with, 130
    mitigation and deflection meeting role
        of, 118

    NEO research by, 312
    Palomar-Leiden Survey role of, 57–58
    photo of, **58**
    polarization curve for Icarus by, 177,
        177n5
    radar research assistance from, 70
    rotation rate studies by, 191
    Snowmass workshop on hazard aspect
        of NEOs, role in, 108
    Spacewatch camera and telescope devel-
        opment by, 10–11, 58–64, 108n5,
        109–10, 130, 141
    Spacewatch Report, The, newsletter, 60, 62
    telescopes for Spacewatch, acquisition
        and funding for, 146–48
Geminids meteor showers, 2
Genesis mission, 205–6n91
geocentrism, 2–3
Geographos asteroid (1620), 322
Geological Society of America
    Gilbert address to, 24–25, 28
    Snowbird I meetings, 85–86, 90
    Snowbird II meeting, 89–90, 93,
        143–44
    Snowbird III meeting, 96
Geological Survey, U.S. (USGS)
    Astrogeological Studies branch at,
        creation of, 49
    chief geologist for, 22, 22n24
    cratering research by astrogeology
        branch of, 31, 49
    Dream Moonshaker skit at, 46, 49
    formation of, 22n24
    Meteor Crater surveys by, 22–26, 27,
        30, 31
    Shoemaker return to, 55
Geological Survey of Canada, 81
geology
    geological markers for terrestrial impact
        sites, 44, 44nn37–38, 46
    impact cratering as geological process,
        36, 42–46, 51
    small solar system bodies as geological
        agents, 13
    time periods, **82**
    uniformitarianism doctrine and, 4–5,
        8–9

George E. Brown Near-Earth Object Survey Act (Brown Act), 230–31, 240, 242, 273–74, 273n1, 280, 281, 286–87, 302
German Aerospace Center (DLR), 180, 288n46, 289
Gerstenmaier, William, 261–62
Giacobini-Zinner, Comet, 75–76, **184**
Gifford, Algernon Charles, 37–38, 37n18, 38
Gilbert, Grove Karl
   lunar craters research by, 23, 25–26n35, 25–30, 26n37, 27–28nn39–41
   Meteor Crater surveys by, 22–26, 27, 30, 31, 32, 33n4
   projectile experiments by, 24–25, 27, 27n40, 32
   USGS position of, 22, 22n24
Ginsburg, Robert, 96
Giotto camera system, images, and mission, 75–76, 76n68, **184**
"Global Catastrophes in Earth History" (Snowbird II) meeting, 89–90, 93, 143–44
glycine, 198, 198n68
Goldin, Daniel, 139
Goldstein, Richard, 70
Goldstone Deep Space Information Facility, 69–74, 189–90, 194, 236, 297
Golevka asteroid (6489), 160–61
Google Lunar XPRIZE, 265–66
Gould, Benjamin Apthorp, 18, 318, 319
Gould, Stephen J., 9, 100
Gravity Recovery and Interior Laboratory (GRAIL) mission, 205–6n91
gravity tractor, 233, 235, 261, 282
Green Bank Observatory, 72
Greenberg, Richard, 204–5
Griffin, Michael, 228–29, 235
Grigg-Skjellerup, Comet, 73, **184**
Ground-based Electro-Optical Deep-Space Surveillance (GEODSS) systems, 130–31, 148–52, 153, 224
Grunsfeld, John, 293
Gunn, James, 10

H
Hahn, Gerhard, 288n46
Hale-Bopp, Comet, 154–55
Halley, Comet
   albedo feature of nucleus, 76
   color of, 65–66
   Comet Halley Watch, 155
   discovery of, 4
   missions to and flybys of, 75–77, **184**, 197
   nucleus observations and characterizations, 75–77, 76n67, 159
   orbital characteristics of, 159
   radar observations of, 74
   size of, 123
   spacecraft mission to and flybys of, 74
Halley, Edmond, 4, 80–81
Hansen, Thor, 92
Hapke, Bruce, 195–96
Harding, Karl Ludwig, 19
Harris, Alan William (German Aerospace Center, DLR), 65n31, 180, **188**, 289
Harris, Alan W. (JPL), 65n31, 116–18, 191, 215–16, 216n8, 301, 301n82
Hartley 2, Comet, **184**, 201, 201n77
Harvard Arizona Meteor Expedition, 43
Harvard University, 43
Hathor asteroid (2340), 161
Hawai'i
   GEODSS system in, 130, 148–52
   Infrared Telescope Facility (IRTF), 59, 68, 179, 180–81
   Keck telescope, Maunakea, 68–69n43
   Pathfinder telescope in, 240–41
Hayabusa and Hayabusa2 missions, 183, **184**, 185, 185n30, 193, 196, 207–9, 258, 313
Haystack radar, 70
Hazards Due to Comets and Asteroids (Space Science Series), 120–21
hazards/impact hazards
   AIAA research and white paper on, 112–13, 276
   asteroid size, hazard risk relationship to, 217–19, **218**, **219**
   automated warning systems for, 142, 166, 167

closest known asteroid approach and move toward NEO research, 111–12nn12–14, 111–13

computation of impact probabilities, 165–69, 166n60, 168n64

congressional directive to discover and track asteroids, 9, 113, 113n21, 128–29, 129n70, 138–39, 230

congressional directive to track and deflect hazardous asteroids, 213–14, 274, 303–4

cost-benefit analysis and risk for, 215, 216, 218–19n16, 218–20, 223, 244, 309–10

credibility of impact-event warnings, concern about, 135, 137, 137–38n93

data release and public announcement guidelines for possible impacts, 138, 138n94, 167, 168–69

death plunge, finding asteroids during, 239–41

detection of short-warning events, **279**

early studies of asteroid hazards, 106–11

identification and detection of short-warning impact, 275–80, 280n21

international conference on Earth-approaching bodies, 113–14

media attention given to, 134–35, 135n84

mortality rates associated with impact events, 110

natural hazard, cosmic impact as, 244

NEOs as, 10, 59, 141

notification of Federal agencies and emergency response institutions of impending NEO threats, 274, 276, 283–85

opinions about impact hazards after Shoemaker-Levy–Jupiter collision, 128

policy development related to, 106

predictions of close-Earth approaches, 105, 165–69, 168n64, 279–80, 314–15

reassessment of, 300–305

residual hazard, 219, **219**

risk analysis and development of NEO hazard reduction goal by SDT, 215–24, 301–2

risk assessment scales for communicating impact hazards, 169–74, **171–72**, 296–97

risk-to-Earth analysis and assessments, 1, 10, 11, 105–6, 141

simulations and modeling of NEOs as, 105–6, 105n1, 285–86, 285n38, 300–302, 300n79

size-frequency-consequences chart on, 110–11, **110**

Snowmass report on detection and investigation of, 108–11, **110**, 143–44

workshops on search programs and interception/deflection strategies, 113–21, 116n28

See also mitigation/deflection

Heidelberg Observatory, 322

Hektor asteroid (634), 68–69, 68–69n43

Helin, Eleanor "Glo"

   asteroid 1997 XF11 observations by, 134

   asteroid follow-up observation by, 63

   asteroid naming by, 151, 224

   Aten asteroid discovery by, 53, 323

   CCD camera for, 149–51

   exhibit to honor, 173

   film, decision about moving away from, 148–49

   JPL transfer of, 55

   NEAT program role of, 55, 148–52, 173

   NEO research by, 312

   PCAS role of, 51–55, **52**, 57, 142–43, 173–74, 178–79

   photos of, **52, 224**

   retirement and death of, 173

   Shoemaker-Levy 9 image taken by, 124n57

   Shoemaker split from, 55

   telepossession claims, opinion about, 265n76

   Teller asteroid naming by, 116

use of Palomar facilities by, 53

heliocentric orbits, 165n59, 321–23, **321**

Hells Creek, Montana, 90

Henbury Station, Australia, craters, 44

Hera (Ceres), 16

Hera mission, 299–300

Hergenrother, Carl, 155, 169

Herger, Paul, 157

Hermes asteroid (69230), 111, 111n12, 154

Herschel, William, 8, 17–18, 318

Herschel observatory, 185

Herschel Space Telescope, 201, 201n77, 210

Hildebrand, Alan, 91–95, 275, 289

Himmelspolizei (Lilienthal Society, Celestial Police), 15–16

Holdren, John P., 283–84, 291, 292

Holsapple, Keith, 192

Holt, Henry, 55, 111

Homeland Security, U.S. Department of, Federal Emergency Management Agency (FEMA), 284, 285, 286, 300

Houston, Kevin, 206

howardite, eucrite, and diogenite (HED) meteorites, 67, 67n37, 163, 163n52, 164, 195, 209

howardites, 67n37

Huang, Su-Shu, 247–48

Hubble Space Telescope
   launch of, 111
   Shoemaker-Levy–Jupiter collision witnessed by, 9, 106, 125, 127, 128
   Vesta observations with, 209
   Wide Field and Planetary Camera, 58

Hudson, Scott, 188

Hughes, David, 20n17

Human Exploration Roadmap, 260

Huntress, Wesley, 139

Hut, Piet, 214, 267

Hutton, James, 4–5

Hyakutake, Comet, 74

I

Icarus asteroid (1566), 70, 106–8, 177, 322

ICE mission, **184**

Ida/Dactyl asteroid (243), **184**, 194, 196, 204, 205, 207

Illustrations of the Huttonian Theory of the Earth (Playfair), 4–5

Imaging Photopolarimeter instrument, 57

impact and explosion mechanics, research on high-velocity, 49

impact events
   credibility of impact-event warnings, concern about, 135, 137, 137–38n93
   dinosaur extinction from, 8–9, 80, 90
   effects on Earth's surface and life on Earth, 8–9, 30, 31, 77, 79–81, 99–104
   explosive event with no crater, 41–42, 41–42nn29–30
   forecasting of, 59
   forest/trees uprooted and flattened by, 2, 40
   geological markers for terrestrial impact sites, 44, 44nn37–38, 46
   Halley research on, 4
   impact process as agent of change in solar system, 50
   keyhole passages to Earth close approaches, 136–37, 228, 235, 236, 278
   mass extinctions from, 9, 79–81, 79n1, 80n3, 81n5, 88–90
   mineralogic marker for, 47, 49, 96–97
   mortality rates associated with, 110
   ocean, asteroid impact in, 86, 91
   public attention given to no-warning and short-warning events, 274–80, **279**
   size of impactor and duration of effects from, 101–2
   sterilization of Earth from, 101–2
   transfer of life between planets through, 102–3
   uniformitarianism and, 4–5, 8–9
   volatilization of impactor, 33–34n6–8, 33–36, 38, 38–39n21
   See also craters/impact craters; Cretaceous–Paleogene impactor and extinction (K–Pg)

impact physics and theory of cratering mechanics, 47
impact winter, 88–90
Infrared Astronomical Satellite (IRAS), 183
infrared observations
asteroid size estimates from, 177–83, 185–87
near-infrared and infrared spectral observations, 180–83, **181**, 182n22
SDT space-based survey telescope recommendations, 220–22
space-based IR observations and data, 183–87, **184**, 185n30, 186n33
thermal models and standard thermal model (STM), 177–80
Infrared Processing and Analysis Center (IPAC), 183
Infrared Telescope Facility (IRTF), 59, 68, 179, 180–81
Inouye, Daniel, 232
Institute for Astronomy (IfA), 67, 231, 277–78, 279, 279n19
Interception Workshop and report (Near-Earth Object Interception Workshop and report), 114, 115–19, 116n28, 119n41, 247, 251–52
International Asteroid Warning Network (IAWN), 10, 295–97, 298, 300
International Astronomical Union (IAU)
asteroid 1997 XF11 circular from, 133–34, 134n78, 137
asteroid 2000 SG344 impact risk announcement by, 168–69
Committee on Small Body Nomenclature (CSBN), 320
dwarf planet reclassification by, 8
minor planet term use by, 318
telegram alerts for follow-up observations from, 54, 63
See also Minor Planet Center (MPC), International Astronomical Union
International Comet Explorer (ICE), 74–75
International Geophysical Year, 46
international space policy, 247

International Space Station (ISS), 256, 266, 287–88, 305
International Sun-Earth Explorer (ISEE)–3 heliophysics spacecraft, 74–75
interstellar object, discovery of first, 307–8, 313
IRAS-Araki-Alcock, Comet, 73–74, 183
Irene asteroid, 319
iridium, 79–80, 81–84, 92, 103, 250
Iris, 177, 177n5
iron-nickel objects, 24, 44, 68, 209, 211, 249–50
iron particles (nanophase iron), 195–96
Itokawa, Hideo, 185n30
Itokawa asteroid (25143), 183, **184**, 185, 185n30, 193, 196, 207–9
Ives, Herbert, 36–37, 37n16, 38

J

Jain, Naveen, 266
James, Joseph F., 29–30
Jangle U nuclear bomb test crater, 47, **48**
Japan
AKARI telescope, 183, 185
Hayabusa and Hayabusa2 missions, 183, **184**, 185, 185n30, 193, 196, 207–9, 258, 313
Jennings, Barbara, 300n79
Jenniskens, Petrus "Peter," 277
Jet Propulsion Laboratory (JPL)
CCD camera for, 148–51
CCD development role of, 10–11, 58
Goldstone Deep Space Information Facility, 69–74
Helin transfer to, 55
Near-Earth Object Program Office (Center for Near-Earth Object Studies) at, 139, 141, 256n38, 280n21, 314
PCAS role of, 54, 54n5
press information sheet release about 1997 XF11 by, 134
Scout system development, 280, 280n20
Shoemaker-Levy–Jupiter impact website at, 128n68
Solar System Dynamics Group, 158

Yucatán Peninsula crater research at, 95

Jiang, Ming-Jung, 92

Johns Hopkins University, Applied
Physics Laboratory, 45, 206,
298–300

Johnson, Brandy, 131n75

Johnson, Lindley N.
Apophis (2004 MN4) impact, discussions about, 226–27, 228–29
asteroid named for, 151
on effects of ARM saga, 271
Gehrels interaction with, 130
GEODSS access for Helin, role in,
150–51
NASA roles of, 132n76
NEOOP role of, 223, 242, 292–93
on NEO program directives, priorities,
and funding, 237
notification about 2008 TC3 from, 276
photo of, **224**
Planetary Defense Coordination Office
(PDCO) role of, 132n76, 293–94
Planetary Defense Officer role of,
132n76, 294
planetary defense program, development of, 285
planetary defense program role of,
129–32
planetary defense term use by, 213
retirement from Air Force of, 132n76,
223
SDT report, commissioning of an
update to, 302
SpaceCast 2020 planetary defense
paper by, 131–32, 223–24
Space Command role of, 130–31,
150–51

Johnson, Lyndon B., 247

Johnson asteroid (5905), 224

Johnson-Morgan photometric system
(UBV system), 66, 66n34

Johnson Space Center
meteorite from Mars research by, 103
planetary defense workshop at, 214

Jones, Terry J., 179

"Journey to Mars" campaign, 270–71

Juno, 16, 19, 176, 319

Jupiter
asteroids between Mars and, 8
comets around obit of, 124–25, 125n59
distance from Sun, 14
Galileo mission to, 9, 58, 125, 204
gravitation effects on main-belt asteroids, 163–64
meteor from, 6
Pioneer 10 mission to, 57, 60, 63
search for missing planet between Mars
and, 13–17, 14n2, 18–19
Shoemaker-Levy 9 collision with, 9,
105, 106, 123–28, 125–26nn61–
62, 204, 273
Trojan asteroid regions, 68, 68–69n43,
187

Jupiter Icy Moons Orbiter, 235

Jurgens, Ray, 71, 187

K

Kaarlijarv, Estonia, crater, 43

Kaiser, Nick, 231, 237–38

Kamoun, Paul, 73

Karin family, 192

Keck Institute for Space Studies (KISS),
257, 257n43, 259

Keck telescope, 68–69n43

Keller, Gerta, 97–98

Kentland, Indiana, cryptovolcanic structure, 44n38

Kepler, Johannes, 4

Kepler mission, 205–6n91, 242

Kepler's third law, 194, 194n58, 205

Kerr, Richard, 96

Kervin, Paul, 149–50

keyhole passages to Earth close approaches,
136–37, 228, 235, 236, 278

kinetic impactor strategy, 117–18, 233,
278, 282–83, 283n29, 287, 298–300

Kirkwood, Daniel, 19

Kirkwood gaps, 19

Kitt Peak Observatory
Apophis (2004 MN4) observations
from, 225–26
CCD–based camera for, 10–11
Shoemaker-Levy 9 confirmation by,
124

Spacewatch camera and telescope at, 11, 59, 60–62, 132
Koehn, Bruce, 154
Koronis family, 205
Korsmeyer, David, 254
Koschny, Detlef, 288
Kowalski, Richard, 275, 279n20
Krakatoa explosion, 85
Krauss, John, 262–63
Kring, David, 291
Kuiper, Gerard, 57
Kulik, Leonid A., 39–40, 43–44

L

L5 Society, 263–64, 264n71
Lagrange, Joseph-Louis, 19
Lagrange points, 221–22, **221**
L'Aigle, France, fall, 7
Large Space Telescope, 58
Large Synoptic Survey Telescope (LSST), 215, 215n5, 231–32, 302, 311
Larson, Steve, 154–56
Lawrence, Kenneth, 54n5, 134, 150, 151
Lawrence Livermore National Laboratory (LLNL), 115, 119, 300
least squares technique, 17
Lebofsky, Larry, 68, 179–80, 182
Leiden Observatory and the Palomar-Leiden Survey, 57–58
Levy, David, 55, 123–24
Lewicki, Chris, 267–68
Lewis, John S., 268
Lexell 1770 I, Comet, 112n16
Lick Observatory, 65
light curve observation and analysis, 64–65, 65n31, 190–91, 193–94, 194n56
Lilienthal Society (Himmelspolizei, Celestial Police), 15–16
"Limits to Growth" report (Club of Rome), 246, 248–49
Lincoln Laboratory
    CCD development role of, 130
    Experimental Test Site (ETS), 153
    Fast Resident Object Surveillance Simulation Tool (FROSST), 222
    Millstone Hill radar, 69

Lincoln Laboratory Near-Earth Asteroid Research (LINEAR) program
    Air Force role in, 130, 153, 156
    asteroids found with, **239**
    Atira discovery by, 323
    data from and MPC data processing and dissemination, 158
    fast-readout capabilities of CCD cameras of, 130, 130n72, 148, 153
    first NEO discovery by, 153, 153n29
    follow-up observations of discoveries by, 153, 153n30, 220
    funding for, 153
    SDT survey systems recommendations, comparison against, 220
    success of, 130, 153, 220
    survey and observations strategies of, 148, 153, 238
LINEAR, Comet (176P), 183n25
LINEAR asteroid (118401), 183, 183n25
Linke, Felix, 21–22
Lipps, Jere, 90
Lo, Martin, 256–57, 257n41
long-period comets, 72–74, 111, 129n70, 145, 151, 217
Los Alamos National Laboratory (LANL), 114, 115
Lowell, Percival, 246
Lowell Observatory, 144, 148, 154
Lowell Observatory NEO Survey (LONEOS), 148, 154, **239**
Lowry, Stephen, 162
Lu, Edward T., 214, 267
Lunar and Planetary Laboratory, 57, 67
Lunar and Planetary Laboratory survey, University of Arizona, 154–56, 297
Lunar Prospector spacecraft, 173, 205–6n91
Lutetia asteroid (21), **184**
Lyell, Charles, 5
Lyot, Bernard, 177, 177n5
Lyttleton, Raymond, 75

M

magnetic properties, 24, 32, 40, 46, 80, 81–82, 94, 207

magnitude/brightness
    absolute magnitude, 18n13, 143n3,
        176
    apparent magnitude, 18n13, 143n3,
        175–76
    concept of, 18n13
    estimates of, 143n3
    light curve observation and analysis, 65,
        65n31, 190–91, 193–94, 194n56
    variations in and rotation of asteroids,
        64–65
    viewing geometry and, 175–76, 176n1
main-belt asteroids, 8, 163–64, 175, 179
Mainzer, Amy, 186–87, 242, 243–44
Manned Orbiting Laboratory, 146
Margot, Jean-Luc, 190
Mariner IV, 50
Mark, Hans, 61–62, 148
Markley, Robert, 246
Mars
    asteroids between Jupiter and, 8
    atmosphere of, research on, 84
    cratering on, 31, 50, 101
    distance from Sun, 14
    gravitation effects on main-belt aster-
        oids, 163–65
    Human Exploration Roadmap develop-
        ment for mission to, 260
    meteorites from, 103, 261–62
    missions to, 255
    planet-crossing asteroid discoveries for,
        142
    priority of missions to, 304, 305
    resources on and planetary evolution,
        246
    search for missing planet between
        Jupiter and, 13–17, 14n2, 18–19
    transfer of life between Earth and,
        102–3
    transfer of life between Venus and, 103
Marsden, Brian
    asteroid 1997 XF11 impact and orbit
        calculations by, 132–37
    model development for nongravita-
        tional forces on comets, 160
    MPC director role of, 121, 132–33,
        157–58

    press information sheet release about
        and media attention given to 1997
        XF11, 133–35, 134n78, 135n84,
        137, 137–38n93
    Shoemaker-Levy–Jupiter impact and
        orbital computations by, 124–25
    Swift-Tuttle comet impact probability
        computations by, 121, 122–23
Mars Pathfinder mission, 205–6n91
Massachusetts Institute of Technology
    (MIT)
    Haystack radar observations, 70
    Lincoln Laboratory, Experimental Test
        Site (ETS), 153
    Lincoln Laboratory, FROSST, 222
    Lincoln Laboratory, Millstone Hill
        radar, 69
    Lincoln Laboratory role in CCD devel-
        opment, 130
    Project Icarus exercise, 106–8
    See also Lincoln Laboratory Near-Earth
        Asteroid Research (LINEAR)
        program
mass driver strategy, 117–18, 247–49,
    252–53, 263
Matese, John, 100–101
Mathias, Donovan, 301
Mathilde asteroid (253), **184**, 205–6
Maurrasse, Florentin, 93
McBain Instruments, 55
McCord, Tom, 66
McDonald Observatory, 132
McKinnon, William, 127
McLaren, Digby, 81
McMillan, Robert S. "Bob," 59, 60, **149**,
    186
McNaught, Robert, 63, 155, 226
Meadows, Vikki, 126
Medium Deep Survey, 237–38
Meech, Karen, 307
Megaton Ice-Contained Explosion
    (MICE) project, Atom Energy
    Commission, 46–49
Meinel, Aden, 146–47, 146n12
Melamed, Nahum, 300n79
Melosh, Jay, 285n38
Mercury, 14, 307

MErcury Surface, Space ENvironment, GEochemistry, and Ranging (MESSENGER) mission, 205–6n91

Merrill, George P., 33–34

Meteor! (film), 107

Meteor Crater (Barringer Meteorite Crater)
  buried remnant of impactor in, 22–24, 31–33, 38, 38–39n21
  coesite in samples from, 47, 49
  impact origin of, 24, 31–39, 34–35nn10–11, 38–39n21, 42, 42n30
  meteoric iron in area of, 23–24, 31–33
  mining iron from, 31–33, 34, 39
  numerical simulation of impact, 49
  Shoemaker and MICE project study of, 47–49, **48**
  steam explosion origin of, 24–25, 35
  surveys by Gilbert and USGS, 22–26, 27, 30, 31, 32, 33n4
  theories about origin of, 23–25, 26, 31–39, 33n4
  volatilization of impactor at, 33–34n6–8, 33–36, 38, 38–39n21

meteoric iron
  fragments found at Meteor Crater, 23–24
  mining of, 31–33, 34, 39, 40
  tools made from, 6

meteors and meteorites
  composition and texture of, 7
  composition of and asteroid spectral class, 66–69, 67n37
  interdisciplinary research on, 43
  interstellar origin of, 43
  Mars, meteorites from, 103
  origins of and explanations for, 5–7, 23
  reports, myths, and religious writings about, 5–6
  velocity research, 43

meteor showers, 2

Metis asteroid (9), 65–66

Michel, Patrick, 193, 193n53

Micheli, Marco, 277, 279n19

Mid-Course Space Experiment (MSX), 183

Milani, Andrea, 166–67

Mildred asteroid (878), 319

mineralogic marker for impact processes, 47, 49, 96–97

Minimum Orbital Intersection Distance (MOID)
  definition and use of term, 321–22

mining. *See* resources/natural resources

Mining the Sky (Lewis), 268

Minor Planet Center (MPC), International Astronomical Union
  assignment of designations, numbers, and names by, 157, 227–28, 319, 320
  asteroid 1997 XF11 impact and orbit calculations by, 132–37
  automated warning systems for potential impact hazards, 142, 166
  computation of orbit and ephemeris by, 158
  data release and public announcement guidelines from, 138, 138n94, 167, 168
  funding for, 157, 312
  location of, 54, 157
  Marsden as director of, 121, 132–33, 157–58
  minor planet term use by, 318
  NEAT program observations for, 151
  NEO Confirmation web page (NEOCP), 280
  observation data processing and dissemination by, 157–58
  Palomar-Leiden Survey role of, 57–58
  PCAS role of, 54
  press information sheet release by and media attention given to 1997 XF11, 133–35, 134n78, 135n84, 137
  Shoemaker-Levy 9 confirmation by, 124
  Spahr as director of, 158
  Swift-Tuttle comet impact probability computations by, 121, 122–23
  telegram alerts for follow-up observations from, 54, 157
  tracklet issues and Isolated Tracklet File, 238–39, 239n59

Minor Planet Circular, 57–58, 157
minor planets, 13, 318. *See also* asteroids
mitigation/deflection
  Analysis of Alternative (AOA) study on
    deflection, 233–35, 274, 280–81
  Apophis (2004 MN4) deflection dis-
    cussions, 228–30, 233, 234
  civil defense and civil defense exercises,
    10, 274, 282, 300
  congressional directive to track and
    deflect hazardous asteroids,
    213–14, 303–4
  death plunge, finding asteroids during,
    239–41
  deflection demonstration missions, 10,
    288–89, 298–300
  deflection or destruction of asteroids
    with nuclear devices, 115–21,
    116n28, 118n38, 119n41,
    233–36, 280–81, 283
  development of strategies for, 245,
    246–47, 251–53, 261–62,
    280–87, 314–15
  funding for research on, 283, 284–85,
    286
  low-thrust propulsion to divert NEOs,
    214, 267
  policy development related to, 10, 106,
    309–12
  Project Icarus exercise, 106–8
  research on strategies for, 1
  rubble-pile asteroids, deflection and
    mitigation of, 108
  SMPAG focus on, 297–300
  workshops on search programs and
    interception/deflection strategies,
    113–21, 116n28
  See also hazards/impact hazards
MODP (Moving Object Detection
  Program), 63
Monte Carlo technique, 127, 136, 136n90
Moon
  Artemis mission to return to, 270,
    270n94
  autonomous rover mission to, 265–66
  circularity of craters on, 35–36,
    35–36n14
colonization of, 248–52, 263–64
cratering on, 101
imaging of far side of, 49–50
impact crater and geological research
  on, 46, 49–50
impact origin of craters on, 26–30,
  27–28nn39–41, 33–38,
  34–35n11, 44–45, 49–50
influence on Earth by, 5
interdisciplinary approach to study of
  craters on, 42–43, 45
meteorites from, 7, 29–30
mining resources on, 266
origin of and research on craters
  on, 23, 25–26n35, 25–30,
  26n37, 27–28nn39–41, 33–38,
  34–35nn10–11, 42–43, 42n31,
  44–45
radar research efforts, 69
samples from Apollo 11 mission, 66
Shoemaker ashes on, 173
spaceflight and impact-related investigations of, 46
spectral characteristics of, 195, 196
terrestrial analogs to craters on, 42–43,
  42n31
volatilization of impactors on, 33–34,
  33–34nn6–7
Moon Express, 266
moonlet theory, 27–28, 28n41, 29
Moon's Face for Science, The (Gilbert),
  29–30
Moon Treaty (Agreement Governing the
  Activities of States on the Moon
  and Other Celestial Bodies), 250,
  263–64, 264n71
Morgan, Thomas H. "Tom," 214–15,
  223, 225
Morozov, Nikolai A., 35–36, 35–36n14
Morrison, David
  asteroid classification system develop-
    ment by, 67
  asteroid size and albedo research by,
    178–79
  Cosmic Catastrophes, 109
  Erice, Sicily, retreat role of, 119
  search program workshops, role of, 114

Spaceguard Survey report role of, 109, 114

talk to congressional staffers about NEO threats by, 113n21

Tunguska event workshop role of, 42n30

mortality due to asteroid impacts, 110

Mottola, Stefano, 288n46

Moulton, Forest Ray, 38, 38–39n21

Mount Lemmon telescope, 68, 148, 155–56, 275

Mount St. Helens eruption, 85

Mount Wilson Observatory, 43, 66, 319

Mueller, Tom, 265–66

Muinonen, Karri, 134, 144–45

Multi-Mirror Telescope (MMT) project, 146–47

mirror for Gehrels telescope project from, 146–147

Musk, Elon, 265–66

N

Nakamura, Tomoki, 208–9

Nakano, Syuichi, 125

nanophase iron (iron particles), 195–96

NASA (National Aeronautics and Space Agency)

Advisory Council workshop on hazards of NEOs, 108–11, **110**

applications program, 303, 310–11, 310n7

applied research as part of congressional mandate for, 303–4, 310–12

Authorization Bill and NEO research by, 113–14, 231

budget and funding for, 60, 111, 129, 174, 237, 241, 253–54, **254**, 292–93, 292n55, 304

congressional directive to discover and track asteroids, 9, 128–29, 129n70, 138–39, 230, 236–37

congressional directive to track and deflect hazardous asteroids, 213–14, 274, 303–4

cratering research by, 31, 49–50

data release and public announcement guidelines from, 138, 138n94, 167, 168–69

founding of, 31, 46, 49

human spaceflight and policy shift for, 241–44

lunar program and scientific aspirations for the Moon, 46

Near-Earth Object program of, 132n76

Office of Program Analysis and Evaluation (OPAE) study, 231–35, 274, 280–81

Office of Space Science, 138–39

planetary astronomy program of, 51, 59, 111, 132n76, 148, 214, 312

planetary science program of, 111, 132, 186, 282, 304

public-private partnerships with, 266

Shoemaker-Levy–Jupiter impact internet resources of, 128, 128n68

Space Shuttle and space station programs, 111

Spacewatch funding from, 148

Vision for Space Exploration (VSE) policy of, 230, 230n38, 241

NASA Innovative and Advanced Concept (NIAC) study, 268

National Academy of Sciences study, 302–3

National Commission on Space, 250–51

National Research Council (NRC), 253, 274, 274n3, 281–82, 303n85

National Science Foundation, 71

National Space Institute, 264n71

National Space Society, 264n71

Near-Earth Asteroid Prospector (NEAP), 265

Near Earth Asteroid Rendezvous (NEAR/ NEAR-Shoemaker) mission

Discovery Program funding for, 198, 205, 205–6n91

Eros visit by, 70–71, 173, **184**, 193–94, 196, **197**, 205, 205–6n91, 207

Mathilde visit by, **184**, 205–6

renaming spacecraft to honor Shoemaker, 173, 207, 207n95

near-Earth asteroids (NEAs)

discovery of first, 8, 19–22, 20–22nn19–20, **21**

evolution of fragments of main-belt
asteroids into, 163–65, 164n58,
175, 179
list of candidates for missions to, 256,
256n38
mission to send astronauts to, 241–44,
253–56, 257
number known, 314
public database for, 154
use of term, 22n23
Near-Earth Asteroid Thermal Model
(NEATM), 180
Near-Earth Asteroid Tracking (NEAT)
program, 55, 148–52, 156, 173, **239**
Near-Earth Object Interception
Workshop and report (Interception
Workshop and report), 114, 115–19,
116n28, 119n41, 247, 251–52
Near-Earth Object Program Office (Center
for Near-Earth Object Studies), 139,
141, 256n38, 280n21, 314
near-Earth objects (NEOs)
ancient interpretations of, 2, 13
changing views on, 312–15
characterization of, 64–69, 174, 312–13
data release and public announcement
guidelines for possible impacts,
138, 138n94, 167, 168–69
definition and use of term, 22n23,
321–22
as exploration steppingstones, 247–62
first fully automated discovery of,
63–64
internationalization of concerns about,
10
international NEO discovery and characterization program, 156
malign intent and warning associated
with, 2, 3–4, 5
media attention given to, 134–35,
135n84
number known, 11, 312–13, 314
orbit determination techniques for,
16–17, 16n7
policy development related to, 9–10,
309–12
sources of, 2, 51, 210–11, 314–15

Near-Earth Object Science Definition
Team (SDT). *See* Science Definition
Team (SDT)
Near-Earth Object Surveillance Satellite
(NEOSSat), 289
Near-Earth Object Survey, 139
"Near-Earth Object Survey and
Deflection Analysis of Alternatives"
report, 213, 213n2
Near-Earth Object Wide-field Infrared
Survey Explorer (NEOWISE),
185–87, 186n33, 237, 238, **239**,
241, 242, 271, 313
Near Earth Objects Dynamic Site
(NEODyS), 166, 167, 169, 173,
173n73, 226, 279–80, 288n46
Near-Earth Objects Observations
Program (NEOOP), NASA
budget and funding for, 139, 146,
213–14, 242, 292–93, 304
establishment of, 139, 146, 273
funding for surveys out of budget for,
153, 154, 155–56, 236–37, 238,
240, 242
IG report on, 292–93
importance of research by, 314
NEO discovery effort to meet
Spaceguard goal, 214
PDCO relationship to, 1, 293
policy and mission to send astronauts
to an NEA, 241–44
SDT findings briefing for, 223–24
telescope renovations to reach
Spaceguard goal under, 146
workshop on 2011 AG5 impact risk,
278–79
Near-Earth Objects Survey Working
Group
budget and funding for, 129
formation of, 128–29
report and recommendations from,
129, 129n70
Shoemaker as chair of, 129
Near Infrared Spectrometer, 207
Nemesis hypothesis (Siva hypothesis), 100
NEOCam (Near-Earth Object Camera),
242, 243–44, 303, 304, 311

NEOShield, 289

New Zealand, iridium identification in, 84

Newton, Isaac, 4, 6

Nicholson, Seth, 319

nickel-iron objects, 24, 44, 68, 209, 211, 249–50

Noble, William, 29

nuclear energy and weapons

deflection or destruction of asteroids with nuclear devices, 115–21, 116n28, 118n38, 119n41, 233–36, 280–81, 283

Meteor Crater and nuclear bomb test craters research, 47, **48**

MICE project, 46–47

nuclear policy and atmospheric effects of nuclear warfare, 87–88

Project Icarus exercise use of, 107–8

SDIO defense systems against incoming nuclear missiles, 115–16, 119, 119n39

nuclear winter, 87–88

O

Oak Ridge Observatory, 322

Obama, Barack, 241–42

Obama administration and Barack Obama, 253, 255, 256, 260, 266, 270–71

Ocampo, Adriana, 95–96

occultation measurements, 176

oceans

asteroid impact in, 86, 91

heat from impact events and boiling of, 101–2

tsunamis, impact-generated, 120, 216–17, 301

Odessa, Texas, crater, 39, 43

O'Keefe, John, 86–87, 161n50

Olbers, Wilhelm, 17, 18–19

O'Leary, Brian, 249, 252

Oliver, Bernard M. "Barney," 62, 108, 108n5

olivine, 208, 209n106

O'Neill, Gerard K., 248–49, 250, 252, 263

Oort cloud, 100–101

Öpik, Ernst J., 35–36n14, 43, 81

Oppolzer, Egon von, 64, 193

Orbital Sciences, 266

orbit determination techniques, 16–17, 16n7

orbits

asteroids, orbits and ephemerides of, 8, 16–17, 16n7, 51–57, 54n3, 54n5, 319

calculation techniques for, 16n7, 157

comets, characteristics of, 4, 159–60, 160n46

computation of orbit and ephemeris through numerical modeling, 158–62, 160n46

early opinions about, 4

elements of orbital motion, 165–66, 165n59

heliocentric orbit classifications, 165n59, 321–23, **321**

main-belt asteroids, orbits of, 8

nongravitational forces on objects and, 159–62, 160n48, **161**, **162**

orbital periods less than a year, 53

PHA orbits and SDT survey systems recommendations, 221–22

Shoemaker-Levy 9 orbital characteristics, 127

Oriani, Barnaba, 15–16

Origins, Spectral Interpretation, Resource Identification, Security, and Regolith Explorer (OSIRIS-REx) mission, **184**, 185, 242, 258, 313

Orion Crew Exploration Vehicle (CEV), 254

Ostro, Steven "Steve," 65n31, 71, 119, 120, 187–89, 188–89n40, **188**

'Oumuamua (1I/2017 U1 interstellar object), 307–8, 313

Outer Space Treaty (Treaty on Principles Governing the Activities of States in the Exploration and Use of Outer Space, including the Moon and Other Bodies), 264–65, 264–65n72, 269

P

PACS (Palomar Asteroid and Comet Survey), 55–57

Paddack, Stephen, 161n50

Paine, Thomas O., 250

Palazzo, Steven, 259

Paleogene period, 79, 79n1

paleontology and extinction hypothesis, 90

Palermo Observatory, Sicily, 13, 15, 15n4

Palermo Scale, 170, 173, 296

Pallas

discovery of, 8, 18–19

effects of impactor the size of, 101

occultation observations of, 176

polarization curve for, 177, 177n5

size estimates of, 177, 180

symbol designation for, 319

Palomar Asteroid and Comet Survey (PACS), 55–57

Palomar-Leiden Survey of minor planets, 57–58

Palomar Observatory

asteroid 1997 XF11 observations at, 134

CCD-based camera at, 10

female observers at, 53

Oschin Schmidt telescope at, 152

Planet-Crossing Asteroid Survey (PCAS) at, 51–57, 54n3, 54n5, 142–43, 173–74, 179

Schmidt telescope at, 51, **52**, 55, 56, **56**, 57, 63, 123, 134, 142–43, 145, 173

Shoemaker-Levy 9 discovery by, 123–24, 124n57

Palomar Sky Survey prints, 54

Panoramic Survey Telescope and Rapid Response System (Pan-STARRS), 186, 231–32, 237–41, **239**, 242, 297, 307

Panoramic Survey Telescope and Rapid Response System 2 (Pan-STARRS 2), 239

Park, Robert L., 119

Parry, R. P., 29

Pathfinder telescope, 240

Patroclus asteroid (617), 68–69n43

PCAS (Planet-Crossing Asteroid Survey), 51–57, 54n3, 54n5, 142–43, 173–74, 178–79, 179

peaceful use of outer space, 263–65, 288, 294–95

Peiser, Benny J., 168

Pell, Barney, 266

PEMEX, 93–94, 95

Penfield, Glen, 93–94, 95

periodicity in extinction events, 99–100, 103–4

Perseids meteor showers, 2

Pettengill, Gordon, 70, 71

Phaethon asteroid (3200), 2, 183

Phanerozoic extinction events, 98

Philae lander, 202–4, 203n81

photography and photographs

advances in and impact theory on crater formation, 36–38

aerial photographs of shell and bomb craters, 36–37, 37n16

CCD technology advantages over, 57

exposure gating technique, 56

images from cameras affixed to telescopes, 19–22, 21–22n20, **21**

lunar craters research with, 25–26, 25–26n35

near-Earth asteroid discovery with, 19–22, 20–22nn19–20, **21**

PCAS examination methods, 53–55, 54n3, 54n5

photograph comparison to find NEOs, 9, 10

photometer and photometric studies, 57, 64–65, 66–68, 66n34, 143n3, 205

Piazzi, Giuseppe, 8, 13–17, 15n4, **15**, 18, 318

Pieters, Carlé, 196

Pilcher, Carl, 138–39

pinball analogy, 164–65, 164n58

Pioneer 10 mission, 57, 60, 63, 248

Pioneer 11 mission, 57, 60, 63

Planet 9, 104

planetary astronomy program, NASA, 51, 59, 111, 132n76, 148, 214, 312

planetary defense

advocacy for, 10

Air Force role in, 129–32, 132n77, 223–24, 273, 285–86, 291–92

budget and funding for, 213–14

development of strategies for, 245, 246–47, 251–53, 261–62, 280–87

increase in attention given to, 213–14

international partners and collaborations, 10, 287–89, 294–300

NASA role in, 129, 132, 132n76, 273–74, 280–87, 291–94, 315

notification of Federal agencies and emergency response institutions of impending NEO threats, 274, 276, 283–85

public support for and priority of, 304–5, 315

simulation of asteroid impact scenario and strategies for, 285–86, 285n38, 300–302, 300n79

Planetary Defense Conferences, 252–53, 300, 300n79

Planetary Defense Coordination Office (PDCO)

budget and funding for, 1, 286–87

establishment of, 1, 274, 286, 293

events leading to establishment of, 1–2, 291–93

focus of work of and research by, 1, 286–87, 293–94

Johnson role in, 132n76, 293–94

NEOOP relationship to, 1, 293

NEOWISE funding by, 186

Planetary Resources (Arkyd Astronautics), 266–67, 268, 269–70

planetary science program, NASA, 111, 132, 186, 282, 304

Planet-Crossing Asteroid Survey (PCAS), 51–57, 54n3, 54n5, 142–43, 173–74, 178–79, 179

planets

asteroids as fragments of, 18–19

asteroids as minor planets, 13

evolution of and resources on, 246

impact events as means to transfer life between, 102–3

orbit and ephemeris of, 16n7

spacing of and distances between, 8, 14

Planet X, 100–101

Playfair, John, 4–5

Pluto, 158

plutonium-244 (Pu-244), 83

Polariscope, 60

polarization measurements and curves, 176–77, 176–77nn4–5

political impacts of comets, 121–28

Pollack, James, 84–85, 87

Pons-Winnecke, Comet, 41n27

Pope, Kevin, 95–96

Posey, Bill, 291

Potentially Hazardous Asteroids (PHAs), automated warning systems for, 142, 166, 167

Potentially Hazardous Objects (PHOs)

data release and public announcement guidelines for, 138, 138n94, 167, 168–69

definition and use of term, 217, 217n14, 321–22

orbits of and SDT survey systems recommendations, 221–22

Potsdam Observatory, 64

Pravdo, Steven, 151

Pravec, Petr, 191

predictions of close-Earth approaches, 105, 165–69, 168n64, 279–80

Principles of Geology (Lyell), 5

Probabilistic Asteroid Impact Risk (PAIR) model, 301, 301n82

prograde rotation, 160, 160n46, **161**

Project Icarus exercise, 106–8

Provin, S., 177, 177n5

Ptolemy, Claudius (Ptolemaeus), 2–3

Almagest, 2–3

Tetrabiblos, 3

pyroxene, 66, 208, 209n106

Q

Q-type asteroids, 196

R

Rabinowitz, David, 63–64, 151

radar

accuracy of observations with, 72, 72n54

Apophis (2004 MN4) radio transponder tracking mission, 228–30, 236
asteroid observations with, 69–72
asteroid shape modeling with, 70–71, 177, 187–90, 188–89n40, 194–95
bistatic radar, 69
comet observations with, 72–74
Doppler and range measurements for orbits and ephemerides, 72, 72n54, 159
function of, 69
Goldstone radar observations, 69–74
limitations on research with, 69, 71–72
list of small-body radar observations, 72n55
satellite search with, 194–95, 194–95nn58–60
Radebaugh, Arthur, 245, 247
Radzievskii, V. V., 161n50
Rahe, Jurgen, 113–14, 115, 148
Ramsden, Jesse, 15
Randall, Lisa, 101
Rather, John, 114, 115
Ratiner, Leigh, 264
Raup, David, 99–100
Reagan, Ronald, 60
Reagan administration and Ronald Reagan, 250, 264, 310–11
Reddy, Vishnu, 297
reflectivity. See albedo/reflectivity
Reinmuth, Karl, 322
Reinvaldt, I. A., 43
Remo, J. L., 118n38
resources/natural resources
asteroid mining, 245, 247, 251, 262–71
exploitation of NEO resources, 245, 246–47
international treaties related to, 263–65, 264–65nn71–72
legal and ethical concerns related to, 247, 263–65, 269
mining of meteoric iron, 31–33, 34, 39, 40
mission to survey extraterrestrial resources, 249–50
private company interests in resource extraction, 266–71
repositories of, NEOs as, 247–62
scarcity of on Mars and planetary evolution, 246
space settlement and resource utilization, 246, 248–52
value of asteroid resources, 268–69
retrograde rotation, 160, 160n46, **161**
Richards, Robert, 266
Richardson, Derek, 191n49
Ries Basin, Germany, 47
risk assessment scales for communicating impact hazards, 169–74, **171–72**
Rivkin, Andrew, 182
Robotics Asteroid Prospector (RAP), 268
Roddy, David, 108
Rohrabacher, Dana, 230, 273, 304
Rosetta mission, **184**, 198n68, 201, 202–4, **202**, 203n81
rubble-pile asteroids
collisional history and family members of, 192–93
concept of and use of term, 108
deflection and mitigation of, 108, 117, 233–36
structure of, 191–95, 191n49, 204–5, 207–9, 211, 314
YORP effect on, 162
Rubin, Vera C., 215n5
Rubin Observatory, 215n5
Russell, Henry Norris, 65n31
Ryugu asteroid (162173), 183, **184**, 185, 313

S
Sagan, Carl, 118, 119, 120
Sakigake mission, **184**
Sánchez, Paul, 192
Sandia National Laboratory, 115, 285n38, 300
Sandorff, Paul, 106–8
satellites, asteroid, 193–95, 194–95nn58–60, 194n56, 204
Saturn, 8, 14, 27, 57, 60, 63, 164
Scaglia Rossa, 81–82
Scheeres, Dan, 192

Schweickart, Russell L. "Rusty," 214,
    228–29, 234, 235, 267, 277–78,
    280–81, 295
Science Definition Team (SDT)
    asteroid size, analysis of hazard risk
        relationship to, 217–19, **218, 219**
    charter for, 214–15, 273
    cost-benefit analysis by, 215, 216,
        218–19n16, 218–20, 223, 309–10
    establishment and leadership of, 214
    goal of NEO hazard reduction of,
        222–23, **223**, 236–37
    ground-based and space-based options
        for systems, recommendations for,
        220–23, 231–33, 302–3
    NEOOP briefing about findings of,
        223–24
    new search/survey systems, develop-
        ment of recommendations for,
        220–23, 230–34, 236–37, 273–74
    report from, 73, 215
    risk analysis and development of NEO
        hazard reduction goal by, 215–24,
        301–2
science fiction, 245, 246
Science Mission Directorate, NASA
    (Office of Science), 258, 293, 303–4,
    303n88, 310–11
Scotti, James V. "Jim," 60–62, 63–64,
    124, 132, 154, 157–58
Scout system, 280, 280n20
Search for Extraterrestrial Intelligence
    (SETI) office, 62
Sekanina, Zdenek, 160
Sentinel mission, 243, 267
Sentry system, 167, 169, 173, 173n73,
    226, 277, 279–80, 280n20, 280n21
Sepkoski, J. John, Jr., 99–100
Serkowski, Krzysztof, 60
Serviss, Garrett P., 245
Shaddad, Muawia, 277
Shaler, Nathaniel, 33–34, 34n7, 81, 81n5
shape of asteroids
    light curve observation and analysis
        and, 64–65, 65n31, 190, 193–94,
        194n56

radar, shape modeling with, 70–71,
    177, 187–90, 188–89n40,
    194–95, 194–95nn58–59
Shapiro, Irwin, 70
Shapley, Harlow, 41, 41n27, 43, 319
shatter cones, 44, 44nn37–38
Shelton, William L., 291, 292
Shelus, Peter, 132
Shoemaker, Carolyn
    comet discovery by, 9
    hazard of Earth-colliding objects,
        Snowmass report on, 143–44
    injury of, 173
    Meteor Crater visit by, 47
    NEO research by, 312
    PACS role of, 55
    photo of, **56**
    Shoemaker-Levy 9 discovery by,
        123–24, 124n57
    talent for NEO discovery and observa-
        tion of, 55
Shoemaker, Eugene M. "Gene"
    comet bombardment time scale, opin-
        ion about, 100
    death of and honors for contributions
        of, 55n8, 173, 207, 207n95
    hazard of Earth-colliding objects,
        Snowmass report on, 108–11,
        110, 143–44
    Helin split from, 55
    impact cratering research by, 9, 46–49
    impact physics and theory of cratering
        mechanics of, 47
    lunar crater and geological research by,
        46, 49
    Mathilde structure, discussion about, 206
    Meteor Crater research and MICE
        project role of, 46–49, **48**
    Meteor Crater visit by, 47
    Near-Earth Objects Survey Working
        Group role of, 128–29
    NEO research by, 312
    PACS role of, 55–57
    PCAS role of, 51, 53–57, 142–43, 179
    photo of, **56**
    return to Geological Survey of, 55

Shoemaker-Levy 9 discovery by, 123–24
Spacewatch proposal to, 59
Swift-Tuttle comet impact probability correspondence with, 122
Shoemaker-Levy 9, Comet, 9, 105, 106, 123–28, 124–25nn57–59, **124**, 125–26nn61–62, 204, 273
short-period comets, 72–73, 129, 159, 201
Siding Spring Observatory, 63, 155, 226, 275
Signor, Phillipp, 90
silica, 47, 49, 96–97
silicates and silicate minerals, 66, 208–9, 209n106, 210
Siva hypothesis (Nemesis hypothesis), 100
Skiff, Brian, 154
Sleep, Norman, 101–2
Small Bodies Assessment Group (SBAG), 255, 258
Smit, Jan, 84, 92
Smithsonian Astrophysical Observatory, 54, 157
Smithsonian Institution, 147
Snowbird I meeting, 85–86, 90
Snowbird II meeting ("Global Catastrophes in Earth History"), 89–90, 93, 143–44
Snowbird III meeting, 96
solar system
clockwork maintenance of, 6
early solar system mixing, 210–11, 210n110
effects of impacts from small objects from on Earth's surface and life on Earth, 8–9, 30, 31, 77, 79–81, 99–104
formation process as source of NEOs, 2, 51, 210–11, 314–15
impact process as agent of change in, 50
motion of heavenly bodies and theory of epicycles, 2–3
partial impact theory for formation of, 37n18

peaceful use of outer space, 263–65, 288, 294–95
planet spacing in, 8, 14
small solar system bodies, efforts to understand, 51
Solar System Exploration program, NASA, 138
solar wind particles (space weathering) and asteroid spectra, 195–96
Soviet Union/Russia
Chelyabinsk fireball and airburst, 1–2, 9, 274, 289–91, **290**
Sputnik and the race to the Moon, 46
Tunguska impact event, 2, 39–42, 41n27, 41–42nn29–30, 43–44, 105, 215, 218–19n16, 282, 291
Tunguska region expedition, 39–41
VEGA 1 and VEGA 2 missions, 75–76, **184**
SPACE Act (Commercial Space Launch Competitiveness Act, U.S.), 269
Space Act Agreement, 242–43
SpaceCast 2020 study, 131–32, 131n75, 223–24
Space Development Corporation (SpaceDev), 265
spaceflight/human spaceflight
budget and funding for, 253–54, **254**
exploration steppingstones, NEOs as, 247–62
Moon and space colonization, 248–52, 263–64
NASA policy and mission to send astronauts to a NEA, 241–44
private, commercial spaceflight companies, 265–68
private asteroid mining, 262–71
space settlement, 246, 248–52, 263–64
Spaceguard/Spaceguard Survey
attitude toward and priority of goal of, 138–39, 273
budget and funding for, 114, 146
commitment to goal of, 138–39
deflection or destruction of asteroids with nuclear devices, 115–21, 116n28, 118n38, 119n41

goal of and achievement of goal, 111–21, 138, 141, 146, 186, 213–14, 215–16, 273, 277

modeling asteroid population and simulation studies for survey system, 142–46

survey network for, 141–42, 146

workshops on search programs and interception/deflection strategies, 113–21, 116n28

Spaceguard Survey report, 109, 114, 141

Space Launch System (SLS), 255

Space Mission Planning Advisory Group (SMPAG), 297–300

space program

impact-related investigations of the Moon, 46, 49–50

policy and mission to send astronauts to an NEA, 241–44

race to the Moon, 46

space settlement, 246, 248–52

Space Settlement workshops, 248–49, 251–52, 263

Space Situational Awareness (SSA) program, 156, 287–88, 288, 288n46

Space Studies Institute (SSI), 263

space tug, 233, 235

Spacewatch

0.9-meter telescope for, 11, 60, 61, 63, 145, 146, 147–48

1.8-meter telescope (Spacewatch II) for, 60, 141–42, 146–48

4K by 4K CCD for, 151–52

2,048- by 2,048-pixel CCD for, 63, 146

asteroid 1997 XF11 discovery with, 132

asteroids found with, 60, 62, 63–64, **239**

automated discoveries with, 61, 63–64, 141–42

control center for, **149**

data from and MPC data processing and dissemination, 157–58

development of, 10–11, 58–64, 108n5, 109–10, 130

drift scanning technique, 61, 145

first camera designed and built, 60–62

first fully automated discovery with, 63–64

funding for, 59–60, 62–63, 147–48, 312

location of, 11, 59, 60–62, 132

NASA role in, 59, 62

observation strategies of, 151–52

Phase A study, 59–60

recovery of known asteroids and comets with, 61–62

Shoemaker-Levy 9 confirmation by, 124

software for and function of system, 60–62, 63–64, 154

Spaceguard survey network role of, 141–42, 146

success of, 147–48

upgrading of, 63

Spacewatch Report, The, 60, 62

space weathering (solar wind particles) and asteroid spectra, 195–96

SpaceX, 265–66

Spahr, Timothy, 154–55, 158, 220

Spain

Calar Alto Observatory in, 126, 288

iridium identification in, 84

spectral observations and asteroid colors, 64–69, 66n34, 68n42, 180–83, 182n22

spectral taxonomy, 181–82, **181**

spectrographs and spectroscopic observations, 65–68, 68n42, 177, 180–81, 207, 208

spectrometer data, 76, 207

Spencer, John, 180

Spencer, L. J., 44

SPeX spectrograph, 180–81

Spitzer Space Telescope, 185

Sputnik, 46

Squyres, Steve, 259

Stanford University, 101–2

Stardust mission, **184**, 198, 201, 205–6n91

Stardust-NExT mission, **184**, 201–2, 202n79

State, U.S. Department of, 276, 284, 286

Steinheim Basin, Germany, 44n38

Steins asteroid (2867), **184**

sterilization of Earth from impact events, 101–2

Steward Observatories, 59, 147

stishovite, 49

Stokes, Grant, 130, 148, 153, 214, 217, 220, 223–24, 302

Strategic Defense Initiative Organization (SDIO), 115–16, 119, 119n39

S-type asteroids, 67–68, 194, 195, 196, 196n64, 205, 207, 208–9, 209n106

substantive uniformitarianism, 9

Sudan, 275–77

Sudbury Basin, Ontario, 44n38

Sugano-Saigusa-Fujikawa, Comet, 73–74

Suisei mission, **184**

sulfur, 97

Sun

asteroids near, search for, 225–26, 237, 238, 238n58

distances of planets from, 14

influence on Earth by, 5

Lagrange points, 221–22, **221**

sunlight reflection. *See* albedo/reflectivity

surveys/discovery surveys

achievement of Spaceguard goals, 186, 213–14, 273

advocacy for discovery surveys, 10

automation of, 11, 141–42, 146–56, 174

budget and funding for, 153, 154, 155–56, 174, 236–37, 238, 240, 242, 271

follow-up observations of discoveries by, 153, 153n30, 156, 220, 238–39, 239n59, 312

next-generation surveys, 237–41, 274

SDT development of recommendations for new search/survey systems, 220–23, 230–34, 236–37, 273–74

whole-sky survey, 145

See also Spaceguard/Spaceguard Survey; Spacewatch

"sweet spots," 237, 238, 238n58

Swift-Tuttle, Comet (109P), 2, 105, 121–23, 132, 135, 137

Szafranski, Richard, 131, 131n75

T

Tagish Lake, Yukon Territory, impact event, 274–75

Tagliaferri, Edward, 112–13, 114–15, 118, 123, 276

Tamkin, Steven, 158

Tancredi, Gonzalo, 125n59

Tantardini, Marco, 256–57

Taylor, Gordon, 176

Taylor, Patrick, 162, 189

Taylor, Theodore, 108

Teapot Ess nuclear bomb test crater, 47, **48**

technology

advances in and NEO discoveries, 10–11, 173–74, 307–8, 312–15

computational and experimental methods for impact research, 49

orbit determination techniques and advances in, 16n7

Tedeschi, William, 120

telepossession claims, 265, 265n76

telescopes

electronic asteroid-hunting telescope, 240–41

Near-Earth Objects Survey Working Group recommendations for, 129

photographic images from cameras affixed to, 19–22, 21–22n20, **21**

SDT ground-based and space-based survey telescopes recommendations, 220–23, 231–33, 302–3

See also Hubble Space Telescope; Spacewatch

Teller, Edward, 115–16, 119, 120

Teller asteroid, 116

Tempel 1, Comet, **184**, 199–201, 199n71, **199**, 200n76, 201–2, 235–36, 283, 283n29

Tertiary period, 79n1

Tetrabiblos (Ptolemy), 3

Texas Instruments, 58

Themis (24) and Themis family, 182–83, 183n25, 192

thermal models and standard thermal model (STM), 177–80

thermal radiation
  atmospheric chemistry and radiative
    transfer research, 84–85
  heat from impact events, 101–2
  size of impactor and duration of effects,
    101–2
  Yarkovsky effect and drift, 160–61,
    160n48, **161**, 162, 164–65, 229,
    236
  Yarkovsky-O'Keefe-Radzievskii-
    Paddack (YORP) effect, 161–62,
    161n50, **162**, 189
Tholen, David "Dave"
  2000 SG344 discovery by, 168
  2011 AG5 recovery by, 277, 279,
    279n19
  Apohele asteroids discovery and
    naming by, 323
  Apophis (2004 MN4) naming by,
    227–28
  Apophis (2004 MN4) spotting by,
    225–26, 237, 238
  asteroid classification system expansion
    by, 68, 181
  follow-up observations by, 238
Thomson, Elihu, 34
Tilghman, Benjamin C., 31–32, 33n4, 47
time domain astronomy, 215
Titius-Bode law (Bode's law), 8, 14, 16, 18
Tonry, John, 239–41
Toon, O. Brian, 85–86, 87, 114
Torino Scale, 170, **171–72**, 173, 226, 296
Toro asteroid (1685), 70
Toutatis asteroid (4179), 70, 72, **184**
Treaty on Principles Governing the
  Activities of States in the Exploration
  and Use of Outer Space, including the
  Moon and Other Bodies (Outer Space
  Treaty), 264–65, 264–65n72, 269
trees. *See* forests and trees
Treysa, meteorite impact at, 37
Trilling, David, 185
triple systems, 195, 195n60
Trojan asteroids
  2010 TK7 asteroid, 187
  Al heating and hydration of, 209n106
  color and classification of, 68–69

Hektor (634), 68–69, 68–69n43
Jupiter regions for, 68, 68–69n43, 187
naming of, 68–69n43
orbital characteristics of, 187
Patroclus (617), 68–69n43
surveys to identify, 57
tsunamis, impact-generated, 120,
  216–17, 301
Tucker, Roy, 225–26, 227–28, 238
Tunguska, Russia, event, 2, 39–42,
  41n27, 41–42nn29–30, 43–44, 105
Tunguska impact event, 215, 218–19n16,
  282, 291
Turco, Richard, 87–88

U

UBV photometric system (Johnson-
  Morgan system), 66, 66n34
uniformitarianism doctrine, 4–5, 8–9
University of Arizona
  CCD–based technology research at,
    10–11
  Lunar and Planetary Laboratory, 57, 67
  Lunar and Planetary Laboratory survey,
    148, 154–56, 297
  Optical Sciences Center and Project
    COLT, 146–47, 146n12
  Steward Observatories, 59, 147
University of California, Davis, 90
University of Chicago, 88–89, 99
University of Hawai'i
  2011 AG5 recovery by, 277–78, 279,
    279n19
  Asteroid Terrestrial-impact Last Alert
    System (ATLAS), 237, 239–41,
    239, 302
  Infrared Telescope Facility (IRTF), 59,
    68, 179, 180–81
  Institute for Astronomy (IfA), 67, 231,
    277–78, 279, 279n19
  Panoramic Survey Telescope and Rapid
    Response System (Pan-STARRS),
    186, 231–32, 237–41, **239**, 242,
    297, 307
  Panoramic Survey Telescope and
    Rapid Response System 2 (Pan-
    STARRS 2), 239

University of New Brunswick, Earth Impact Database, 103–4
University of Rhode Island, School of Oceanography, 96–97
Uppsala Observatory, 155
Uranus, 8, 14, 18, 318
Urey, Harold C., 45, 80n3, 81, 209n106
Urias, John M., 132n77

V
Valsecchi, Giovanni, 167
van der Hucht, Karel, 288n46
van der Waals forces, 192, 192n52
Van Flandern, T. C., 194n56
van Houton, Cornelis Johannes, 57
van Houton-Groeneveld, Ingrid, 57
Vega mission, 75–76
    VEGA 1 and VEGA 2 spacecraft, 75–76, **184**
Veillet, Christian, 187
Venus
    atmosphere of, research on, 84
    distance from Sun, 14
    greenhouse effect and heat on, 101
    missions to and flybys of, **184**
    radar research efforts, 69
    transfer of life between Mars and, 103
vermin of the sky, asteroids as, 19–20, 20n17, 314
Very Large Array (VLA), 72
Very Long Baseline Array (VLBA), 72
Vesta
    composition, color, and structure of, 66–67, 67n37, 163, 195, 209–10
    discovery of, 19
    effects of impactor the size of, 101
    evolution of fragments from into near-Earth space, 163–65, 164n58, 175
    HED meteorites from, 67, 67n37, 163, 163n52
    mission to, **184**, 209–10
    polarization curve for, 177, 177n5
    reflected sunlight from, 66n33
    rotation period and color variations of, 65, 65n32
    size estimates of, 177, 178
    symbol designation for, 319

Veverka, Joseph, 177, 177n5
Vision for Space Exploration (VSE), 230, 230n38, 241
Vokrouhlický, David, 160–61
volcanoes
    climate, effects of eruptions on, 84–86, 98
    crater formation from steam explosions, 24–25
    crater origins associated with, 23, 23n26, 26, 26n37, 28–29, 35, 37n18, 38
    impacts related to as extinction mechanism, 84–86, 98
    meteorites from lunar eruptions, 7, 29–30
    stratovolcano, 26
von Zach, Franz Xaver, 14–15, 14n2, 16, 17
Voyager spacecraft missions, 68n42
Vredefort Dome, South Africa, 44nn37–38, 45, 103–4
Vsekhsvyatskij, Sergey, 75

W
Wabar, Arabia, craters, 44
Wainscoat, Richard, 238, 238n58, 277, 279n19
Ward, Steven "Steve," 217, 285n38
War of the Worlds, The, 245
wartime innovation and theories on crater formation, 36–38
Watson, Fletcher Guard, Jr., 43
Weber, Robert, 153n29
Wegener, Alfred, 37, 38
Weidie, Alan, 94–95
Weiler, Edward J. "Ed," 169, 215
Weiss, Edmund, 20n17
Weissman, Paul, 120
Weryk, Robert, 307–8
Westland, Charles J., 37–38
Westphal, James, 10, 58
Wetherill, George, 108
Whipple, F. J. W., 41, 41n27
Whipple, Fred L., 41, 43, 75–76, 77, 159
White House Office of Science and Technology Policy, 274

Whitely, Robert, 168
Whitmire, Daniel, 100–101
whole-sky survey, 145
Wide-field Infrared Survey Explorer (WISE) project, 185–86, 237, 242
Wild 2, Comet, **184**, 198–99, 201
wildfires/fires, 2, 88, 89
Williams, Gareth, 157–58, 307, 319
Williams, James, 54, 164
Wisdom, Jack, 163–64
Witt, Gustav, 8, 21–22, 21–22n20
Wolbach, Wendy, 88–89
Wolf, Frank, 259–60
Wolf, Max, 20
Wolfe, Ruth, 143–44
Wood, Lowell, 115, 116n28, 119
Worden, Simon "Pete," 119–20, 148, 215, 216, 262
Worthington, A. M., 34–35n11
Wright, Edward, 185
Wright, Frederick E. "Fred," 42–43, 42n31

X
X-ray Spectrometer, 207
Xu, Shui, 164

Y
Yarkovsky, Ivan, 160, 160n48, 161n50
Yarkovsky effect and drift, 160–61, 160n48, **161**, 162, 164–65, 229, 236
Yarkovsky-O'Keefe-Radzievskii-Paddack (YORP) effect, 161–62, 161n50, **162**, 189
Yeomans, Donald K. "Don"
    advocacy for Comet Halley visit by, 74
    AMOS telescope, trip to discuss access to, 150
    Apophis (2004 MN4) impact, discussions about, 226–27, 278
    asteroid 1997 XF11 impact and orbit calculations by, 134
    asteroid 2000 SG344 impact risk announcement by, 169
    asteroid 2008 TC3 statement from, 276
    comets, analysis of impact risk from, 217
    Mathilde structure, discussion about, 206
    model development for nongravitational forces on comets, 160
    Near-Earth Object Program Office role of, 139
    on orbit determination techniques, 16n7
    press information sheet release about 1997 XF11, concerns about, 137, 137–38n93
    Science Definition Team role of, 214
    Shoemaker-Levy–Jupiter impact and orbital computations by, 125, 125–26nn61–62, 127
    simulation of asteroid impact scenario, role in drafting, 285n38
    Spaceguard goal, report on achievement of, 277
    Swift-Tuttle comet impact probability computations by, 121–23
YORP (2000 PH5) asteroid (54509), 162, 162n51
Yucatán Peninsula, 80, 93–95, **94**, 97–98, 97n50, 101, 104, 128

Z
Zellner, Ben, 67